Disaster Studies and Management

Janki Andharia · Pankaj Anand
Editors

Shifting Lands, Flowing Waters

Transboundary Water Governance and Risk-Informed Planning

Springer

Editors
Janki Andharia
Jamsetji Tata School of Disaster Studies
Tata Institute of Social Sciences
Mumbai, India

Pankaj Anand
Oxfam
Dili, Timor-Leste

ISSN 2730-5635 ISSN 2730-5643 (electronic)
Disaster Studies and Management
ISBN 978-981-96-7690-3 ISBN 978-981-96-7691-0 (eBook)
https://doi.org/10.1007/978-981-96-7691-0

This Springer imprint is published by the registered company Springer Nature Singapore Pte Ltd.
The registered company address is: 152 Beach Road, #21-01/04 Gateway East, Singapore 189721, Singapore

Preface

This volume is titled "Shifting Lands, Flowing Waters" to denote the altering of entire topographies in the aftermath of desolating disaster events. The inherent planetary vulnerabilities cause a variety of dramatic changes to the Earth's surface, which are compounded even as human-induced activities manufacture newer risks and hazards that reconstruct the contours of the land. Further, overflowing rivers, flooded by severe rains or melting snow, cut new courses, and reshape riverbanks. Floodplains, previously stable environments, are now dynamic zones of change as water exerts its authority over the land in the face of more severe hazards. Human settlements are frequently caught in the crossfire, as houses and infrastructure are washed away, leaving behind a trail of destruction, changed landscapes, and shattered populations. The scars on the land often caused by water and the collective memory of communities survive, acting as permanent reminders of the fragile balance between human civilisation and the forces of nature. In contemporary times, understanding the interaction of shifting lands and flowing waters becomes critical in negotiating this complicated and dynamic relationship for limiting disaster impacts even as each nation-state works towards developing greater resilience in the face of an ever-changing geography.

This volume collates multi-faceted inquiries, reflections, critical scholarly probing of transboundary rivers and their impacts on life, on land and landforms, frameworks of governance both within and across countries. Shifting lands and flowing waters contextualises the visceral nature of changing topographies owing to land-water interactions, and their marked influence on riverine communities.

The fact that planet earth operates as a single, complex adaptive system is the contribution of earth system sciences which studies and integrates the interplay of bio-physical processes and human dynamics. As the idea of planetary boundaries and tipping points assume contemporary significance in environmental sciences, climate discourse, risk assessments, geo-politics, transboundary resources and in global growth theories, different disciplines converge today to address complex problems that confront the sustainability of the planet. Despite advancement of scientific understanding, the frequency and intensity of disasters are rising at an alarming rate and is a cause of concern. In 2022, 387 natural hazards and disasters were recorded on the global Emergency Event Database: costing 30,704 lives and affecting 185 million others (CRED, 2023). It also recorded economic losses of approximately 223.8 billion USD (CRED, 2023).

According to the IPCC's Special Report on the Ocean and Cryosphere in a Changing Climate[1], (2019) the projected global mean sea level (GMSL) rise, extreme sea level events that are historically rare, (today's hundred-year events) will become common by 2100 under all RCPs[2]. Thus, many low-lying cities and small islands at most latitudes could experience such events annually as early as 2050—a grim warning of humanity being dangerously close to the tipping point after which some climate change impacts would be irreversible (IPCCa, 2021). The report findings also warn of more intense and frequent heatwaves, erratic weather, droughts, and melting glaciers. This has been corroborated by the annual reviews published by CRED[3] in the last decade, which suggest the frequency and intensity of disaster events is on the rise, and so are the resultant human and economic costs of disasters. The UN Secretary-General has flagged this as "the code red for humanity" (IPCCb, 2021) emphasising the urgency for climate action, which cannot be further delayed.

Fresh water is called the bloodstream of the biosphere and is at the centre of the changes taking place at a planetary scale. Water fluxes and stores regulate the Earth's climate and are essential for thriving aquatic and terrestrial ecosystems, as well as water, food, and energy security. Since the beginning of civilisation, water has been the essence of life on earth. It has sustained activities related to agriculture, energy, industry, and several modes of livelihoods. However, water as a resource is threatened by the increasing population across the world, burgeoning domestic priorities, and multiplying developmental demands which place growing pressure on water sources. Degradation, pollution, adverse changes to water bodies—all have far-reaching impacts on adjoining lands and human societies living there. Added to the already existing challenges of managing the water resources are the risks posed by climate change that impacts earth systems processes and the water cycles.

Around 50% of the earth's land surface is covered by shared rivers and lake basins. Around 276 river basins all over the world are shared by two or more sovereign states and 40% of the world population lives surrounding such river and lake basins which cross international borders. Nearly 2 billion people are dependent on groundwater from 300 transboundary aquifers (Jagerskog, 2013). As a consequence, transboundary water governance is an area that requires significant attention and intervention.

---

[1] The SROCC is one of 3 Special Reports published by the IPCC during the 6th Assessment Cycle; the other two being the SR1.5 (on Global Warming of 1.5 degrees Celsius) and SRCCL (on Climate Change and Land).

[2] Representative Concentration Pathways signify the expected future trends in the concentration of greenhouse gases in the atmosphere, subject to change in the future on account of anthropogenic factors; chiefly categorized into 4 RCP levels ranging from very high (RCP 8.5) to very low (RCP 2.6) projected future concentrations.

[3] Centre for Research on the Epidemiology of Disasters, Belgium is a research unit that publishes and promotes research, training, and technical expertise on humanitarian emergencies; and maintains the open access international Emergency Database (EM-DAT).

In specific regional contexts of South Asia, the management of water resources and water governance are interlinked with geopolitics, demography, and challenging socio-cultural and economic constraints. Understandably, the past, present and future of the countries of South Asia are rooted in the equitable sharing of resources. For example the governance and utilisation of transboundary rivers of South Asia pose growing risks for water scarcity, degrading quality of water, floods, and inequitable access to water for the riparian communities. It is to be noted here that transboundary rivers are complex to govern because they often involve concerns related to national security, sovereignty, and territoriality. The decision-making with regard to transboundary waters is dominated by a few people consisting largely of leaders of the countries involved, politicians, bureaucrats, and engineers who often tend to lose perspective of communities whose lives and livelihoods are dependent on these rivers.

From 2017 to 2022, Oxfam in collaboration with four regional and 15 national and local partners implemented the Transboundary Rivers of South Asia (TROSA) programme in four regionally significant river basins shared between Bangladesh, India, Myanmar and Nepal, with funding from the Swedish Government. The overall goal of TROSA was to reduce poverty and marginalisation of vulnerable river basin communities through increased access to, and control over the riverine water resources on which their livelihoods depend. The efforts focused on strengthening the capacities of river basin communities and civil society. This included amplifying the voices of women to influence government and private sector policies and practices and promoting respect for communities' rights in relation to water resources management across the region. The programme was designed to contribute directly to Sustainable Development Goals (SDGs) 1: End poverty, and 6: Clean water and sanitation.

Over the years, TROSA created positive momentum for transboundary cooperation on shared water resources in South Asia. In a challenging regional context, where governance of transboundary water resources is deeply politicised, and a notable lack of multilateral cooperation mechanisms, or structures for this purpose is observed, Oxfam and its partners made significant progress in laying the foundation for inclusive and accountable regional cooperation.

One of the programme's key successes was bringing together a diverse range of community and civil society actors across the region and supporting them to engage in water governance processes with a collective voice. TROSA also fostered community-to-community cooperation on shared water resources, which resulted in several unique collaborative initiatives, including the establishment of cross-border flood early warning communication systems, and the signing of joint declarations and agreements by civil society leaders and local governments committing to more inclusive and sustainable water governance. The empowerment of women and their strengthened voice and participation in decision-making processes around the water and other natural resources that they, and their communities depend on, was another notable achievement of the programme. These achievements of TROSA are only illustrative, not exhaustive.

During one of the multi-stakeholder engagements under TROSA, the idea of publishing an edited book based on experiences around Transboundary River engagements was mooted. Many organisations and practitioners around the world evinced interest in this book. All contributors to the book have brought deep insights based on their unique experiences and understanding. Janki Andharia, the then Dean of the Jamsetji Tata School of Disaster Studies at The Tata Institute of Social Sciences, Mumbai, was working on risk informed planning and through a series of meetings and discussions between TISS and Oxfam, it was decided to organise a Roundtable conference on the theme of transboundary water governance and risk informed planning. A call for papers followed and we got an excellent response. The conference titled, "Shifting Lands, and Flowing Waters: Transboundary Cooperation for Water and Related Issues in South Asia" was organised between 16th and 19th March 2021 The National Institute of Disaster Management (NIDM), New Delhi also collaborated and hosted this online event. The conference was supported by Sverige, Sweden and was widely attended by scholars, government officials, and representatives of grassroot organisations, who are deeply committed to improving transboundary water governance. This book is a compilation of select papers presented at the Roundtable, Each paper was systematically peer reviewed, carefully edited and revised. It offers emerging debates, contestations, challenges and interventions aimed at enhancing regional cooperation in transboundary river basins.

The introductory chapter by **Janki Andharia and Lavanya Shanbhogue Arvind** lays the canvas of issues covered in the volume, reflects on the neo-liberal reading of the transboundary water issues and argues for the urgent need for risk informed planning.

Dipak Gyawali had offered the keynote address during the Roundtable conference which was later converted into a paper. Using the cultural theory he critiques the top down, hegemonic approaches of institutions, development programmes and water treaties and suggests that grassroots conditions are often ignored. He also argues that in water treaties, atmospheric "sky rivers" and underground water issues remain ignored. He proposes pluralizing the discourse, by ensuring that three social solidarities—of market individualism, agency hierarchism, and activist egalitarianism; with their different perspectives on risk and justice, are given adequate voice at the policy table. He proposes a spirit of constructive engagement to foster resilience and preparedness for future disasters.

The next paper by **Mukunda Upadhyay** emphasizes that regional transboundary geo-political actors today are facing struggles at various levels in managing hazards, risks and vulnerabilities due to lack of adequate commitment in ensuring the basic principles and values of inclusive governance. He demonstrates models of people centric co-operation and dialogues between nations especially in the South Asian region as pragmatic solutions to manage shared risks. Initiatives taken by community led institutions to strengthen people to people ties across South Asian nations have marked a sustainable impact on the risk governing systems by strengthening community level capacities to manage uncertainties generated by shared risks. He presents seven principles which states can adopt to ensure comprehensive transboundary risk governance. Melita Grant and Cameron Holley in their paper offer a perspective on how collaborative water governance can be strengthened by learning from international experiences including failures. They make the argument that Australia has experienced "failures" in transboundary water governance through political interference, breaches of compliance as well as inadequate stakeholder engagement.

Amit Ranjan's paper highlights hydro-politics between India and China, China as an upper riparian, covering nature of disputes, and the efficacy of co-operative arrangements. He suggests that contrary to the popular perceptions of "threat", such concerns do not exist beyond the Indian state of Arunachal Pradesh on the eastern side. The MoUs and water treaties have not been very successful in satisfactorily addressing India's waters concerns.

Nitin Sathe adopts a security perspective on transboundary water issues in South Asia, focusing on geopolitics in the region. He offers a complex and shifting idea of "boundaries" from a security perspective of a nation and how they shape international relationships between India, Bangladesh, Pakistan, and China. He argues for a deeper understanding of cross-border "attitudes" and relationships which help appreciate how laws, policies, judgements and actions by governments would pan out in future. He suggests that cross-border security issues are also linked to water and can potentially help create better cross-border dialogues around key water challenges at hand.

Pointing to the reality of climate change which will have a huge impact on disaster risks, **Animesh Prakash and Pankaj Anand** argue that preparedness is key. For communities to be prepared, especially at the last mile, early warning systems play a significant role. While reflecting on their study that analysed gaps in early warning systems, they also designed intervention programmes. They present Oxfam's experience with democratising early warning systems through active engagement with communities in disaster prone regions of India, through three case studies where they worked with local civil society groups. Their work included riparian communities residing along transboundary river basins. Their work offers valuable lessons and insights that riparian communities across India and the South Asian region could use as they face similar challenges of governance with regard to disaster risks and vulnerabilities.

Saran Prakash dwells on the Indus River Treaty between India and Pakistan The contentions of both countries are presented within a historical and geo-political context. The paper highlights the role of democracy in fostering cooperation in transboundary river treaties and stresses the idea that treaties must also be broad enough to respond to changing social and economic issues of the countries involved.

Rajan Subedi and Mukunda Upadhyay offer a citizen science framework to address transboundary water pollution. The author provides a case study of engaging local people and community volunteers in monitoring riverine pollution through systematic data collection and analysis over a period of three years at multiple spots along the Mahakali basin along India-Nepal border.

Saumya Kumar's paper reflects on the role of International Court of Justice (ICJ) invoking the idea of "transboundary harm" in dealing with contestations around water issues between two countries. The paper cites several case laws from the ICJ to trace the development of legal principles in environmental law. She examines some bilateral treaties on water sharing in South Asia to explore the scope and possibility of applying these principles for creating more robust governance mechanisms to address the increasing impact of climate change in the region.

Aparna Unni, Eklavya Prasad, Ajaya Dixit, and Rakesh Mishra in their paper argue for the need to systematically identify and manage water-related risks posed by smaller transboundary river systems. They focus on the region along the Indo-Nepal Border in the state of Bihar (India) a landscape impacted by regular flash floods from small transboundary rivers and also affected by arsenic contamination. Most of these challenges remains unmapped and therefore unattended to, in policy frameworks. They, advocate for the need to gather data and evidence systematically, engaging the local communities that are directly impacted by these seasonal and localised floods and contamination. They offer a framework for management to address these issues of water-based risks and argue that such a framework can be extended to similar landscapes in other parts of the India-Nepal border in a carefully calibrated manner.

Shobhit Chepe and Jacquleen Joseph point out that while the three larger rivers the Kosi, the Gandak, and the Mahakali are governed by agreements between Nepal and India, there are numerous smaller rivers and streams that flow between Nepal and India, which are not governed by any formal agreements. These rivers are very significant to the local economies at the border. The authors study one such river—Pandai at Bhikhna—Thori in India/ Thori in Nepal. Water sharing took place through a traditional system of distribution. The river faces flash floods and riverbank erosion and also dries up during certain months. Over the years a conflict has erupted over the Pandai, and the region faces collapse of the local water-sharing arrangement as the upstream villages have started to divert the streams of the Pandai for their own use. The paper argues for community level engagement with formal government level dialogue.

The final paper by **Bateshwar Das and Meghna Guha** pertains to the Sundarban deltas that share a porous border between India and Bangladesh. Deltaic regions "have played an important role in societal development since the last ice age, offering flat, fertile lands with abundant freshwater which are ideal for agriculture" (Scoun and Dunn, 2023). Globally, deltas and their roughly half a billion inhabitants are in peril especially due to rising sea levels and ongoing historical pressures from upstream rivers. Das and Guha argue that the Sundarban area, rich in biodiversity and spread over 10,000 kms, is designated as a "Ramsar site". However, the people living here are among the poorest and most vulnerable. Transboundary conflict around complicated international legal procedures affect the both the population and the management of natural resources. The paper argues that there is need for sustainable resource management in the region, requiring co-operation between the two countries.

A final note on acknowledgements. The book was made possible through the efforts of many. The Tata Institute of Social Sciences graciously sanctioned a sabbatical to Dr. Janki Andharia, for completion of this volume. Ms. Saran Prakash, Saumya Kumar, and Lavanya Shanbhogue, Tania Mathews —and several colleagues had provided support in organising the Roundtable, following up with paper presenters and also in reviewing some of the papers. Their contributions are gratefully acknowledged.

In conclusion, Oxfam is delighted that the publication of this book coincides with the start of TROSA Phase 2 titled "Rivers, Rights and Resilience" which is being supported by SIDA for another four years (2023 to 2027). TROSA Phase-2 has an explicit focus on further improving cooperation in governing transboundary water resources to strengthen resilience of riparian communities to the impacts of climate change. The program will work in selected transboundary sub-basins within the GBM river basin system, across Bangladesh, Bhutan, India, and Nepal. We hope that this volume will be of use to those engaged in transboundary water governance, risk assessment, and planning in various capacities.

Mumbai, India — Janki Andharia
Dili, Timor-Leste — Pankaj Anand

References

CRED (2023) 'Disasters Year in Review 2022', Cred Crunch Issue No. 70, April 2023

IPCC (2021a). Climate Change 2021: The Physical Science Basis. Contribution of Working Group I to the Sixth Assessment Report of the Intergovernmental Panel on Climate Change [Masson-Delmotte, V., P. Zhai, A. Pirani, S. L. Connors, C. Péan, S. Berger, N. Caud, Y. Chen, L. Goldfarb, M. I. Gomis, M. Huang, K. Leitzell, E. Lonnoy, J. B. R. Matthews, T. K. Maycock, T. Waterfield, O. Yelekçi, R. Yu and B. Zhou (eds.)]. Cambridge University Press. In Press https://www.ipcc.ch/report/ar6/wg1/#FullReport

IPCC (2021b). 'Code red' for human driven global heating, warns UN chief' UN News ', Climate and Environment, 9 August 2021. https://news.un.org/en/story/2021/08/1097362 Accessed April-May 2023.

Jagerskog, A. (2013, January). Transboundary water management-why it is important and why it needs to be developed. UNESCO. Retrieved 2021, from https://www.researchgate.net/publication/262916482_Transboundary_Water_Management_why_it_is_important_and_why_it_needs_to_be_developed_in_Free_Flow_Reaching_Water_Security_Through_Cooperation_UNESCO_Publishing_and_Tudor_Rose_ Paris

Scown M., and Dunn F. E (2023). River deltas-are-threatened-by more than climate change leaving hundreds of-millions of people at-risk, The Conversation 12th Dec, 2023 Availabe at:https://theconversation.com/river-deltas-are-threatened-by-more-than-climate-changeleaving-hundreds-of-millions-of-people-at-risk218304?utm_source=whatsapp&utm_medium=bylinewhatsappbutton

Contents

About the Editors

Janki Andharia is Senior Professor with the Jamsetji Tata School of Disaster Studies, Tata Institute of Social Sciences (TISS), Mumbai. Her doctoral degree is from the School of Environmental Sciences, University of East Anglia, Norwich, UK. With over 40 years of experience in the field of community organisation and social development, Prof. Andharia's work focuses on disaster studies, gender issues, environmental concerns, democratic governance, and on addressing diverse forms of marginalisation and vulnerability. She has been involved in national and state-level policy-making work and has had a long association with grassroots organisations. She served as the Head, Department of Urban and Rural Community Development at TISS (1994–2006) and having led TISS teams to respond to several disasters, in the year 2006, she was entrusted the responsibility of creating and establishing a pioneering academic entity on Disaster Management at TISS. She laid the foundation for a new and unique multidisciplinary Masters' programme in Disaster Management at TISS which was the first of its kind in the Asian region and in India, and the third, globally. Through her leadership, this Centre has now evolved to become the Jamsetji Tata School of Disaster Studies (JTSDS).

She was awarded the Association of Commonwealth Universities Scholarship to pursue her Ph.D, invited to the Rockefeller Foundation's Residency programme at the Bellagio Centre in Italy, and was the recipient of the "Best Professor in Disaster Management award", instituted by the National Education Leadership Awards, Lokmat, Mumbai. This was in recognition of the critical leadership she provided to the establishment of a multidisciplinary Centre for Disaster Management.

Over the years Professor Andharia has conducted research, trainings and consultancy assignments for governments, industry, and NGOs and international agencies. She has travelled abroad to several countries to deliver special lectures, establish institutional linkages with Universities and also on research collaborations. She has been appointed as Honorary Professor at the University of Manchester's Humanitarian and Conflict Response Institute until 2026. She has published extensively and is the series editor for Springer's book series on *Disaster Studies and Management*. She also serves on editorial boards of several International journals.

Pankaj Anand is currently the Country Director of Oxfam in Timor-Leste. Prior to this, Pankaj had served Oxfam India for over seven years first as Director of Programme and Advocacy and later as Interim Chief Executive Officer (CEO). In his current role as Country Director, he heads Oxfam's work in Timor-Leste on justice issues relating to Climate, Gender and Economy. Overall, he has 25 years of rich experience cutting across humanitarian and development work.

In addition to experiences in project planning and management, he is a strong hand in advocacy, knowledge management, development communication and resource mobilisation. His thematic experiences cut across WaSH, Disaster Risk Resilience, livelihood, public health, gender, SRHR and local self-governance, among others. His work across themes have been guided by the basic tenets of human rights—civil, political, social and economic—manifesting in equality, dignity, justice and non-discrimination for all.

He is a strong hand in organisational development and strategic planning. He has devised long-term programmatic strategy including programmes adaptable to emergencies like Covid-19 programme. In the area of Project Management, he has long years of experience of multi-country projects in Asia cutting across Water Resources, Gender, Fair Finance, and Responsible Business.

Pankaj believes that transformational changes can happen only through people-centred approaches where communities are not treated as passive recipients but as equal partners in making the change possible. He also believes that sustainable changes do happen only when collaborative relationships are built and consolidated over time with government, citizen collectives, CSOs, coalitions/alliances, businesses, academia, media, and development partners. It is Pankaj's strong belief that humanitarian or development frameworks and actions must address the central question of inequalities and discrimination based on caste, gender, ethnicity, and religion.

Pankaj has published many articles and opinion pieces on topical issues in the area of development and humanitarian work in India. He has been a member of the Executive Committee of Sphere India and has played a key role in strengthening the humanitarian coalition and support to Inter-Agency Groups (IAGs). He has also been a Governing Body member at Gargi College and Swami Shraddhanand College (both University of Delhi).

Chapter 1
Building Bridges over Troubled Waters: Towards Risk-Informed Transboundary Water Governance

Janki Andharia and Lavanya Shanbhogue Arvind

1.1 Introduction: Transboundary Governance of Water and Disasters

Transboundary rivers basins traverse across and connect multiple nation-states inextricably creating pathways for shared ecological futures that demand cooperation. These complex riverine systems aid agriculture, livelihoods, energy production and are the primary sources of freshwater for riparian communities on all sides of the human-made borders. Going beyond economically deterministic imaginations of benefits from transboundary rivers, it is important to recognize that these river basins house rich biodiversity that provide significant ecosystem services that support the socio-economic development and well-being of humanity.

The political calculations of capitalism combined with the colonial and post-colonial legacy that privileges structural measures or "engineering fixes" on rivers for irrigation and creation of hydro-powers, need attention. The governance architecture of the various nation-states that share common hydrological resources such as aquifers create policies and laws that serve the interests of capitalists but remain apathetic to the dependence of the poor and marginalized on these water bodies for livelihoods and food security. On account of this, regulatory systems are top-down, technocratic and contested spaces where communities struggle for recognition of their rights. Such contestations by communities coupled with institutional ennui create great complexities that can sometimes turn into geopolitical risks.

Transboundary water governance therefore extends beyond traditional methodologies of state-initiated diplomacy efforts and includes a range of both formal and informal network of non-state actors including civil society organizations, rural action groups, academic, and research institutes, developmental agencies, and other community-led advocacy groups that work in the areas of river basin conservation

J. Andharia (✉) · L. S. Arvind
Jamsetji Tata School of Disaster Studies, Tata Institute of Social Sciences, Mumbai, India

J. Andharia and P. Anand (eds.), *Shifting Lands, Flowing Waters*, Disaster Studies and Management, https://doi.org/10.1007/978-981-96-7691-0_1

and rights of riparian communities. This evolving and dynamic arena requires policy measures to be cognizant of interests of diverse groups. A thorough understanding of these dynamics is necessary to ensure equitable and effective water governance in a complex, multi-actor environment.

Transboundary water basins are fraught with a range of issues. Riparian communities of different nations sharing these "troubled waters" experience devastating consequences of diplomatic deficits and extreme weather events. More than ever before, it is necessary to strengthen governance of transboundary water basins by building collaborative and cooperative bridges between and among water actors, all stakeholders, communities, local, national, and international governments. Such proactive diplomatic efforts to build bridges over troubled waters will pave the way for risk-informed transboundary water governance.

A quick look at the sources of water on the planet (see Fig. 1.1) will help contextualize the arguments made not only in this paper but also by the various authors of this volume. Water constitutes 71 per cent of the planet's surface; 96.5 per cent of which is found in the oceans. While freshwater constitutes only about 2.5 per cent of the available water resources on Earth, it may be argued that there is sufficient fresh water on our planet to meet basic human needs throughout the world. However, this water is unevenly distributed due to geophysical reasons as well as socio-economic ones. Several factors are combining to aggravate the freshwater crisis across the world in contemporary times.

As neoliberalism takes newer forms, capital seeks new terrains for profitable investment. As Boelens et al. (2022) argue, "River systems worldwide are subjected to multiple forms of domestication, enclosure, erasure, and pollution on an unprecedented planetary scale;" (p.2). How then do power asymmetries influence inequitable outcomes among riparian communities, the ones that are most vulnerable and most dependent on water for survival? This raises the question of rights and water justice. Further, unbridled pursuit of environmentally insensitive, water-wasteful mining and agricultural practices, exploitative industrial practices have brought us closer to the sixth mass extinction (Andharia 2020). These, combined with the newer aspects of global climate change, has defined the "water challenge" of the twenty-first century. We are also clear that developments in water sciences and engineering alone have not delivered adequately to engage this challenge, and there is a strong need to consider the social dynamics, cultural, and heritage issues towards more holistic responses and solutions, those which combine local knowledge and scientific advancements in hydrology and biodiversity and also address diverse transboundary challenges.

Water risks are diverse and varied and therefore water responsibilities must be shared through prolonged and sustained cooperation between communities, regions and nations that share common water bodies and sources. It is in this context that community-led cross-border sharing of water assumes significance. A nation's sovereignty is intrinsically embedded in questions of resources shared across the border. This brings into question water rights of communities settled along national borders. What should be the role of the international community in responding to such rights? How do changing climate and disasters affect water bodies and sources

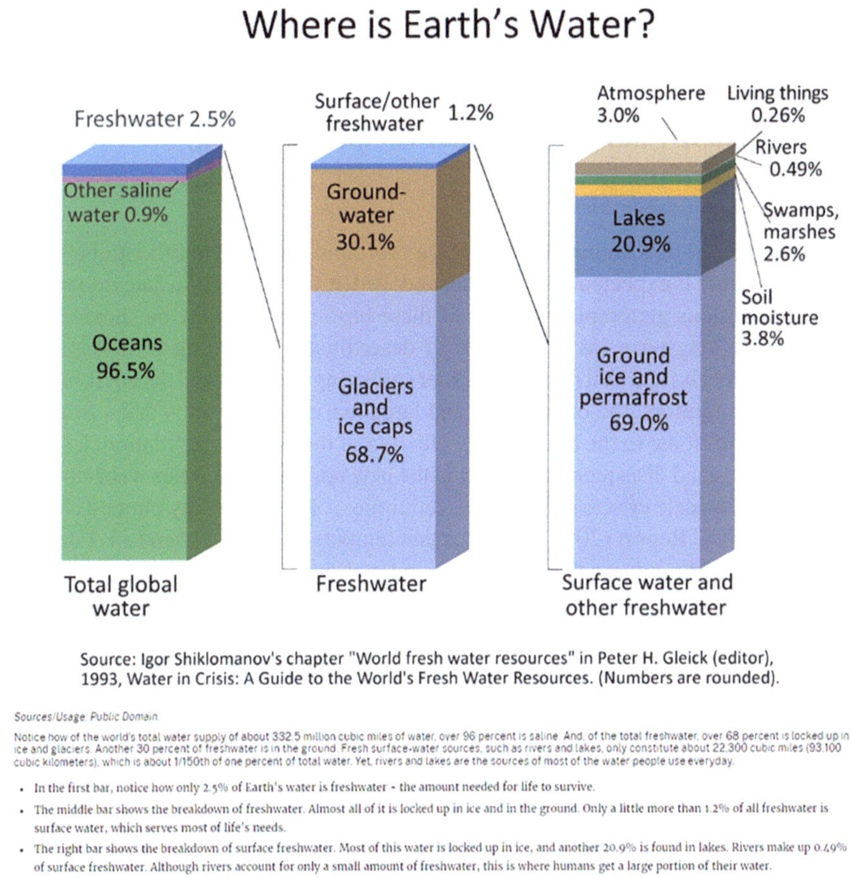

Fig. 1.1 Where is Earth's Water *Source* United States Geological Survey https://www.usgs.gov/special-topics/water-science-school/science/where-earths-water; UNECE https://unece.org/fileadmin/DAM/env/documents/2017/WAT/05May_16-18_Workshop_Kisumu/4.1_Fauconnier___Demilecamps___Azza_Transboundary_Water_Governance.pdf

and river basins? What kind of governance and planning are needed? The paper will discuss aspects.

The chapter further interrogates the neo-liberal frame of free markets where water is treated as a commodity, a resource, and not valued as source of life itself. Perspectives that help create pathways to social and environmental justice are discussed. It argues that transboundary water governance, must integrate an iterative risk-informed approach that identifies and mitigates water risks at all stages of policy planning and implementation while being cognizant of the roles of water users and water actors in the region. An analysis of risk framed within processes of transboundary water governance will strengthen community resilience and positively contribute to building adaptive capacities of the community in a changing world. It will also serve to alleviate the woes of the water poor.

1.2 "The Canvas": Transboundary River Basins, Disasters, and Risk-Informed Planning

1.2.1 Climate Change, Water Crisis, Species Facing Extinction:

It is well- known that climate change has a profound influence on the global water cycle. The future is expected to be warmer and wetter, which will translate into natural hazards in various geographies. Many of these hazard events may be short term or protracted or long term (as in the case of desertification, melting of icecaps with cascading impacts) and will result in disasters affecting societies and the environment. Rivers, lakes aquifers, the open ocean and large marine ecosystems that sustain the biosphere, contribute to the health and wellbeing of the world's population. Extreme weather events and disasters are becoming the new normal as societies wake up to the reality of large-scale crises, risks, and uncertainties, often caused by climate change. The recent IPCC Report (2023) indicates that temperatures have risen to 1.1 degrees Celsius above pre-industrial levels. Water related disasters are already on the rise as they increase significantly with higher frequency of intense storms and floods, rainfall patterns and extremes in discharge of water in rivers, prolonged droughts, variability and shifts in precipitation patterns. These water-related disasters account for 90 per cent of all natural disasters. Various climate change models expect these to rise which will challenge the capacity of many nation-states to meet their domestic water demand (Intergovernmental Panel on Climate Change 2007).

Around one million animal and plant species are facing extinction. Freshwater species have suffered the greatest decline, falling by 84 per cent since 1970. According to the 2020 UN Water Development Report, around four billion people currently experience severe physical water scarcity for at least one month per year, a situation that has been exacerbated by the climate crisis. As the planet warms, water cycles are impacted which in turn affects food security, energy production, economic development and efforts at poverty reduction.

1.2.2 Transboundary River Basins

Many of these water systems are shared by two or more nations. The situation gets more complex because climate change, disasters, and water flows do not respect national boundaries and adaptation in water management needs to be seen as a transboundary water governance issue (Timmerman et al. 2017). The transboundary waters, which stretch over 71% of the planet's surface, in addition to the transboundary subsurface aquifers, and the water systems entirely within the boundaries of the individual countries, comprise humanity's water heritage.

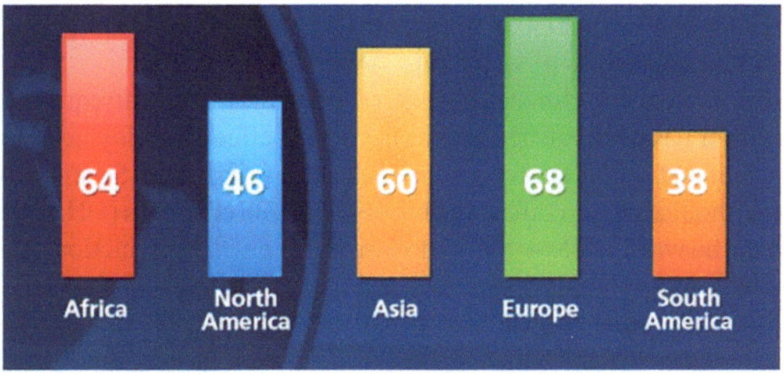

- 276 transboundary river basins (TRB) and 200 transboundary aquifers
- 148 countries include territory within one or more TRB
- 39 countries with more than 90% of their territory within 1 or more TRBs
- 21 countries entirely within one or more TRBs

Fig. 1.2 Geographical spread of transboundary river basins *Source* Fauconnier, et al. (2017).

Approximately 60 per cent of the world's fresh water supply is located in 276 transboundary rivers, such as the Nile, Rhine, Danube, Indus, Columbia, and Mekong basins (Fig. 1.2). These transboundary rivers cover 45 per cent of the earth's land surface and provide a home for 40 per cent of the world's population (Wolf et al. 1999). About 90 per cent of rivers and lakes in Africa are transboundary, while in the Middle East transboundary basins provide over 60 per cent of the region's domestic water supply (United Nations World Water Development Report 2015).

River basins across international borders are home to over 70 per cent of the world's population and supply water for roughly 60 per cent of global food production. (IGRAC and UNESCO-IHP 2015). Clearly, the world's transboundary river basins link countries in a common future; and their significance for sustaining life cannot be over-emphasized.

1.2.3 Uneven Distribution of Fresh Water and Scarcity

It may be argued that there is sufficient fresh water on our planet to meet basic human needs throughout the world, however, this water is unevenly distributed. For instance, regions containing large populations, such as the Middle East, North Africa, western portions of the United States, and northern portions of China all confront extreme shortages of fresh water. Other nations across the world, such as Pakistan, Afghanistan, and India experience what is referred to as economic water scarcity. This arises due to insufficient investment in hydrological infrastructure to enable the capture of water in the monsoon period for use during the remaining dry period. As a result, across the globe, 1.6 billion people live under conditions of economic water scarcity (FAO 2007). By 2030, it is expected that half of the world's population will be living under conditions of water stress (World Bank 2016).

As mentioned earlier, climate change will affect the availability, quality, and quantity of water for basic human needs, threatening the effective enjoyment of the human rights to water and sanitation for potentially billions of people. Seventy-five per cent of people without access to safe water reside in sixteen countries (UNFPA 2004). The alteration of the water cycle will also pose risks for energy production, food security, human health, economic development, and poverty reduction, thus seriously jeopardizing the achievement of the Sustainable Development Goals (UN Water 2020). Research shows that transboundary river basins with high economic dependence on water resources, low levels of societal well-being and high exposure to floods and droughts have the highest climate-related risks.

In summary, the exhaustion and degradation of water supplies create tremendous risks to communities dependent on them. Repercussions sometimes also result in social strife and conflict. This broadens the canvas of questions around transboundary water governance. What is the diverse nature of risk? How is it perceived, understood and analyzed in different regions? Are risks acknowledged differently at local, regional sub-national and international levels? These are important questions and issues that several authors try to explore in their respective contributions in this book.

1.2.4 Water Scarcity and Conflict

Research shows that environmental, human, and agricultural water stresses often occur in the same transboundary river basins, resulting in competition for water between sectors and between countries (UNEP-DHI and UNEP 2016). In the absence of people-to-people institutional ties and arrangements, Geo-political tensions often heighten pressures and regional instability, and require years or, more commonly, decades, to resolve. For instance, the Indus water treaty between India and Pakistan took almost 10 years to negotiate, the bilateral Ganges water treaty between India and Bangladesh took over 30 years to negotiate, while the Jordan treaty between Israel and Jordan took over 40 years to negotiate. Even the inter-state waters within the same nation can be a source for conflict rather than cooperation. The dispute between two south Indian states with respect to the Cauvery water sharing has been a prolonged battle. The tussle between the downstream Tamil Nadu that uses the Cauvery waters for irrigation and the upstream Karnataka that seeks to increase irrigated agricultural land has been a source of conflict in the region. Conflicts also arise if water quality is polluted and degraded so as to be detrimental to the health of riparian lives (Priscoli and Wolf 2009).

Over the years, more than 450 transboundary water treaties have been signed. These instruments offer an important means to avoid, and where necessary resolve disputes between states. However, over 60% of TB river basins worldwide still lack a framework for any form of cooperative management (International Union for Conservation of Nature 2023).

Thus, data and evidences from the past reveal that acute and persistent problems persist hindering effective and just management and governance of transboundary waters giving rise to hydro-political tensions. Given the scope of the problems and the resources available to address them, avoiding water conflict is vital as they can be expensive and destructive. To ensure long-term sustainable and equitable cooperation and peace, there is a need to extend water governance beyond the state-actors to include non-state actors who play an important role as well.

In order to establish regional cooperation, bilateral modes of collaboration that India prefers is only part of the solution and not a substitute. The questions on collaboration at what level, between whom, and for what purpose remain unclear (Iyer 2007) and need to addressed in situated contexts.

1.2.5 Transboundary River Basins and Disasters

Further, transboundary river basins pose significant disaster risks for riparian communities on different sides of the border. For instance, the Koshi River Basin (KRB) is a transboundary basin shared by three countries—China, India, and Nepal. The basin is (in)famous for its recurring history of flooding. Between 1954 to 2015, Nepal experienced almost 41 flooding events, while studies by the Bihar Disaster Management Authority revealed that almost 9500 lives have been lost between 1979 and 2017.

Apart from mortality losses, there have been huge socio-economic impacts such as loss of livelihoods, mass displacement, housing and food insecurity, erosion of coping capacities, among others factors that exacerbate vulnerabilities of communities. Climate change in the KRB reveals a potential 4 °C increase in temperatures with expected increase in monsoon precipitation, changes in evapotranspiration that can alter hazards both upstream and downstream areas (Shreshta and Gupta 2019).

Similarly, flooding in the Gandak River Basin, another Indo-Nepal transboundary basin too has posed significant challenges to both nations. Cloud bursts, precipitation, glacial melts and landslides are common occurrences in the wake of climate variability and changing spatiality of these terrains. While, engineering fixes such as dams and embankments are being attempted, the persistence of calamity in these regions has produced distinct and unmistakable problems that demands our critical and urgent intervention. In other parts of Asia, such as the Mekong River Basin, recurrent floods have devastated the lower Mekong countries—China, Laos, Vietnam, Cambodia, and Thailand, among others.

Transboundary disaster events are not unique to Asia. Both disasters and conflicts are common across several shared water geographies. The Nile Basin sees routine friction between the eleven riparian countries: Egypt, Rwanda, Tanzania, Uganda, Burundi, the Democratic Republic of Congo, Eritrea, Ethiopia, Sudan, and South Sudan. Damage as a result of flooding is estimated at over US$ 25 million in the rural settlements alone. However, formal cooperation in the region through the Nile Basin Initiative has enabled transboundary flood preparedness and early warning alerts for over two million people in the region (World Bank 2015).

There exists routine water distress along the Euphrates-Tigris river basin shared by Iraq, Syria and Turkey. Several irregularities along the diverse irrigational systems across the basin have proved costly for the communities dependent on the waters as recurrent floods devastate the regions. These are only a few instances of the universal character of the disaster risks in shared water geographies.

Various hazards, such as the rise in relative sea levels (SLR) and human-induced changes, pose significant threats to tropical delta areas. This makes these regions more susceptible to severe flooding, cyclones, storm surges, intrusion of salinity, and other dangers, which could potentially escalate in both intensity and frequency due to climate variations (Rahman, et al. 2021). Deltas are home to over half a billion people —with over 250 million people living in the three deltas of the Ganges-Brahmaputra Meghna, Mekong, and Nile rivers alone (Foufoula-Georgiou et al. 2013).

Over the last decades, flooding risks in transboundary basins and deltas have been on the rise worldwide. Fatalities caused by excessive rainfall are amplified by overflows of lakes, rivers, and dams leading to devastating losses of life, livestock, crops, livelihood, property, critical infrastructure, wildlife, and other personal losses for people all around the basin. Years of development gains are often wiped out by a single instance of inundation.

In addition to hazards and disasters, in many transboundary river basins the risk of pollution is high and projected to increase. Relative risk for either nutrient pollution or wastewater pollution is high or very high in 218 out of 286 basins (UNEP-DHI and UNEP 2016).

1.2.6 Transboundary River Basins and Risk Informed Water Governance and Planning

As stated earlier, although two-thirds of our planet is covered by water, but only 2.5 per cent of this water is fresh. Nations across the world sharing common water sources have been struggling to maintain cooperation over shared commons. This is because water management regimes vary greatly across countries, making transboundary water management more complex and at times very challenging. The growing threats of climate induced disasters is bound to have unprecedented and cumulative impact across the transboundary regions which in turn may aggravate fragilities in peripheral and vulnerable geographies. The interconnectedness among countries due to social economic and cultural factors create a cascading effect in the aftermath of a natural hazard or any emergency across the region. Therefore, a collective approach and regional solidarity are needed to maintain access to freshwater supplies. These then are broader issues of water governance which include infrastructure choices and management, decisions on land-use, reginal, and spatial planning. Water governance also needs to be risk informed, i.e. cognizant of extant risks as also future of future dangers in view of the vagaries of the changing climate. Creating a robust data base on basin wide impacts would help feed into water management policies (Timmerman et al. 2017).

Several factors are combining to aggravate the fresh water crisis across the world. Some argue that water has become one of the main ways we experience climate change and opportunities that improved water management offers in terms of adaptation and mitigation need to be explored systematically (Zawahri 2018).

Climate change is also expected to influence the management of transboundary water sources, both in basins governed by treaties and in those lacking such contracts. Climatic variability is affecting the timing, quantity, and quality of water flowing in the basin, which can challenge a state's capacity to comply with treaty commitments. There is also potential tension between treaties, which typically fix rights and responsibilities, and climate change, which seems to demand flexibility and resilience.

Water from great river systems such as the Indus, Mekong, Ganga, and the Brahmaputra are shared by different nations and have also been the source of tension and disagreement at different levels, affecting the vulnerable communities the most. There is considerable evidence of riparian communities being caught in the cycles of poverty induced by risk and vulnerabilities. Risks are projected to increase in the next 15–30 years, particularly for four hotspot regions: the Middle East, Central Asia, the Ganges-Brahmaputra-Meghna basin, and the Orange and Limpopo basins in Southern Africa. Action should be taken now to reduce future costs and impacts (UNEP-DHI and UNEP 2016). Such action would include improving the governance capacity in basins with high climate-related risks.

Risk-informed planning and programming (Andharia 2021), a relatively newer and emerging body of thought, calls for the integration of risk consciousness into developmental and policy planning of river basins. It should aim to fortify resilience against shocks and stresses by recognizing and addressing the root causes and drivers of risk and uncertainties (Andharia 2020). These include pre-existing vulnerabilities, lack of capacity, and exposure to various hazards that can amplify poverty. It requires a strong risk analysis of the multiple hazards faced by households and communities in the context of shared water sources. Perpetuating unmitigated risks threaten equitable preservation of water bodies.

In the transboundary context, the water governance framework therefore must integrate an iterative risk-informed approach that identifies and mitigates water risks at all stages of policy planning and implementation while being cognizant of the roles of water users and water actors in the region. In other words, an analysis of risk that informs transboundary water governance will strengthen community resilience and positively contribute to building adaptive capacities of the community in a changing world. Further, it will serve as a tool to alleviate the woes of the water poor.

Risk-Informed Water Governance as a concept not only provides national government the scope to deal with vulnerabilities in a cooperative manner but also boosts economic opportunities which otherwise remain relegated and insignificant in national policies. A comprehensive transboundary river governance can have components of sustainability only when policies and programmes are risk informed and also viewed through the kaleidoscope of cooperation.

1.2.7 Transboundary Water Governance for Equity and Regional Cooperation

Transboundary water sources are of two types: ground water reserves and surface water sources. Since water is continuously in motion: issues of control, jurisdiction and sovereignty are much more complicated as compared to static land resources and this task is even more difficult when water sources are shared by a number of political entities (Shamir et al. 2001). The Register of International Rivers listed 214 international transboundary river basins in the world in 1978 (UN 1978). In a later inventory of transboundary river basins, 261 international rivers, covering 45.3% of the land surface of the earth were identified and listed (Wolf et al. 1999) while others extended the number of transboundary river basins to 280 (Shamir et al. 2001). The need for regional cooperation has been unlined for decades.

According to the UN Water, the interagency mechanism that coordinates the work of the United Nations surrounding sanitation and water, there are a total of 276 transboundary river basins and 200 transboundary aquifers. More than 148 countries include territory within one or more transboundary basin. About 39 countries include more than 90% of their territory within 1 or more transboundary basin. Besides, 21 countries are contained within one or more transboundary basin (Fauconnier et al. 2017).

A Report by UNEP-DHI and UNEP (2016) states, "The construction of dams and water diversions is in progress or planned in many transboundary river basins, sometimes without adequate international water cooperation instruments. While many transboundary agreements exist, more effort is needed to update them to reflect modern principles of transboundary water management, such as the obligation not to cause significant harm and the principles of cooperation and information exchange" (p.1). It argues that improved guidelines for siting new dams, and their designs. It requires cooperation and commitment in order to ensure that they serve multiple purposes, and optimize the operation of dams to maximize human benefits and minimize negative ecosystem impacts and sediment trapping. This is particularly relevant in transboundary river basins, where dams are often in upstream countries.

Further, at regional and local levels, water has historically served as a traditional medium of exclusion that leads to further disenfranchisement of poor and marginal populations including riparian communities. Governance of water thus becomes a critical tool to not only regulate water management within and across countries but should also operate to ensure equity in the society. This calls for sensitivity to issues of equity and water justice that must be enshrined within policies and governance frameworks.

Water management planning and implementation should be based on clear legal and regulatory framework offering guidance and pointers along with mechanisms that are flexible, coherent and in consonance with local priorities within a river basin approach. The mechanisms should include sharing of costs and benefits which in turn would enhance mutual trust and confidence and foster collaborative learning. For example, sharing of hydrometeorological data, climate change predictions, alternate pathways towards sustainable management of waters, or even early warning systems can support policy convergence across countries and help enhance transboundary equity and justice. A political economy approach ensures a sharper focus on the struggles of river-dependent communities and points to the need to go beyond the international relations framework and acknowledge the heterogeneity of actors at multiple levels. This is discussed in the next section.

1.3 The "Stage": Contextualizing Transboundary Issues through a Political Economy Approach

Transboundary river basis are complex geographies around which the lives of riparian communities are organized. Nation-states on either side of these riverine borders regulate socio-economic arrangements for river-dependent communities. Households that are directly dependent on river waters utilize a range of subsistence and economic activities including cooking, drinking, livestock rearing, fishing, irrigating farmlands, household chores, tourism and transportation. Further, rivers have deep ritualistic and ceremonial meaning and may embody divinity. Traversing an array of cultural, economic, eco-social and political dimensions, river regulation, and governance surrounding water development must strike a balance between social development, ecological health, and geo-political equilibrium in the region. Thus, a political economy approach that recognizes the interconnectedness of systems, people, habitats, culture, rights, and law is required.

An intersectional and multi-pronged inquiry that is cognizant of compound socio-political identities burdened by caste, class, gender, disability, ethnicity, indigeneity, race, and language among others in the context of geo-polity and international relations must inform governance in transboundary river basins.

1.3.1 Heterogeneity of Actors and Inadequacy of International Perspectives

The perspective offered by international relations however remains tardy and inadequate, for geo-politics is a complex endeavour. Frameworks are either absent or ineffective and do not consider the heterogeneity of actors. The "theatre" of transboundary issues is not acquainted with the on-ground realities of people's lived experiences. Even as one attempts to respond to older crises, newer debates emerge. Most of the key so called "global policies" on transboundary governance are limited in their scope and geographical acceptance. With many scholars and even the International Court of Justice arguing that the UN Water Treaty now enjoys the status of international custom, see limited application in the South Asian context.

Possibly due to these pressing challenges, today, numerous transboundary river basins worldwide lack formal cooperative arrangements. Moreover, the existing transboundary water treaties and institutions are often weak in terms of their mandate, design, sources, and enforcement mechanisms. Only around one-third of multilateral basins have treaties signed by at least three states, only 11 basins have treaties that include all riparians, and only about a quarter of all treaties cover the entire basin to which they apply. The situation is even worse for transboundary aquifers where very few agreements exist: only six aquifer agreements have been adopted at the international level and two agreements regarding the sharing of transboundary ground waters between sub-national/local authorities. There exist more than 450 transboundary water treaties that have been signed since 1820. These instruments may transform into catalysts for regional pacifism and which provide necessary resolution disputes between states. However, around 60% of transboundary river basis in the world today still lack any form of cooperative arrangement (UN Waters 2016).

While the state-actors play a very crucial role in bringing about stability and enhancing security among nations that share waters, through treaties, or agreements, ensuring long-term cooperation and peace can be very challenging. State-driven interactions require strong political will and commitments from the riparian nations—which can be hard to generate and sustain. Analysis of previously signed water treaties are testament to the fact there exists significant time lag between when nations first start to impinge on each other's water planning and when agreements are finally, arduously, reached. This time lag often heightens tensions and regional instability, and requiring years or, more commonly, decades, to resolve.

Most of the treaties provide for a Commission to ensure better management of the issues arising under the treaty. While this could be an excellent forum for a participatory engagement to bring in multiple stakeholders to the forefront, these Commissions continue to be controlled by state-based actors. Some commissions allow for scientific advisors to be a part of its fold. However, in most of these treaties the Commissions were either never constituted or have been utterly unsuccessful. The Indus Water Commission was functional and has been instrumental in addressing some disputes diplomatically but it is often influenced by the geo-political tensions between the countries involved.

Also, when water becomes a political issue, political tensions between countries can impede equitable water-sharing agreements. There always exists a possibility for this state-driven cooperation to be skewed, benefitting only the powerful riparian state. In fact, in certain cases, cooperation through asymmetrical treaties (like for the Nile, Jordan, and Ganges) has become a source of conflict rather than cooperation. Most common reasons for conflicts include divergent priorities or conflicting policies or incomplete or biased understanding of the benefits that could be derived from cooperation.

The treaties especially in the South Asian context have been formulated after multiple political backlashes between the parties. Every treaty has become a matter of sensitivity owing to flaring political tempers. Thus, even a basic process of updating, modifying, or enforcing key aspects of the treaty has the potential to escalate into a full-blown international issue or a tricky domestic issue. Most of the treaties are in dire need of updation as they focus only on the percentage of water sharing during the monsoons or the dry period based on the amount calculated decades ago. The water-use patterns and the dependency of population on the sources has changed drastically. No efforts have been made to ensure that the treaties reflect issues such as pollution, environmental degradation, development efforts, or recurrent hazard assessments as topographies change or even equity and social justice.

Earle and Neal (2017) map the heterogeneity of actors around water governance and argue for inclusivity, focusing on water resource community. Effective transboundary water governance requires cooperation and commitment from all parties involved—governments, corporations, researchers, civil society organizations and most of all, the communities—in order to ensure that transboundary governance of water resources is and remains adaptive and continues to uphold community rights and social justice.

1.3.2 Need for Enhanced Local Participation and co-Operation

Asthana and Shukla (2014) have argued that in the South Asian context, water resource development has not followed a path of sustainable development. South Asia needs to move forward with a cooperative and participatory approach on river basins and water sharing. There needs to be a regional awareness that rivers can be better harnessed through collective efforts and recognition that cooperation is essential to alleviate threats to water security. Their suggested approach for fostering cooperation is that of inclusion and dialogue that moves beyond a top-down approach. They advocate dialogue between state and non-state actors to improve transboundary water interactions (Asthana and Shukla 2014). Effective and sustainable transboundary water management is an imperative to not only support the existing population on a long-term basis, but also for the significantly larger population that will have to be fed in the remaining decades of this century. The question of sustainability is discussed in greater detail in 4.

Specific to the river Ganges that traverses India, Nepal, Tibet, and Bangladesh and is one of the most populous and ecologically vulnerable basins of South East Asia, Sharma (2014) reasons that government and non-government agencies, the private sector, civil society groups, and the public at large need to be jointly engaged in the effort to improve water governance in the basin. Collaborative pathways that include all significant actors must move beyond business-as-usual approaches

(Sharma 2014). In other words transboundary water governance therefore must transcend water diplomacy involving governments as it is known to be a vexed process mired in geo-politics. It must extend to include other non-state actors and large corporations that are often legal entities in their own right.

The next two sections examine theoretical responses and conceptual debates on transboundary water governance, critiquing the neo-liberal frame of approaching the subject of water.

1.4 "The Contestations": Neo-Liberal Frames and Theoretical Responses

The political calculations of capitalism combined with the colonial and pre-colonial legacy that privileges structural measures or techno-engineered approaches, have dominated water management across the globe. Such approaches have neglected the inherent interdisciplinarity in global water management for decades and this very negligence has led to dire consequences, especially the riparian communities.

Colonial approaches, treat all ecological systems as "resources" within the borders of their colonies, which were to be appropriated, harnessed, making it an extractive process rather than deploy a conservatory and protective framework. The extractive relationship continues well into the post-colonial era with colonial powers replaced by corporations and their political lobbies. The unbridled pursuit of environmentally insensitive, water-wasteful mining and agricultural practices, exploitative industrial practices have brought us closer to the sixth mass extinction. For instance, construction of dams affects the livelihoods of the population living along the dammed river and downstream. The livelihoods of river-dependent population, particularly farmer and fisherfolk, are most affected. Dams are claimed as symbols of development for their multipurpose utility and contribution to the welfare of mankind. But compulsory displacement caused by dams has raised major issues of social justice and equity. The oustees not only lose their income and wealth but also social goods (liberty, opportunity, and the very basis of their self-respect) raising the question of equality and justice. Initiatives to rehabilitate them by the state proved to be a total failure because of poor planning and a callous attitude.

For instance, a recent arbitration between India and Pakistan on the construction of the Kishanganga Dam is very telling on the importance given to the exploitative approach of the governance mechanism with respect to transboundary river. While Indus Water Treaty formulated in 1960 provided a range of technical details on the kind of hydroelectric plants allowed on some parts of the river there is barely anything on environment protection in the treaty or in the communication of the Commission. The Arbitration Panel had to rely on contemporary international law on environment protection to identify that there is a need to protect the environment of the lower riparian state and therefore consider the environmental impact of the project. The emphasis was mostly on the loss of livelihood due to the reduction in the level of water.

Neo-liberalization of developmental planning in river basins has presented a problem of erosion of water rights of communities on both sides of the border. This lies on a shaky foundation of the histories of displacement, devastation, and disenfranchisement of river-dependent communities.

As economic liberalism and free-market capitalism join hands, capital seeks new terrains for profitable investment (Klein 2007). National calamities and emergencies are often exploited for profit and to push dubious policies, while citizens are too distracted (emotionally and physically) to engage and develop an adequate response. For example, India's rivers and riparian ecology have been badly affected by the alarming rate of unrestricted sand mining, and industrial pollution that impacts downstream communities that are often from a different nation. These, combined with the newer challenges of global climate change, has defined the "water challenge" of the twenty-first century. There is ample evidence ((Boelens et al. 2022) to suggest that developments in water sciences and engineering alone have not delivered adequately. There is a strong need to consider the social dynamics, culture, and heritage issues towards more holistic and solutions which combine local knowledge and scientific advancements in hydrology and biodiversity while also addressing diverse transboundary challenges.

In an increasing neo-liberal world, the interdependencies between water, energy, food, and the environment require considerable thought and strategic planning. The idea that livelihoods of the poor and their food security are reliant on ecosystem services is often missing in policies. Existing power hierarchies and systems of oppression on the basis of caste, class, ethnicity, religion, race that create marginalization and discrimination of specific groups of people affect access and bring about alienation. These include women as the primary users of water in many parts of the world, for domestic consumption, subsistence agriculture, and health as well as the youth of today that constitute the future generation that will need to solve issues related to water. Intergenerational equity or the concept of fairness and justice between generations have to be thought out in economic, psychological, political, and sociological contexts.

Justice outcomes including gender justice, climate, and environmental justice that address questions of access, opportunity, equity, participation, agency, and decision-making of historically disadvantaged communities must inform public policy and programmatic perspectives.

In doing so one must also reject hegemonic knowledge that claims universality while acknowledging that knowledge rooted in the local ways of being and knowing must be regarded as legitimate and valid.

1.5 Emerging Debates

1.5.1 State-Driven and People-Centric Approaches

Conventionally, transboundary interactions in the context of shared water sources (rivers, lakes etc) confined to state-actors and has inherently been a political process. Cooperation and conflicts have largely been state-driven. Formal diplomacy and government-to-government interactions dominate the transboundary water negotiations and decisions. Concepts such as hegemony, power, power asymmetries, and domination etc., are terms used to describe the engagement of state-actors, dominate the academic discourse on transboundary water governance. While state-actors play a very crucial role in bringing about stability and enhancing security among nations that share waters, through treaties or agreements, ensuring long-term cooperation, and peace can be very challenging. State-driven interactions require strong political will and commitments from the riparian nations—which can be hard to generate and sustain. Besides they could merely remain on paper, without the active participation and engagement of local communities and other non-state actors.

However, in the last decade or so, newer debates have emerged that demonstrate the need to confront the heterogeneity of issues and actors in transboundary river basis. These range from state officials, communities themselves, civil society organizations, and other decentralized bodies on both sides of the border. Communities often form people-to-people ties across the borders owing to cultural and linguistic similarities.

It is necessary to induct and include non-state actors to participate in water governance to increase pathways to cooperation. Benefits derived in such an approach are three pronged: a) this would help ensure greater efficiency and effectiveness of policies and project outcomes b) would help improve empowerment of individuals, particularly of the marginalized groups and, c) would lead to greater legitimacy of governance measures. Furthermore, scholars also argue that lack of public support can hinder state-led implementation of policies or water-related agreements and even lead to opposition and protests. They also argue that with participation comes increased feelings of ownership and responsibility, which would in turn increase the efficiency of certain water governance measures. Lately, increasing efforts at involving citizens and a broad range of actors from the private and non-profit sectors, more regularly and directly, in policy and decision-making processes can be observed in numerous parts of the world.

1.5.2 Limits of Structural Measures

Traditionally, science and technology has played the dominant role in devising solutions to overcome the daunting problems of management of transboundary water. GFDRR in 2013 presented some of the challenges or disadvantages of relying solely on structural measures and heavily engineered interventions such as dams, dykes, reservoirs etc. for water management. It stated that such heavily engineered measures generally require very high investments and cannot evolve with the ever-changing goals and needs. Structural measures also often have major side-effects that are typically overlooked. For instance, in the 1960s, India built a barrage at Farakka, diverting a portion of the Ganges flow away from its course into Bangladesh, in an effort to flush silt away from Calcutta's seaport, some 100 miles to the south. In Bangladesh, the reduced upstream flow resulted in a number of adverse effects: degraded surface and groundwater, impeded navigation, increased salinity, degraded fisheries, and endangered water supplies and public health. Migration from affected areas further compounded the problem. Ironically, many of those displaced in Bangladesh found refuge in India (United Nations Environment Programme 2005). This case demonstrates that science offers limited help in determining the best decision to make in the face of multiple conflicting and evolving goals held by multiple stakeholders. An integrated and balanced perspective is imperative to the success of sustainable, basin-wide water management strategies. Technology needs to join hands with the social sciences. Implementation of hydrogeological science-based protocols that incorporate gender, social, and governance issues need to be developed. Expertize from some of the social science disciplines such as anthropology, psychology, sociology, and geography that should be involved have not been systematically engaged or incorporated into interdisciplinary investigations. Efforts will need to be made to encourage contributions from these disciplines to ensure the much-needed holistic approach (Jury and Vaux Jr 2005).

1.5.3 Water Justice and "Riverhood"

The larger philosophical underpinnings guiding such articulations is that technological interventions cannot be made on a standalone basis, or worse still in a way that "tinkers around at the margins." Therefore, guiding philosophies must be rooted in socio-ecological justice, and particularly water justice. In order to do so, various factors that endanger rivers worldwide require interrogation. Mega-damming, pollution, depletion, and socially blind technological fixes disrupt hydro-social worlds and intensify water injustices. Globally, anthropogenic activity in riverine belts have brought about unparalleled destruction. The river therefore becomes a site of contestation.

Ontologies and epistemologies of water justice requires newer forms of criticality and interrogation. Remaking rivers through extensive anthropogenic intervention destroy and disrupt cultures that have emerged around water bodies. The daily sociality of riverine communities is thus deeply and adversely affected.

To fully understand the extent of the problem, there is a need to look at rivers as a distinct socio-natural entity. To this extent, "riverhood" is a theoretical and conceptual register to frame the ontological question surrounding riverine ecologies. In framing the river as an entity that has material, symbolic and social historiographies, newer ways of thinking about rivers will enable pathways to social and environmental justice. One of the primary arguments for conceptualizing riverine existence through the lens of riverhood is to promote water justice. Riverhood looks at rivers as subjects with rights, as eco-societies, as territories that require protection and as movements towards egalitarianism (Boelens et al. 2022).

In TRBs how might "riverhood" be applied and operationalized? When rivers are considered as common property, enclosures, dams, embankments, and other "fixes" by one country will impact other countries. These interventions require participation and consensus by all actors who have rights to access commons. Use of commons by one group diminishes availability to other, often more vulnerable groups. This ontological position provides impetus to grassroots movements' claims on commons and demands against state hydrocracies.

Secondly, thinking of the river as a subject in itself with rights provides both legislative and social protection to the river. In 2017, the Whanganui River of New Zealand was granted same legal rights as human beings. This verdict was an outcome of years of struggle by the Māori tribe of Whanganui who fought to get the river recognized as an ancestor. This status means that, in the eyes of the law, harming the river is akin to harming the tribe in itself (Roy 2017). In the Indian context too, the Ganga and the Yamuna were declared "living entities" by the Uttarakhand High Court in 2016 thereby granting them the legal status of a person. Although this ruling was subsequently overruled by the Supreme Court of India, the intention of the lower court was to protect the river as a subject in itself (Sen 2019).

Viewing rivers as eco-societies is a form of biocentric approach in which all forms of life are considered to have intrinsic value. However, modernist developmental approaches are "top-down" and stem from perspectives of ownership, management, and control and view rivers as resources. Therefore, a reversal of an ontology rooted in anthropocentricism to one rooted in bio/eco-centrism will benefit cooperative efforts in transboundary river basins.

1.5.4 Perspectives on Hydrodiplomacy

When viewed as territories, rivers are often placed within the bureaucratic lens of control and safeguarding of territorial sovereignty of States. Deemed as sacrosanct in the post-Westphalia world, territory is considered sovereign and exclusive to nation-states. In transboundary water issues, there are tensions between national sovereignty and TR water cooperation. According to Sindico (2021), a nation has two kinds of statuses: a) hydrological (upstream versus downstream) b) hydropower (economic and political dominance in the region) and based on these two statuses, nations can assert their sovereignty over the waters flowing through their territories. Upstream states often decide how much flow they require and regulate their wates accordingly while downstream states do not tolerate downward revisions to their water needs imposed by upstream states.

These different perspectives present valuable ontological positions to view hydro-diplomacy. While biocentric approaches may seem too puritan for a world at the brink of violating planetary boundaries (Rockström et al. 2009), the need for cooperation and democratic deliberations should not be lost in the brambles of overpoliticization.

1.6 Conclusion

The vexed picture of the future outlook of water and increasing water stress in Asia warrants a shift in the existing paradigm. We need to clearly embrace an effective multi-agency, people-centric approach that is cognizant of historical disadvantages of the people in the context of water.

Among these multi-faceted concerns, transboundary water governance assumes significance and includes the management of various risks and hazards. Water related vulnerabilities must be read along with the various underlying socio-political, economic, and administrative systems which influence the general well-being of the riparian communities across nations. Only then can transformation be achieved in a desired direction.

This important compilation of deliberations and thoughts on transboundary cooperation in river basins also calls for newer tools to enable such affiliations. Risk-informed planning and programming, a relatively newer and emerging body of thought, calls for the integration of risk consciousness into developmental and policy planning in river basins. Risk-informed planning and programming should aim to fortify resilience, against shocks and stresses by recognizing and addressing the root causes and drivers of risk (Andharia 2020). These include pre-existing vulnerabilities, lack of capacity, and exposure to various hazards that can amplify poverty. Risk Informed Planning requires a strong risk analysis of the multiple hazards faced by households and communities in the context of shared water sources. Perpetuating unmitigated risks threaten equitable preservation of water bodies. On the other hand, mechanisms for distribution can help create the context for good, robust governance. Risk-informed transboundary water governance requires governments, civil society organizations, and other non-state actors to be involved in the design and adjustment of programmes to ensure that they make a proactive commitment to reducing risk and enhance wellbeing of those most vulnerable. A participatory process that includes community engagement, and one that is built on the foundations of regional cooperation and solidarity will ensure a system of water governance that protects water rights and leaves no one behind.

References

Andharia J (2020) Thinking about disasters: a call for intersectionality and transdisciplinary in disaster studies. In Andharia J (ed) Disaster studies: exploring Intersectionalities in disaster discourse. Springer Nature, Singapore

Andharia J (2021) Disaster management: institutionalising risk-informed planning. In Ravi Agarwal, Omita Goyal (eds) The Crisis of Climate Change: Weather Report, Routledge, Taylor and Francis. https://www.routledge.com/The-Crisis-of-Climate-Change-Weather-Report/Agarwal-Goyal/p/book/9781032106953

Asthana V, Shukla A (2014) Water security in India; Hope, despair, and the challenges of human development. Bloomsbury (Academic), New York, USA

Boelens R, Escobar A, et al (2022) Riverhood: political ecologies of socio-nature communing and translocal struggles for water justice, *the journal of* peasant *studies*, 15 November. https://doi.org/10.1080/03066150.2022.2120810

Earle A, Neal M (2017) Inclusive Transboundary Water Governance. https://doi.org/10.1007/978-3-319-43350-9_8. https://www.researchgate.net/publication/309441042_Inclusive_Transboundary_Water_Governance

Fauconnier I, Demilecamps C, Azza N (2017) Session 4. Transboundary Water Governance (Workshop paper). International Union for Conservation of Nature (IUCN) & United Nations Economic Commission for Europe (UNECE). https://unece.org/fileadmin/DAM/env/documents/2017/WAT/05May_16-18_Workshop_Kisumu/4.1_Fauconnier___Demilecamps___Azza_Transboundary_Water_Governance.pdf

Food and Agriculture Organization (2007) Coping with water scarcity: world water day 2007. Food and agriculture organization of the united nations, on behalf of the 24 agencies and programme members of UN Water

Foufoula-Georgiou E, Overeem I, Saito Y, Dech S, Kuenzer C, Goodbred S, Harrison I, Anthony E, Brondízio E, Hutton J, Nicholls R, Matthews Z, Dearing J, Lazar A, Baschieri A, Newton A, Ramachandran R, Renaud F, Sebesvari Z, Twilley R (2013) A vision for a coordinated international effort on delta sustainability

Intergovernmental Panel on Climate Change (2007) Climate change 2007: synthesis report. Contribution of working groups I, II and III to the fourth assessment report of the intergovernmental panel on climate change. IPCC

IGRAC (International Groundwater Resources Assessment Centre) UNESCO-IHP (UNESCO International Hydrological Programme). 2015. Transboundary aquifers of the world [map]. Edition 2015. Delft, Netherlands: IGRAC, 2015

IPCC (2023) AR6 Synthesis Report Climate Change 2023. https://www.ipcc.ch/report/sixth-assessment-report-cycle/

International Union for Conservation of Nature. (2023). Transboundary water governance and diplomacy [Web page]. https://iucn.org/our-work/topic/transboundary-water-governance-and-diplomacy

Iyer R (2007) Towards water wisdom: limits, justice, harmony. Sage, New Delhi

Jury WA, Vaux H Jr (2005) The role of science in solving the world's emerging water problems. Proc Natl Acad Sci USA

Klein N (2007) The shock doctrine: the rise of disaster capitalism. Knopf Canada

Priscoli JD, Wolf A (2009) Managing and transforming water conflicts. Cambridge University Press, Cambridge

Rahman KF, Pal I, Szabo S, Pramanik M, Udmale P (2021) Transboundary water risk governance frameworks in deltaic socio-economic regions: a case study of river deltas in Bangladesh, India, and Vietnam, chapter 3 in disaster resilience and sustainability. Adaptation for Sustainable Development, Pages:49–72. https://www.sciencedirect.com/science/article/abs/pii/B9780323851954000111

Rockström J et al (2009) A safe operating space for humanity. Science 326(5954):472–475. https://doi.org/10.1126/science.1185786

Roy EA (2017, March 16) New Zealand river granted same legal rights as human being. The guardian. https://www.theguardian.com/world/2017/mar/16/new-zealand-river-granted-same-legal-rights-as-human-being

Sen S (2019, April 25) Of holy rivers and human rights: protecting the ganges by law. Yale University Press. https://yalebooks.yale.edu/2019/04/25/of-holy-rivers-and-human-rights-protecting-the-ganges-by-law/

Shamir U, Kliot N, Shlomi D (2001) Institutions for management of transboundary water resources: their nature, characteristics and shortcomings. Water Policy 3:229–255

Sharma, B. (2014, July 12). Cleaning the ganga, step by step. Retrieved from The Hindu: www.thehindu.com/opinion/op-ed/cleaning-the-ganga-step-bystep/article6105068.ece

Shreshta K, Gupta N (2019) A knowledge hub to address transboundary disaster risk and water security issues in the Koshi River basin. International Centre for Integrated Mountain Development

Sindico F (2021) National Sovereignty Versus Transboundary Water Cooperation: can you see international law reflected in the water? AJIL Unbound 115:178–182. https://doi.org/10.1017/aju.2021.24

Timmerman J et al (2017) Improving governance in transboundary cooperation in water and climate change adaptation. Water Policy 19:1014–1029

UN (1978) The register of international rivers department of technical cooperation for development. Natural Resources and Energy Division, UN

UN Waters (2016) High level panel on transboundary water governance. https://sustainabledevelopment.un.org/content/documents/hlpwater/04-TransbounWaterGovernance.pdf

UNEP-DHI and UNEP (2016) Transboundary River basins: status and trends, Summary for Policy Makers. United Nations Environment Programme (UNEP), Nairobi

UNFPA (2004) State of world population 2004: The cairo consensus at ten—Population, reproductive health and the global effort to end poverty. United Nations Population Fund.

United Nations (2015) World water development report. Water for a sustainable world. United Nations Education, Scientific, and Cultural Organization, France. http://unesdoc.unesco.org/images/0023/002318/231823E.pdf

United Nations 2020 World Water Development Report:Water and Climate change Available at: https://unesdoc.unesco.org/ark:/48223/pf0000372985.locale=en

United Nations Environment Programme (2005) Hydropolitical vulnerability and resilience in international river basins. UNEP-DEWA; Universities Partnership for Transboundary Waters. https://waterpartners.geo.orst.edu

Wolf AT, Natharius JA, Danielson JJ, Ward BS, Pender JK (1999) International river basins of the world. Int J Water Resour Dev 15(4):387–427

World Bank (2015) The Nile story: reducing flood devastation in the Nile Basin Retrieved from https://openknowledge.worldbank.org/bitstream/handle/10986/23591/Reducing0flood0on0in0the0Nile0Basin.pdf?sequence=1

World Bank (2016, June 3) A third of people in low-income countries struggle to access clean water (S. Dahan & H. Kashiwase, authors). https://blogs.worldbank.org/en/opendata/sdg-6-water-and-sanitation-essential-sustainabledevelopment

Zawahri N (2018) The Governance of Transboundary Rivers Across the World in Europenow. Available at: https://www.europenowjournal.org/2018/12/10/the-governance-of-transboundary-rivers-across-the-world/

Janki Andharia is Senior Professor with the Jamsetji Tata School of Disaster Studies, Tata Institute of Social Sciences (TISS), Mumbai. She has her doctoral degree from the School of Environmental Sciences, University of East Anglia, Norwich, UK. With over 40 years of experience in the field of community organization and social development, Prof. Andharia's work focuses on disaster studies, gender issues, environmental concerns, democratic governance, and on addressing diverse forms of marginalization and vulnerability. Prof. Andharia is one of the editors of this book. Currently she is working on a research project with the Ministry of External Affairs, Government of India and also engaged in a unique training programme with the Indian Army on Climate and Disaster Risk Assessment of Infrastructure Projects. For more details, see note on editors.

Lavanya Shanbhogue Arvind is an educator, researcher, and author, exploring the intersections of gender, disaster management, climate change, sustainability, and women's labour. Holding a PhD in Women's Studies from the Tata Institute of Social Sciences, her academic journey includes an institutional gold medal in Gender Studies and a rich tapestry of interdisciplinary expertise. She has led impactful projects such as "Infusing Gender Consciousness in Disaster Pedagogies," supported by the GRRIPP (Gender Responsive Resilient Intersectionality in Policy and Praxis) consortium, University College and London and the University of Dhaka. This initiative integrates gender, intersectionality, and social inclusion into disaster management frameworks, reflecting her commitment to socially responsive scholarship. Her doctoral work critically examines water rights and spatial justice among Adivasi women in Maharashtra, contributing to feminist geographies and decentralized governance discourses. With nearly 12 years of overall experience across risk management, gender studies, disaster management, teaching and curriculum development, Dr. Shanbhogue has designed courses on gender, vulnerability, and climate risk for adult learners. Her writings on disaster settings, gender-sensitive DRR, and sustainable development appear in publications like the Economic & Political Weekly, Springer, Routledge as well as the South Asia blog of the London School of Economics. Most recently, her research paper on gender, tribal women, and the environment will appear in the prestigious Oxford Research Encyclopaedia of Natural Hazard Science.A recognized voice in academia and creative spaces, she has received numerous accolades, including the Commonwealth Short Story Prize and invitations as a TEDx speaker and as a panellist at the Melbourne Writers Festival. She was an invited awardee of the Writers in Immersion and Cultural Exchange (Wrice) Fellowship by the Royal Melbourne Institute of Technology (RMIT), Melbourne and travelled to Indonesia and Australia to write and speak about her work. Her work in national conferences and collaborations underscore her dedication to inclusive and transformative education and policy.

Chapter 2
Disasters as Unfinished Business of Development: Why Pluralizing the Natural and Social Aspects of Transboundary Waters May Help

Dipak Gyawali

2.1 Duryog Nivaran's Rethinking Disasters

In the mid-1990s, academics and activists of South Asia got together to set up a network to champion an alternative perspective on disasters, which I had the honour of being the founding chair. Even the name selected for the network was unconventional in that it would not be known by an Anglo-Saxon acronym but words that reflected South Asian languages: it was called *Duryog Nivaran* (1995). Duryog (the opposite of Rajyog for ill effects of disaster) and Nivaran (to get rid of) had resonance in Sinhala, Nepali, Bengali, Gujarati, Punjabi, Sindhi and other South Asian tongues. The network put forth two fundamental propositions to promote this alternative perspective:

1. *Disasters are cyclical*; and what societies do in between disasters is as, if not more, important than the necessary post disaster activities in their aftermath such as rescue and relief. Floods and droughts have their own seasonal and long-term cycles. Earthquakes and volcanic eruptions, though near impossible to anticipate with any degree of meaningful certainty, do happen with cycles of 10/15/20 years, and so on. Other natural disasters such as glacial lake outburst floods (GLOFs) are one-time events and not cyclical at that place; and even less predictable are the much more frequent landslide-dammed lake outburst floods (known in Nepali as *bishyaris*—see Bandyopadhyay and Gyawali (1994)). These are essentially large-scale slope failures in the Himalaya adjacent to rivers that temporarily dam them. Unlike GLOFs that can be aerially identified and some measures taken to prevent their outburst (e.g. Tsho Rolpa on the Tama Kosi, see Gyawali and Dixit (1997)), these *bishyaris* can occur anywhere on a river bank. The mass wasting land slip blocks the river, temporarily dams it but the unstable rock and

D. Gyawali (✉)
Nepal Academy of Science and Technology, Kathmandu, Nepal

J. Andharia and P. Anand (eds.), *Shifting Lands, Flowing Waters*, Disaster Studies and Management, https://doi.org/10.1007/978-981-96-7691-0_2

mud dam eventually burst with river overtopping it, taking the entire mass of rock and debris with it, and causing much damage downstream. Between past disasters and new ones coming, it is important that we get to doing the necessary institutional tweaking that is necessary to better prepare society to face the other, similar big ones that will be coming in the future. This is about learning the right lessons from a disaster, and preparing for better resilience the next time. This was the first proposition of this network.

2. *Disasters are the unfinished business of development.* Not all extreme natural events and hazards translate into disasters: they become so when society is less prepared than it should or could be. "Development" or *bikas* is the new religion of modernity in the post-colonial era; and this enterprise is what should have prepared society to better cope with natural hazards and their unexpected visitations. Unfortunately, many proposed development activities do not do that: on the contrary, as mal-developments (or *ku-bikas*) they contribute to depletion of resilience and increased vulnerability where so-called development projects are implemented without wider consultations and reviews. If development projects are to be apt-developments (or *su-bikas*), they need to build into their design and implementation the lessons from past disasters and what could be done differently to ensure that when they strike again, their impacts are less harmful. This "always learning, never quite getting it wholly right" is the ever-present challenge before *su-bikas* if it is to contribute to a society more resilient in the face of disasters.

2.2 Kosi Breach of August 2008

It is with these perspectives that a recent disaster can be re-examined to understand what a better approach might be to transboundary Himalayan Rivers and the diasters they wreak—the 18 August 2008 Kosi breach. Even as Maoist leader Pushpa Kamal Dahal "Prachanda" was taking his oath of office as prime minister of Nepal, the Kosi river sliced its left embankment near the Nepali Tarai village of West Kushaha. This breach allowed the river to afflux to a new eastward channel affecting some 65,000 families in Nepal and three and a half million in downstream Bihar (in India) along its path from the Nepal border to the Kosi's confluence with the Ganga at Kursela in Katihar district.

However, it needs to be argued that this was not a flood, even though there were pictures in the media of thousands of people moving through neck-deep waters with their belongings on their heads towards safer high grounds! It was also not a natural disaster but a man-made one with interesting transboundary implications and occasions for finger-pointing. The reason is, when the breach occurred, the Kosi was flowing at 130,000 cusecs, which when compared with the average monthly flow for the month of August of 320,000 cusecs, means that the river was at only 40% of the average flow for the month of August. That can hardly be called a flood even though millions of people were affected by the Kosi waters inundating their villages (Ojha and Bista 2008).

So, what really happened? The answer is much more mundane and simpler: it was "garden variety" contractor corruption (DtE 2008 and Bharti 1991). The catastrophe resulted from an unholy marriage of three things: wrong technological choice, bad institutional arrangements and half a century of political misconduct. Looking at the section of the failed embankment that had been sliced clean, it was obvious that it had no compacted clay core that would prevent or at least reduce seepage through it. What the embankment consisted of, was mostly very porous sand and mud mixture that allowed water to seep through and begin the process of what civil engineers would call piping—a leakage that grows with time, eventually weakening the mud embankment leading to its collapse. (In the immediate aftermath of the breach, Nepali media did manage to interview old men nearby who had worked as young labourers in the Kosi project some 50 years earlier, and who recounted how the contractor only required them to carry the sand from the river bed and pile it on the embankment.)

The massive Kosi Project of Independent India (which the British had feared undertaking) basically ignored the sedimentation aspect of Himalayan rivers such as the Kosi where, along those drainage channels, matter is a co-equal passenger with water (Gyawali 2003). Sapta Kosi (as the river is known in Nepal after the confluence of seven upstream mountain tributaries), acts as a massive conveyor belt taking sediment from the ever rising and ever eroding Himalaya to the Bay of Bengal. Indeed, geologically speaking, if this process did not exist, much of today's Uttar Pradesh, Bihar and Bengal would be the prehistoric (Mesozoic to Cenozoic eras') Tethys Sea! Some 100 million cubic metres of gravel, sand and mud flow out of the Chatara gorge every year at the point where the river debouches onto the flat Tarai plains that are the northern limits of the Indo-Ganga plains.

And this flow cannot be blamed on deforestation. Nepal, thanks to community forestry and other such measures (Ojha 2017) has more forest cover in the Kosi catchment today than ever before in its recent history. It is caused by Himalayan geo-tectonics coupled with the monsoon regime's heavy precipitation. As the river slows down in the flat plains beyond Chatara, it deposits this sediment it is carrying, filling up the river's main channel until it overflows and begins a new sideways course. A map of the Ganga-Brahmaputra courses prepared by Col. Rennell for Governor General Warren Hastings in 1789 compared with the course of these rivers in the 1980s shows how Kosi has shifted westwards almost 115 kms in 200 years (Thompson and Gyawali 2007). This natural process produced the large inland delta that lies across southern Nepal and the Indian state of Bihar before the river's confluence with the Ganga at Kursela. Traditional means of coping with this meandering nature of the river have been technologies such as houses built on stilts that allowed flood to pass below without harming people or their possessions.

But, for the last half century and more, the "Kosi Project" has used embankments to restrict the river's sideways course with the aim of confining it within the narrow limits between the embankments that extend from Chatara gorge all the way across the Kosi barrage at the Nepal-India border to Kursela, a distance of some 120 kms. This has kept sediment deposits within the main canal, raising the riverbed some four metres above the surrounding land, and correspondingly perching the river to make it a disaster waiting to happen any moment as earthen embankments are not designed to withstand high water pressure sideways. In Nepal it happened in August 2008 but in Bihar it has happened at least eight times over the last 50 years. In all these cases, no lessons of geo-tectonics and river morphology were ever learned: the breach in Nepal was "repaired" at massive cost in the following years and it was back to business as usual, waiting for the next one and similar expensive "repairs".

Another factor was the fact that this type of flood control measure is suited to arid zones, not ones in regions with heavy monsoon precipitation. Good embankments do prevent a river from spilling over to adjacent land; but they also prevent, when river level declines after the heavy flow event, the draining of water that has accumulated outside of the embankments into the river, especially in a monsoon-dominated regime where massive rainfall occurs outside of the embankments, aggravating flooding, unlike in more desert climates. Such myopic embankment building causes massive year-round water logging and loss of otherwise fertile agricultural land which gave three crops a year, now unable to give even one. It is estimated (Mishra 2003) that more land has gone out of production in the Kosi plains from water logging than have been brought under irrigation from the government's Kosi project, causing immiseration to millions of Biharis who are forced to migrate to major Indian metropolis in search of meagre labour-based livelihood. This certainly makes the Kosi project an "ideal" example of *ku-bikas*!

Rethinking development to avoid the negative impacts of *ku-bikas* is also becoming imperative in this age of climate change when there is a tendency to just blame climate change for everything, ignoring the negative consequences of our ill-considered actions. Indeed, the press secretary of the president of Nepal was quick to blame the August 2008 Kosi breach-induced flood as caused by climate change. As argued above, it was anything but that! This is not to say that climate change will not make the cost of past *ku-bikas* much worse in the future, if corrective measures are not taken today: they indeed will. An institutional aspect of *ku-bikas* is the very 1954 Kosi treaty itself between Nepal and India (HMG/N 1975). While the Kosi Barrage (and its upstream embankments) is within the territory of Nepal, the agencies responsible in Nepal government have absolutely no hand in its maintenance or operation. Interestingly, they are not even within the purview of the government of India but with that of Bihar's water agencies.

This unsatisfactory transboundary river management arrangement has been the source of much finger-pointing and has caused bad blood between the two countries, especially when it comes to taking responsibility for tackling governance issues like corruption and communications' failure. Bharti (1991) cited earlier estimates that as much as 60 per cent of approximately three billion Indian rupees (INR) spent annually by the Bihar government on construction and repair works was pocketed by an iron triangle nexus of politicians, contractors and government engineers. It is said that there is a perfect system of percentages in which a share exists for everyone who matters in the hierarchy—from the minister to the junior engineer. The actual expenditure never exceeds 30 per cent of the budgeted cost. Contractor's bills are paid without being verified—many of the desilting and maintenance works allegedly completed, are never done at all, and yet payments are made. Indeed, engineers are said to make so much money from the percentages that they do not bother to collect their salaries (Bharati 1991).

2.3 Apt Versus mal-Development

Kosi project is only one glaring example of *ku-bikas*: the development landscape of the last seven decades is strewn with many more examples, some of which are discussed in this section. They would help in distinguishing between *su-bikas* and *ku-bikas*, and lead to any analysis of why the former is sidelined by the latter to the detriment of regional and transboundary cooperation.

2.3.1 Roads Versus Ropeways

It is well accepted that access to markets is important for development, especially for rural areas. Unfortunately, it has been defined for the hills only in terms of building roads, which is not a most mountain-friendly technology. First, it relies on expensive, imported climate-unfriendly fossil fuel such as diesel and petrol, instead of clean hydroelectric power that mountain regions are blessed with. Second, it requires road building as a priori infrastructure, which is often ecologically very damaging in geologically fragile mountain areas. Well-designed hill roads with sufficient environmental safeguards are very expensive, compared to similar length roads in the plain areas.

Unfortunately, political pressure for road access is very high and Nepal Himalaya especially is seeing a wild spree of road building, mostly by local entities without adequate technical studies or designs to address issues of stability of slopes and mountains. It is said that the bulldozer driver is the primary engineer deciding where and how to gouge the hillside for a road track!! The result is destabilization of fragile hill slopes that, coupled with monsoon downpour, lead to massive landslides, drying up of traditional springs and destruction of terraced farms below. Further, the debris from these exacerbated landslides end up in the river below, raising the riverbed level and causing more floods downstream, both in the hill valleys as well as in the Tarai plains and cascading on to Bihar and Uttar Pradesh in India. This *ku-bikas* or maldevelopment is contributing to increased instances of what can only be called "human-induced" flooding that has no relation to climate change. In Nepal, villagers often refer to this as "*dozer aatanka*" or bulldozer terrorism. Fortunately, there are alternative mountain-friendly and climate-friendly technology of ropeways (cable cars) available. And they have a very small environmental footprint since they are supported only by a limited number of towers and do not require gouging out hundreds of kilometres of fragile hillsides. They also run, not on imported (expending scarce foreign currency) fossil fuel but on locally produced hydropower. Furthermore, this is not a new technology: ropeways came to Nepal way back in 1922–24, at least 100 years ago, and just a decade after Nepal's first hydroelectric power plant was built in Pharping south of Kathmandu in 1911.

Unfortunately, ropeways as a means of *su-bikas* have been completely marginalized in development policy and planning in favour of more roads, more trucks, more diesel spewing vehicles. So has inland water transport which is the cheapest form of bulk goods transport. Its development would have much positive benefits both for regional economy as well as ecological health of the river. Navigation requires water in the river, unlike an irrigation-only view that sees any water not diverted as "wasted to the sea". However, inland navigation too has been ignored nationally and regionally in South Asia (Gyawali 2016).

Studies have shown (Gyawali et al. 2004) that goods carrying ropeways between valley bottom (*besi*) to villages on hill tops (*gaon*) are three times cheaper to build than equivalent, cheapest gravel "green roads", eight times quicker to install (in one dry season construction versus minimum of four years for road tracks), and two times more energy efficient (34 MJ/ton of goods carried for ropeways versus 53 MJ/ton for road). The example of Bhattedanda Milkway is instructive. Initially it required NRs 34,000/month[1] of diesel not including the cost of transporting the fuel whereas when it shifted to community (hydro) electricity, it took only NRs 7000/month with transport cost included! The reason lies in technological "locked-in-ness" and its constricting political economy that will be discussed in the section further below.

2.3.2 *Traditional Technologies of Marginalized Rivers*

Transboundary water dialogue between upstream Nepal and downstream India is dominated by large rivers such as Sapta Kosi, Sapta Gandaki, Karnali and Mahakali as well as their many snow- and glacier-fed tributaries. This is natural because Nepali rivers contribute upto 70 percent of the dry season flow in the Ganga (and 45% on an annual average), even though snow and glacier melt contribute only 9% to the Ganga flow (unlike in the Indus river where it is up to 40%). However, there are many dozens of smaller transboundary rivers—termed "marginalized rivers" because they are not on either government's radar—which provide much economic benefits but also cause significant harm during floods to the people in their basins. To harness their waters for irrigation, farmers have used centuries-old traditional technologies but in recent times have also turned to modern pumps, while government agencies have built cement dams over them with mixed results.

One such "marginalized river" is the Tinau in central Nepal and its tributary, the Rohani, which carries the distinction of being the first river in history to have had water conflict mediation by none other than the Buddha himself some 2500 years ago. The rivers lie between his mother's home in Dev Daha of Rupandehi district and his father's in Tilaurakot in Kapilvastu and Buddha is recorded in Buddhist literature as having mediated between the two clans. Farmers of the Tinau use traditional technology of brushwood dams, which is what farmers must have used even in

Buddha's time, to divert water from the streams to their fields for dry season irrigation (February to May).

2.4 100 Indian Rupees (INR) Is about 160 Nepali Rupees (NRs)

It works is as follows: during the monsoon (except for rare dry periods of what is called "break monsoon"), there is no need for water. Indeed, the problem is how to get rid of it by draining it off. On the other hand, in the dry spring season, there is a need for irrigation to grow *chaite dhan*—spring rice—when the rivers are at very low flow. Historically, flow variation between flood and dry seasons in snow-fed large rivers is between 25 to 30 times whereas for non-snow-fed rivers like Tinau, dry season flows are 90 to 170 times lesser than monsoonal flows (Gyawali et al. 2019). Hence, traditionally village communities have banded together to build small brushwood dams to divert the low stream flows to their fields. Comes the monsoon when no water is needed, the brushwood dam is washed away—a perfect traditional example of a seasonal, "disposable" dam!

Unfortunately, this community technology is not taught in any engineering school in South Asia: what is taught are "modern" technologies using ferro-cement and concrete that are called "permanent" structures without first asking, (in a hydro-ecological context of flood flows being 170 times or more than the dry season one), how permanent is permanent. Ignoring traditional technology of "disposable" brush-wood dams along the Tinau, agency hierarchism of India and Nepal built a modern, permanent dam on the Tinau in 1961 just south of the town of Butwal. Unfortunately, tragedy struck in 1962 and during that year's monsoon floods, the river moved away, westwards, leaving the dam as a monument that bears testimony to man's stupidity.

In the process of building this concrete dam, *pukka* modern canals were built below it in a prosperous agricultural area known as Marchawar that ignored the traditional, community-driven equitable *warabandi* system of water sharing. Thus, when the river moved away from the dam, with the old hierarchic agency-developed system destroyed, and the new one not functioning, there was a total social collapse when the farmers lost their agricultural livelihood to so-called modern, technological intervention. It became a "dacoit-infested" area simply because people had to do whatever they could to survive. Two decades later, thanks to some policing presence but mostly the appearance of another technology—the humble market-led, individualist "Kirloskar pump", that agriculture-based livelihood was restored. Marchawar it is no longer a "dacoit-infested" area, but is one of the most prosperous agricultural regions in southern Nepal *Tarai* (Gyawali 2004), a case of disaster to normalcy brought about by unplanned re-thinking of development.

2.4.1 "Nexused" Water

The previous section introduced the phenomenon of groundwater, an easy but invisible resource that is fairly ubiquitous as the largest freshwater storage in the world but increasingly mismanaged everywhere (WWDR 2022). Despite groundwater accounting for approximately 99% of all liquid freshwater on earth, with half the water used for domestic purposes and a quarter of irrigation globally, it lies in a policy blind spot of most governments, especially local ones, who are the first victims of the negative impact from its misuse. It is a common-pool resource affecting future generations that requires governments acting as custodians to develop adequate and effective framework for its governance, especially of transboundary aquifers. As it has nexused relationship with food production, health and sanitation, energy as well as industry, its governance structure needs revamping for better sustainability.

Groundwater was an easy resource, a saviour when cheap pumping energy was available, and the water table sufficiently high as in the case of Marchawar. It has also been treated as an infinite resource, requiring only the cost of drilling a well and cheap energy for pumping. Indeed, no thought was given to its recharge characteristics. However, with uncontrolled drilling and pumping, water levels in aquifers are declining, leading to a negative feedback loop of competitive deeper drilling making the situation even worse. This has been the case in India (with the situation being much worse in the southern hard rock areas than in the alluvial Ganga flood plains, but here too conditions are deteriorating), and increasingly in urban and peri-urban areas of Nepal as well.

In Nepal's case, there are two types of groundwater crises, both of which have interlinked transboundary concerns (transboundary being not only international but also sub-national and local administrative). The first is in towns and peri-urban areas as well as in the Tarai plains. Due to unchecked, deep boring and pumping of groundwater, not only are traditional wells drying up but new ones are also facing water availability decline. This is because of competitive drilling by neighbours forcing them onto the same unsustainable path. Groundwater is assumed to be a private good, available for the taking by anyone who can afford the technology. This state of affairs was brought about by the inadequacy or failure of the local or public (i.e., municipal) piped water system that also saw the simultaneous rise of another poorly regulated water-tanker market. Given that groundwater aquifers in the Indo-Ganga plains are contiguously interconnected, the much more massive (and subsidized) volume of pumping out of the ground water, in adjacent Bihar and Uttar Pradesh states of India, has also contributed to water table decline in the Nepal Tarai.

The second is in the hills where mountain population depends upon adjacent springs for its daily domestic water use. Rivers are hundreds of metres below settlements, making their waters impractical to carry up and down for daily domestic needs. Essentially, spring water is groundwater and mid-hills (mostly without snow) are massive water-towers with the groundwater within them filled by monsoon and winter westerly rains that leak into the aquifers from geological fissures. These spring waters are the primary source of stream flows of Himalayan rivers that are also fed by groundwater backflow from riverbanks all along their stretch. However, springs are drying across the Himalaya (Sharma et al. 2016), not so much from climate change as yet as from *ku-bikas* such as over-pumping and desecrating fragile hills with "bulldozer terrorism". In fact, climate change is expected to make matters worse if the impacts of mal-development are not taken care of adequately. Sometimes even *su-bikas*, (such as increase in commercial, local vegetable farming especially by women-headed households in a migration-remittance economy with men folks abroad), can lead to springs drying up with massive, deep boring groundwater pumping in the valleys below.

If we take a more holistic view of the overall water cycle and the kinds of water available to us, we will need to adopt a more interlinked view—the nexus approach (Allouche et al. 2019)—of the development process that provides early warning of unintended cost externalization to weak social groups and unprotected parts of the environment. "Integrated approach" to water management has a long history since the 1980s and resulted in the program of Integrated Water Resources Management (IWRM). However, despite some success in policy pronouncements and academia, it saw little traction in practice because it failed to answer two basic questions: who does the integrating (what kind of organization and through which forms of agency) and how (through what process). To overcome this shortcoming, a new nexus approach to water governance has been proposed which requires some shift in conventional thinking (Gyawali 2015).

First, in light of climate change, it is necessary to expand the spectrum of water, especially transboundary waters (Gyawali 2024). It is not just Blue (river, lakes etc. surface) waters but also Green (soil moisture) waters that all forests, grasslands and terrestrial wildlife as well as dryland agriculture and pastoralism depend upon. Brown (underground) waters is actually most of terrestrial freshwater that also feeds all rivers and streams during the dry season. Water that is produced after their use is either only organically polluted (Grey water) or is heavily laden with industrial chemical pollutants and is much more difficult to recycle (Black water). The ultimate source of all terrestrial waters is White (atmospheric) water, mostly from evaporation of the vast Magenta (saline sea and ocean) waters. Except for Blue water, all other forms of water are ignored or insufficiently studied. Second, how each one of these waters is valued, depends very much on the social management styles of particular types of water users. Pastoralists would be more concerned with Green water whereas city managers more worried about the state of Brown water that their cities require to run. Environmentalists on the other hand would be concerned with the Grey and Black waste waters that pollute Blue and Brown waters. What such a widening of the water and water users' spectrum does is to allow water management both to avoid blind spots and open up wider options for difficult political trade-offs.

2.5 Cultural Theory of Disaster Risks

What each of the examples above point to is the mal-adjusted dynamics between different social solidarities, their worldviews, management styles and technology choices, which ultimately contribute to, and exacerbate, existing natural hazards into becoming a major disaster. The new and evolving social science of Cultural Theory (see: Douglas 1987; Thompson et al. 1990; and Thompson 2008)) argues that a stable policy terrain occurs only when it is not monistic (bureaucratic socialism or market liberalism) but pluralistic—one which also includes activist egalitarianism—not only public or private, nor even public-private partnership (PPP) but public-private-civic partnership (PPCP). More theoretically, these institutional styles emerge from the permutation of two dimensions of sociality: *transactions* that are symmetrical or asymmetrical (nature of group affiliation and incorporation); and *competition* that is fettered and accountable or unfettered and unaccountable (the degree to which an individual's life is circumscribed by externally imposed prescriptions). What these two parameters raise are fundamental but age-old questions of philosophy in human life: who am I (group affiliation)? And what should I do (the context of conforming or not conforming with pre-ascribed rules)?

They give rise to four organizing styles depending upon whether the two dimensions are weak or strong:

1. *Hierarchism*—strong group cohesion, strong prescribed rules,
2. *Egalitarianism communards*—strong group cohesion, but weak on prescribed rules),
3. I*ndividualism*—weak on both group cohesion and prescribed rules.
4. *Fatalism* of the conscripted—weak on group, but strong on ascribed rules pressing down on them.

In common language, they would be bureaucratic agency, activist NGOs, market players and fatalistic voters or consumers, respectively. Agencies such as hydrocracies have as their overriding concern, *control* and are characterized by unequal roles for unequal members. Activist groups are bound together by their perception of threats from outside and the moral outrage they provoke, leading them often to *alarmist* causes. Market players, on the other hand, consider all boundaries provisional and subject to *free* negotiations. Fatalists find themselves so strongly subject to external rules with no commensurate group support for collective action that they lapse into passivism. Because they themselves do not actively cognize and strategize, unlike the other three that are called "active" social solidarities, they are often excluded from analysis. What Cultural Theory argues is that the way in which each of these social solidarities are organized, determines their position on a range of issues from risk to justice to technology and more. Verweij (2017) has summarized the different perceptions on various topics (see Table 2.1). Take risk, for example, that lies at the heart of disaster preparedness (Douglas and Wildavsky 1982). Individualism considers risk as an opportunity; and as is well-known, many market players use disasters like floods, pandemics, earthquakes and even wars to make money, to profiteer. For hierarchism, it is about how you manage that risk, keeping matters under control so that the situation does not go out of hand. Egalitarian activists are risk-sensitizers or risk-amplifiers, using the occasion to highlight issues of social justice, inequities, human rights violation and more to highlight the shortcomings of the system that led to the calamity. Fatalism is about risk absorption, absorbing the strategies of all other (three) active social solidarities.

Similarly diverse is the attitude towards nature (it is robust, robust within limits, fragile or capricious), and technology (whatever is efficient and profitable, capital- and expertize-intensive, manageable by the community or too confusing). Coupled with the fact that ideas of governance, human nature and justice are similarly plural, one can appreciate how single-minded pursuit of solutions, ignoring other views can lead to conflicts. What is important to realize is that these different perspectives cannot be merged into a single neat consensus but need to be engaged with, in order to hammer out a compromise—perhaps even a "clumsy solution" where no one solidarity gets everything it asks for but gets just enough to continue the constructive engagement (Verweij and Thompson 2006).

Most of our inadequate response to disasters and their aftermath lie in the fact that water scarcity or floods, energy shortage, city planning, preventing pandemics or climate change are what are called "wicked" problems (Verweij and Ney 2015).

Table 2.1 Cultural theory's plural perspectives on human development

Social Solidarities à	Individualism	Hierarchism	Egalitarianism	Fatalism
Risk	Opportunity: risk taker	Manage: risk manager	Minimize: risk sensitizer/ amplifier	Cope: risk absorber
Nature	Robust	Robust within limits	Fragile	Capricious
Human nature	Egocentric but intelligent	Sinful unless guided wisely	Altruistic but easily corrupted	Unpredictable
Technology	Anything efficient and profitable	Capital/ knowledge intensive	Simple, manageable by community	Confusing
Governance	Unfettered but minimal protective	Guided technocracy (Guardian)	Participatory democracy	Benevolent dictatorship
Justice	Opportunity equality	Noblesse obligé	Outcome equality	Survivable
Poor	Inventive entrepreneur	Deserving of helping hand	Oppressed and marginalized	Unreliable
Foreign aid	Distorting and corrupting market functioning	Global public good with moral imperative	Fig leaf hiding exploitation and inequities	Rich getting richer anyway

Source Verweij (2017) in Gyawali et al. (2017) *Aid, Technology and Development*

Unlike "tame" problems which are amenable to textbook solutions, "wicked" problems are those that we cannot even agree on their definition, let alone arriving on a solution. Hence, they require "uncomfortable knowledge" that is ultimately able to bring forth stable and sustainable "clumsy solutions", unlike "neat solutions" that are rigid and prone to inciting conflicts leading to an impasse. But how is uncomfortable knowledge generated? Cultural Theory would argue that each social solidarity, seeing problems very differently from others, would generate knowledge that is uncomfortable for the others. And if and when they collectively engage with each other constructively, they would then produce a clumsy solution broadly (but not wholly) acceptable to all.

Unfortunately, most hierarchic agencies are comfortable in what Rex Tillerson (former chair of Exon and former US Secretary of State) calls "exnovation", the exact opposite of innovation (Gyawali 2021). Exnovations are things that have worked and do not need to be tinkered with, let alone completely replaced with new, disruptive innovations. Technologies come with both massive supporting infrastructure and their politically powerful social carriers that leads to a certain technological locked-in-ness that becomes too costly to tamper with. Traditional technologies for water management of brushwood dams and houses built on stilts (mostly maintained by community volunteerism) that have worked for centuries fall in this exnovation category. On the other hand, modern technologies and new innovations introduced by the market, while efficient, productive and profitable for those who can afford it, often lead to unforeseen mishaps such as groundwater levels declining or embankments breaching (Fig. 2.1).

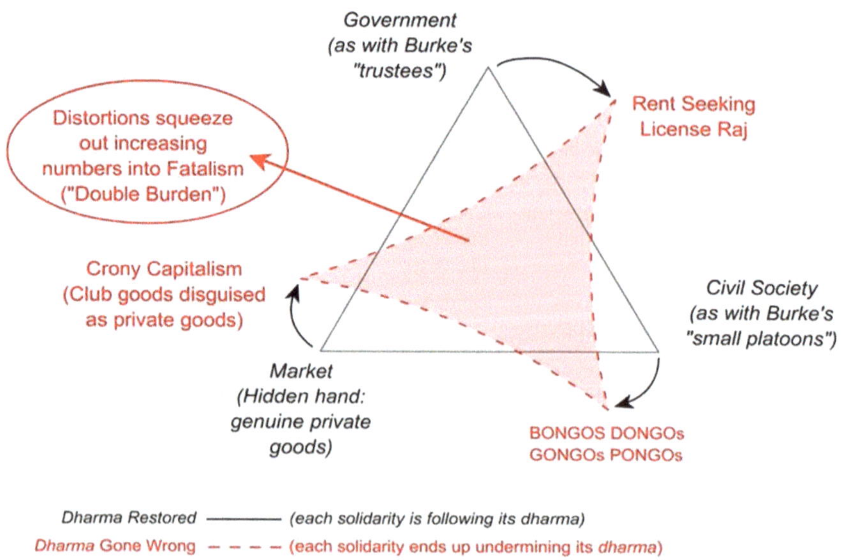

Fig. 2.1 Dharma and its dynamics *Source* Gyawali & Thompson (2016).

The dynamics of this distortion essentially entails seeing policy making as a plural three-legged stool with, in the best of circumstances, space for all three voices—government, market and civic—that are not only heard but also responded to in a constructive manner. Unfortunately, and it is even worse in Nepal currently, with shaky coalitions of ideologically incompatible partners, there is massive distortion of ideology and mission of political parties, market players and NGOs, what we call *dharma* distortion (Gyawali and Thompson 2016).

As can be seen from the above diagram, the governments are not maintaining their *dharma* as trustees but engaged in rent-seeking license raj; market which should be competitive are mostly into crony capitalism where club goods are disguised as private goods; and NGOs, having given up on the essential quality of volunteerism of civic movements, have become business and politics by other means. There are traditional NGOs in Nepal called *guthis* mostly with religious and cultural orientation that do maintain their *dharma*; but many modern NGOs are fronts for family business or political party causes. So, unless this *dharmic* triangle is restored and each social solidarity lives up to its inherent mission, it is unlikely that much would be achieved in disaster mitigation, leaving it as an unfinished business of apt-development that is unable to stop mal-development.

References

Allouche J, Middleton C, Gyawali D (2019) The water-food-energy nexus: power, politics, and justice. Routledge EARTHSCAN, London

Bandyopadhyay J, Gyawali D (1994) *Himalayan Water Resources: Ecological and Political Aspects of Management*. Mountain Research and Development vol. 14, no. 1 pp. 1–24. Himalayan Water Resources: Ecological and Political Aspects of Management on JSTOR

Bharti I (1991) Fighting the irrigation mafia in Bihar. Bombay: EPW vol. 26 no. 38, 21 September

Douglas M (1987) How institutions think. Routledge & Keegan Paul, London

Douglas M, Wildavsky A (1982) Risk and culture: an essay on the selection of technological and environmental dangers. University of California Press, Berkeley

DtE (2008) Engineered flood: not excess water but neglected embankment caused the flood. Down to Earth (DtE), Delhi. 30 September.https://www.downtoearth.org.in/coverage/engineered-flood-5038

Gyawali D (2003) Rivers, Technology and Society: learning the lessons from water management in Nepal. London: Zed Books. (Originally published as Water in Nepal in Kathmandu by Himal Books and Panos South Asia with Nepal Water Conservation Foundation to include the 1989 eponymous monograph from Environment and Policy Institute of East West Center, Hawaii taking an interdisciplinary look at resource uncertainties, evolving problems and future prospects)

Gyawali D (2004, Feb to July) Water, sanitation and human settlements: crisis, opportunity or management? Kathmandu: water Nepal 11(2):5–18

Gyawali D (2015) *Nexus governance: harnessing contending forces at work*. IWA/IUCN nexus dialogue synthesis papers. IUCN, Gland, Switzerland. https://iwa-network.org/wp-content/uploads/2016/03/1448008531-Nexus-Governance-Harnessing-Contending-Forces-at-Work_Dialogue-Synthesis-Paper-2015.pdf

Gyawali D (2016) *Will Inland Navigation Shift South Asia's Water Discourse Positively?* Kathmandu: Spotlight Nepal vol 10 no 3 September 02. https://www.spotlightnepal.com/2016/08/31/will-inland-navigation-shift-south-asias-water-discourse-positively/
Gyawali D (2021) *A Cultural Theory Approach to Innovation: pluralizing the policy terrain to understand what it is and is not.* Chapter in Imtiaz Ahmed and Liyan Zhang (eds) *Innovation in Education.* Bangladesh: Dhaka University Center for Genocide Studies, Bangladesh and Tianjin University Center for Innovation and Entrepreneurship, China
Gyawali D (2024) Reframing wicked water problems from an eco-centric perspective under climate change uncertaintics. Spotlight Nepal, Kathmandu (16 September) www.spotlightnepal.com/2024/09/16/reframing-wicked-water-problems-from-an-eco-centric-perspective-under-climate-change-uncertainties/
Gyawali D, Dixit A (1997) How distant is Nepali science from Nepali society? Lessons from the 1997 Tsho Rolpa GLOF panic. Kathmandu: Water Nepal vol. 5 no. 2. pp. 5–43
Gyawali D, Dixit A., Upadhya M (eds) (2004) Ropeways in Nepal: context, constraints and co-evolution. Kathmandu: Nepal Water Conservation Foundation with Kathmandu Electric Vehicle Alliance. Gyawali, D. (2004). *Water, Sanitation and Human Settlements: Crisis, Opportunity or Management?* Kathmandu: Water Nepal vol. 11 no 2. (Feb-Jul), pp 5–18
Gyawali D, Sharma B, Pokharel G S (2019) *Water Scarcity: Nepal country snapshot.* Report prepared for the International Water Association and the Food and Agriculture Organization of the United Nations during August–December 2018. Kathmandu: Nepal Water Conservation Foundation, NWCF Working Paper 1/2019
Gyawali D, Thompson M (2016) Restoring development dharma with toad's eye science? IDS Bull 47 no 2A. https://bulletin.ids.ac.uk/index.php/idsbo/article/view/2822/ONLINE%20ARTICLE
Gyawali D, Thompson M, Verweij M (eds) (2017) Aid technology and development: the lessons from Nepal. Routledge and Earthscan, London
HMG/N (1975) Revised agreement between his majesty's government of Nepal (HMG/N) and the government of India (GoI) on the Kosi project. Kathmandu. https://www.moewri.gov.np/storage/listies/May2020/revised-agreement-on-nepal-and-india-koshi-river-1966.pdf
Mishra D K (2003) Life within the Kosi embankments. Kathmandu: Water Nepal, 10(1) pp. 227–301
Nivaran D (1995) Disasters and vulnerability in South Asia: program of work for Duryog Nivaran, a south Asian initiative on disaster mitigation. Published by ITDG Sri Lanka, Colombo for *Duryog Nivaran*
Ojha, H. (2017) *Community Forestry: thwarting desertification and facing second generation problems.* In Gyawali et al (2017). Aid Technology and Development: the lessons from Nepal
Ojha A, Bista D (2008) India-Nepal: whither Kosi River? *An Interview with Dipak Gyawali* Kathmandu: The Kathmandu Post, 1st September. http://www.sacw.net/article3.html
Sharma B, Nepal S, Gyawali D, Pokharel G S, Wahid S, Mukherji A, Acharya S, Shrestha A B (2016) *Springs, storage towers, and water conservation in the Midhills of Nepal.* Kathmandu: ICIMOD and Nepal water conservation foundation, working paper 2016/3. https://lib.icimod.org/record/32016
Thompson M (2008) Organising and disorganising: a dynamic and non-linear theory of institutional emergence and its implications. Triarchy Press, Axminster/UK
Thompson M, Ellis R, Wildavsky A (1990) Cultural theory. Westview Press, Boulder
Thompson M, Gyawali D (2007) *Uncertainty Revisited or the Triumph of Hype over Experience.* New introduction to Thompson, M., Warburton M. and Hatley T. "Uncertainty on a Himalayan Scale: an institutional theory of environmental perception and a strategic framework for the sustainable development of the Himalaya". Kathmandu: Himal Books for International Institute for Applied Systems Analysis (Vienna) and James Martin Institute for Science and Civilization (Oxford University). Originally published in 1986 as Milton Ash Editions of Ethnographica (London)
Verweij M (2017) *A Cultural Theory of How to Aid Development.* In Gyawali et al (2017). Aid Technology and Development: the lessons from Nepal, Earthscan

Verweij M, Thompson M (2006) Clumsy solutions for a complex world. Palgrave Macmillan, New York

Verweij M, Ney S (2015) Messy institutions for wicked problems: how to generate clumsy solutions. Environment and Planning C: Government and Policy 33(6): 16791696. (PDF) Messy institutions for wicked problems: How to generate clumsy solutions (researchgate.net)

WWDR (2022) Groundwater: making the invisible visible. World Water Development Report (WWDR) of UN UNESCO World Water Assessment Program, Paris. https://www.unesco.org/reports/wwdr/2022/en

Dipak Gyawali is academician of the Nepal Academy of Science and Technology, trained as a hydroelectric power engineer (Moscow Energy Institute, USSR) and a political economist (UC Berkeley, USA). His research focuses on the interface between technology, society and resource use issues from the perspectives of Cultural Theory (neo-Durkheimian theory of plural rationalities and new institutionalism). He was chair of the non-profit Nepal Water Conservation Foundation till 2018, and in the 1990s of the poverty alleviation focused NGO, Rural Self Reliance Development Center (Swabalamban). During his time as Nepal's Minister of Water Resources in 2002/03, he initiated reforms in the electricity and irrigation sectors focused on decentralization and promotion of rural participation in governance through farmer-managed irrigation systems as well as community electricity. He was the first Chair of Duryog Nivaran, a South Asian network of scholars and practitioners advocating for an alternative perspective on disaster resilience. In 2006 he chaired an international EU panel that reviewed EU's 12 years of international water research cooperation across the globe. In 2011 he chaired UNESCO's review of its International Hydrological Program. He was co-chair of Climate Action Network South Asia (CANSA) till 2021. He has been the lead author on water and the nexus for the global assessment of nexus and biodiversity by the Intergovernmental Science-Policy Platform on Biodiversity and Ecosystem Services (IPBES). He has also served as advisor to The STEPS Center/University of Sussex. He has published extensively both academically and in the media on water, energy, environment and development issues.

Chapter 3
The Symphony of Co-operation: Principles, Values and Ethics of Transboundary Risk Governance

Mukunda Upadhyay

3.1 Introduction

Water is the very basis of human existence and survival. Since antiquity, states have always based their existential foundation around this scarce resource. Great powers across countries have emerged because of the strategies involving the use of water. As a matter of fact, the "*Harmon Doctrine*" translated as the "*absolute territorial sovereignty*" empowers states to exploit the water resources flowing through their territories in any manner they want (Baranyai 2020). In the absence of any regulatory authority, states have been interpreting such doctrine in a manner that does not always evince co-operation. Regulation has always been perceived through the sovereign perspective of nation states within which action around disaster risk is often based on an obligation which translates into an immediate response to a singular hazardous event. States, in the recent past have generally moved from one disaster to another, operating the problem in a co-operation constrained environment, generally assuming that these are disconnected and disparate events. Such unorganized shifts and isolated response centric operations were the product of the scant understanding around the lacunae in the overall disaster governance framework, the cost of which was generally borne by communities living in the hinterlands and state frontiers; unattended and unnoticed.

However, with a shift in the approach to transboundary co-operation observed in international conventions, the dominant nature of the state seems to be shifting, fostering greater participation and negotiation which in turn has given a new direction

The author would like to gratefully acknowledge the editorial suggestions made by Prof. JankiAndharia to improve the quality of the paper.

The author is a Disaster Management Professional Working in India.

M. Upadhyay (✉)
Disaster Management Professional, Guwahati, India
e-mail: mukundaupadhyay@gmail.com

J. Andharia and P. Anand (eds.), *Shifting Lands, Flowing Waters*, Disaster Studies and Management, https://doi.org/10.1007/978-981-96-7691-0_3

to the idea of the need for governance of shared risks (Ebi 2010). The *United Nations Economic Commission for Europe* (UNECE) in 2000 appealed for a paradigm shift in risk governance from the existing defensive mode to a more holistic river basin approach instead of a localized intervention.[1] While people can be controlled through borders, disasters such as floods and earthquakes are beyond any national, regional or institutional boundaries (UNECE 2009). It is imperative therefore, that states co-operate with each other for planning better risk mitigation and cost-effective strategies across borders. The West-Philian[23] perspective on natural resources often allows states to maintain an environment of anarchy beyond their usual jurisdictional spaces which often blocks opportunities of potential co-operation and dialogues over shared resources and risks. With rising discontent, an environment of ever-lasting conflict pervades resulting in a perennial failure to heed to the needs of the governed living in the peripheries (transboundary spaces). In the *Universal Declaration of Human Rights, 1948*[43] the will of the people is seen as the fundamental basis of authority. Self-governance of people however does not merely mean control of political processes but also control of the means and resources of production.

Varis et.al (2008) points out that the discussions around transboundary water governance is mostly confined to conflicts and visible risks but fails to recognize the socio-political and cultural interaction which influence hazards. Nincic and Weiss (2016) argue that the institutional capacity, legitimacy and stability of states have a huge role in determining domestic policies on water. They also draw attention to the fact that governance around water hangs on a very delicate balance of power wherein both co-operation and conflict have equal chance of occurrence. Andharia et al. (2019) emphasize the need for taking an egalitarian approach within which communities, particularly the poor and vulnerable who bear the impacts of climate change, pollution and disasters are involved in decision making involving the management of natural resources. They analyze the intertwining relationship between common property rights and environmental justice, essential as a combination for achieving the goals envisaged under different national and international commitments on disaster risk reduction, sustainability and climate change adaptations. Ideally, it is expected that people engage with their representatives in a continuous consultative process to bring laws and conventions to secure the idea of sustainability wherein each and every state intervention is implemented and evaluated in the light of growing threats

---

[1] Meeting of the parties to the convention on the protection and use of transboundary watercourses and International Lakes, Second meeting, The Hague, Netherlands, 23–25 March 2000. Refer to Item 6 (b) of the provisional agenda Sustainable Flood Prevention, Submitted by the Chairman of the task force on flood prevention and protection, led by Germany, and prepared with the assistance of the secretariat, page 4, para 1, Available at https://unece.org/DAM/env/water/publications/documents/guidelinesfloode.pdf

[2] Refers to system of governance defined under the saga of political dialogues in Europe in the seventeenth century which finally evolved into the Peace of Westphalia in 1648 marking the beginning of the modern system of national identities which allowed absolute jurisdictional freedom within a state's defined territory. Refer to Filho (2007).

[3] Article 21(3) of the Universal Declaration of Human Rights, 1948, Available at https://www.ohchr.org/en/udhr/documents/udhr translations/eng.pdf

of climate change and disasters. Here the autonomy of private stakeholders particularly in transboundary areas should be regulated because privatization often focuses on profit maximization even at the cost of environmental sustainability. Also, more often than not assessment mechanisms within the environmental governance regimes are often inconspicuous (Olagunju and Blakely 2017) which creates opportunities for private capital to mobilize and exploit state systems for their own benefit.

Soliev and Theesfeld (2020) offer a critical argument that no single state can ideally claim exclusive rights over common resources in transboundary areas because of jurisdictional, administrative and ethical limitation. They also argue that it cannot also belong to any private individual or entity because of the same jurisdictional limitation, which in principle prohibits any state actor to take unilateral decisions on these resources. Because of complexities of ownership of systems and its apparatuses, it becomes more important that any state initiative taken within its jurisdictional area or with any private actor, should be risk informed, ensuring the well-being of the people living in those areas. It is also important to understand that international law on water co-operation is possible only under three essential conditions:

(a) the equitable distribution of resources,
(b) taking appropriate risk informed measures and,
(c) preparedness from hazards (Baranyai 2020).

This is however possible only if states agree to take joint steps to achieve shared results towards fostering resilience of transboundary ecosystems.

Risk informed governance is an integration of continuous collaborative and consultative engagement undertaken by relevant stakeholders to measure, analyze and optimize the process of production and development promising sustainability and preventing risks. Clearly, it is important that risks are understood and managed through a multi-stakeholder engagement where no single actor can take advantage of the resources and the loopholes in the systems of governance for their own benefit. But how can progressive, people-centric transboundary risk governance be ensured? This is where the symphony of co operation takes a crescendo in the form of diplomacy and solidarity which not only involves the relevant states but also other stakeholders including members of the civil society, representatives from communities living in transboundary areas as also the private actors.

3.2 Drawing Principles from Field-Based Actions

Transboundary basins often get neglected because of the lack of organized governance apparatus to manage risks. The general provision within article 2(a) (b) and (c) of the Convention on the protection and use of transboundary watercourses and international lake held at Helsinki in 1992 emphasize the protection of shared waters from any kind of harm. It appeals to the parties signing the convention to take appropriate steps to reduce pollution and the use of shared water resources in an equitable manner[45]. Owing to mismanaged information, lack of accountability, integrity and more importantly limited risk solidarity, managing risks has been very difficult in the past. This calls for some guiding principles which are based on the ethics of solidarity and justice and which can also become the foundation for a comprehensive Disaster Management plan.

This paper draws its learning from the field to critique the current approach to transboundary risk governance along with the principles, values and ethics embedded in the arguments. Most of the observations in the form of arguments are taken from the transboundary regions in the Ganga-Brahmaputra Basin. The principles herein could be understood in alignment with the previous introductory analysis as to why the current approach of nation states is exclusive in nature. Governance in the post industrialized era cannot occur in an isolated manner and pragmatic approaches to transboundary co-operation need to be people centric rather than state focused. There is a need to move from an engineering approach which is mostly confined to the usage of water by state parties, to a more community-centric approach where the idea of human dignity is given precedence over the volatile state egos. Such an analysis may also be helpful for the various "stakeholders" to engage in a more participatory dialogue and protect the rights of the people at the margins from hegemonic and predatory tendencies. These principles are focused on:

1. managing a meaningful information system,
2. ensuring accountability to-at-risk population,
3. facilitating localization and cultural sustainability,
4. adopting innovation in structural and non-structural measures in transboundary areas,
5. strengthening dialogues over shared resources,
6. protecting community's right to the river and,
7. promoting regional and global solidarity on risks.

Each of the principles and the ethics governing them are explained in details below with examples from the South Asian experience.

3.3 The Principles

3.3.1 Managing Transparent and Meaningful Information System

One of the fundamental aspects of transboundary risk governance is the dissemination of meaningful and timely information of impending risks, catering to the needs of the most marginalized on both sides of the border. A major sub principle is to estimate the

[4] Convention on the protection and use of Transboundary water course and International Lakes done at Helsinki, on 17 March 1992, Available at https://unece.org/fileadmin/DAM/env/water/pdf/watercon.pdf

timeliness of information dissemination because after the optimum time threshold, such information can be meaningless. In transboundary areas, state actors should build systematic mechanisms for timely early warning dissemination in order to ensure that there are no gaps or breaks in the complete information channel. An ideal channel is transparent in flow not only to all the state departments on both sides of the border, with adequate information of the probable impact that a hazard can generate, and it should also reach those who are responsible for responding to vulnerabilities on both sides. For instance, states can agree to co-operate on an effective transboundary flood information management system which facilitates the flow of information from the grassroots to the relevant departments responsible for maintaining and acting on information on loss and damages. State actors can also take the support of the civil society groups and community volunteers to disseminate information on early flood warning reaching the last mile. A well-organized communication channel should have a pre-established network working as early warning governance collective. Once information is received, the possible damage to critical infrastructure, loss of human life and migration triggered due to risks can be avoided through the execution of a well-designed preparedness and response plan.

Dissemination of early warning information without a proper response plan would not be effective because even if the community receives the information on time, in the absence of provisioning for evacuation, such information would not be helpful. In principle, a hazard becomes a disaster because of the lack of preparedness and the latter is a product of a weak understanding of the vulnerability of any geo-spatial location (Andharia 2020). In the Mahakali Basin, shared by both India and Nepal, it was found that there is huge gap in the mechanism of dissemination of early warning information primarily due to two reasons (a) lack of co-ordination between the state actors in both the nations and (b) myths and misinformation associated with hazards. The downstream communities have a preconceived belief that any water-logging in their villages is due to the negligence of the upstream local governments in Nepal. The case however is that when water is being released from the Sharda Barrage in Tanakpur Uttarakhand, none of the downstream communities receive information in a timely manner. Upadhyay et.al (2020) have highlighted some of these critical issues with regard to the blockage in the early warning information channel in the Mahakali and how it has affected the communities of both the nations. With the burden of responsibility passed on to the other stakeholders, a scene of subtle discontent is prevalent among transboundary communities with no one taking responsibility for the loss of lives and property. Box 1 provides an example of Oxfam's initiative in India.

Box. 3.1 Technology Enabled Transboundary Early Warning System

Oxfam India along with some of the civil society members is one of the pioneers in India that has taken a lead to improve the dissemination of early warning in the transboundary areas of South Asia especially in the Ganga-Brahmaputra Mega Basin. Their well-established basin level network has a proven efficiency in transmitting information on disasters. During the COVID 19 pandemic, it was through such established four network that the community was provided information on the safety precautions to be taken. A web-based technology known as *Adhikarvani* (translated as "Voice for Rights") was installed and implemented to disseminate information on early warning to communities living in the last mile including transboundary areas.

3.3.2 Ensuring Accountability to at Risk Population

Since the state is the common thread which binds people in a contract, the primary responsibility and accountability of progressive risk governance lies on it. To structure a goal-oriented risk governance framework, there is no alternative to co-operation with other states especially on understanding the felt needs of communities living near and around such shared resources. Since there is no entity above the state, there can be no other governing authority beyond them except the bodies created under various agreements and treaties. Therefore, it is suggested that if any entity is more responsible for effective transboundary risk governance, it is the State because of its intrinsic role and mandate to participate in agreements, conventions and treaties. Without a proper framework explaining the nature of accountability of each state, any transboundary dialogue would make little sense. Civil societies can merely facilitate dialogue with support of the people but unless states sharing transboundary resources comply with their cardinal responsibilities to govern their borders in a co-operative manner, their efforts would not be very fruitful. Further, states should engage its citizens on a more polycentric form of governance because of the changing nature of borders itself. Such an approach is under realization in South Asian nations where locally elected bodies (*panchayats*) are playing a relatively better and proactive role in managing risks than the national governments that sign treaties. As compared to earlier times, these bodies through the support of the civil society have been trying to embrace greater accountability for disaster governance. In India, Nepal and Bangladesh such accountability is seen reflected in community level plans which are imbibing the principles of community-based risk informed governance (Upadhyay and Prakash 2019). However, community-based institutions have limitations and since laws on diplomacy lie in the hands of the larger bureaucratic apparatus, it is time that decision makers at the macro level take the initiative to consult with the community directly and create models of responsible governance in peripheral, transboundary areas. In this, the state has to ensure that accountability is not imposed on one another, because as argued earlier, the river does not obey human laws of governance and hence it is axiomatic that we begin to frame our policies according to changing climatic and resource scenarios. Since the burden of risk lies more on people and since the state is formed through a social contract by the people, it is one of the most fundamental duty and obligation of the state to protect people from any disruption in the nature-human relationship equilibrium caused by natural hazards.

3.3.3 *Facilitating Localization and Cultural Sustainability*

Riparian border areas should have their own local plans which may not contradict the state policy but adhere to the principles mentioned in the United Nation's Agenda 21. These plans should be based on local risks which influence human lives in general. For this it is very important for states to provide autonomy of governance to the administrative divisions in border areas so that local governments on both sides of the border can hold continuous dialogue, discussions and negotiations. In fact, hydro-diplomacy should begin from the grassroots at the borders rather than from the centre, in order to develop the understanding of risks more organically from the local community. On the other side of spectrum however, the potential capacities of such community level institutions are being harnessed by the civil societies who have been able to bring communities across borders in South Asia for dialogue, to foster harmony and co-operation. The Indian civil society organizations like Oxfam have taken such initiatives to bring the local panchayats and the people together to understand their risks and then discuss the nature of local intervention to be carried out. For example, the Lohaghat Consultation held in January 2020 here the grass root community-based institutions and experts from Oxfam's water resource pool came together and mapped various vulnerabilities in the basin with specific focus on women, children, especially abled and the elderly. One of the most important aspect of the consultation was to conceptualize water risks considering the local dynamics and then links them to existing policies on disasters in both India and Nepal.

Various networks formed under the banner of Water Governance Collective Action Network (WGCAN)[5] have been able to fill the vacuum on risk governance through co-ordinated efforts. In Nepal, the local government of Bhimdatta Municipality devised its own sand mining rules[6] to protect the river ecosystem from predatory private actors. This action in fact emerged out of a declaration made in Dhangadi. These initiatives although led by the civil society are examples of co-operation with the state to ensure accountability and hold great significance for the future of risk governance in South Asia. One of the most important points in the Dhangadi declaration[7] is the

---

[5] Water Governance Collective Action is a network of grassroot civil society organizations committed to strengthen transboundary risk governance. Refer to https://wgcan.in/About.aspx for more information

[6] River Sand Mining Guideline for ensuring environmentally sustainable and socially responsible mining in transboundary Mahakali river, Oxfam Briefing Paper, published by Oxfam in Nepal and supported by the

Government ofSweden. Availableat: https://cng-cdn.oxfam.org/nepal.oxfam.org/s3fs- public/file_attachments/Sandmining%20Policy%20Brief.pdf

[7] Dhangadi Declarations in Action, 2020 published by Oxfam in Nepal, Available at https://nepal.oxfam.org/latest/policy-paper/dhangadhi-declaration-actions-2020

formation of Federal and Provincial Level Expert Commissions which would aim at the conservation, utilization and development of the Mahakali River. The declaration also emphasized the formation of the Transboundary Mahakali Commission as envisaged in the Mahakali Treaty of 1996.[8] This declaration is considered a landmark because it demonstrated how pragmatically the principles of cultural sustainability could be implemented in a very coordinated and efficient manner.

In 2021 in another consultation in Pithoragarh in India, the local government representatives from Nepal appealed to the local governments in Uttarakhand, India to have a dialogue with them on shared risks so that the concerns of the marginalized population could be addressed in an efficient and co-ordinated manner. If such local hydro-diplomacy would continue with the local government of Nepal, the already strengthened ties between the two nations would help resolve several socio-economic challenges emerging out of shared risks. In the light of the recent catastrophic events in Uttarakhand, it is very important that local governments along with the civil society prepare their own local agenda and set an example of an ideal hydro-diplomacy to the world.

Sustainable development without the involvement of people in decision making process may prove to be mere textual adventure. If the United Nations Agenda 21 is initiated and led by the state, such an approach could revolutionize risk governance in critical geographies. In Chap. 18 of Agenda 21,[9] the significance of transboundary water resources is emphasized strongly with the special focus on the mobilization of riparian resources and the need for effective co-ordination between the states in the areas of integrated water resources development and management, water resource assessment, sanitation and drinking water, impacts of climate change on water and protection of aquatic eco-systems etc..[10] States as a custodian of these resources must take into consideration the involvement of people in decision making by protecting the sustainability of local autonomy over shared resources.

3.3.4 Adopting Innovation in Structural and Non-Structural Measures in Transboundary Areas

For effective risk governance, it is imperative that the state take appropriate structural and non-structural measures so that the shared invisible risks, detrimental to the community's social, economic and psychological well-being can be reduced. Structural measures may include establishment of transboundary emergency response

[8] Mahakali Treaty, 1996, Treaty between his majesty's government of Nepal and the Government of India concerning the integrated development of the Mahakali River including Sarada Barrage, Tanakpur Barrage and Pancheswar project.

[9] Agenda 21, Chap. 18. Protection of the Quality and Supply of Freshwater Resources: Application of Integrated Approaches to the Development, Management and Use of Water Resource, United Nations Conference on Environment and Development, Rio de Janeiro, 1992, "Earth Summit", Available at https://www.earthsummit2002.org/ic/freshwater/reschapt18.html

[10] Ibid, Chap. 18, Agenda 21, United Nations

systems, hazard resistant constructions in border areas as per agreed standards, joint investment on critical social infrastructure and construction of embankments in critical areas etc. To compliment such measures some of the Non-structural measures may focus on improving institutional mechanisms for emergency response in transboundary areas, developing regulatory framework for shared resilience, establishing standards for risk preparedness, mitigation, response, recovery and rehabilitation and ensuring systems reinforcement essential for risk informed transboundary governance. For this, states sharing borders should aim to establish functional well-operated systems with border area emergency operation centres especially for those geographies which are periodically devastated by recurring hazards. States on both the sides should co-operate and plan hazard resistant settlements and devise mechanisms for retrofitting of critical infrastructures including those which are lifesaving. Structural measures in the political sense is a state subject and due to lack of resources, civil society can only support the state in exploring the best ways to ensure a joint structural intervention. Within this it is also important that the civil society advocate for risk informed development initiatives so that the goal of the latter is not achieved at the cost of ecological, economic or cultural sustainability.[11] Since culture has a deep-rooted impact on human traditions, it is very important that local autonomy in transboundary risk governance be given due consideration.

Without ensuring and respecting the sustainability of existing social systems of the community, stewardship of shared resources cannot be ascertained. That is why all structural and non-structural measures are to be aligned with local interest and context and not modelled based on external expectations. Cultural sustainability assures the upholding of community interests and encourages promotion of local policies for the protection of natural resources and risk governance. For instance, nations sharing water resources may invest in the construction of embankments and levees through the respective local governments on both sides of the border. The administrative government at the local level should come together to develop joint evacuation plan for population and livestock that are the most susceptible to the vulnerability posed by specific hazards.

One of the best ways to ensure a comprehensive transboundary structural intervention is through social housing schemes for border communities in which planners and policy makers can imbibe the principles of cultural sustainability bringing local governments to design flood resistant structures catering to the needs of the people sharing common risks. As a matter of emergency preparedness, local government should be provided the autonomy to plan multi-purpose flood shelters for transboundary communities.

that face the risk of hydrological disasters. Other aspects of the holistic integrated transboundary risk governance could be the structural intervention to ensure de-silting of rivers and improvements in the overall drainage system especially in transboundary urban settlements. For this urban planning should be carried out in a joint manner encouraging towns across two borders to hold policy level dialogues and discussions.

[11] See Soini & Dessain (2016) for cultural sustainability in the broader perspective.

According to Lucaciu (2006) some of the important structural measures to be considered as "flood defense measures" are based on soil conservation, inundation reduction techniques, maximum water level reduction techniques and protection of people's lives and properties from hydrological hazards. Besides, non-structural measures also need important consideration. This includes investment in techno-legal regimes and reforms in institutional preparedness by the state. The National Disaster Management Plan, (GoI, NDMA 2019) of government of India is an excellent document which elaborates the roles and responsibilities of each state department in flood preparedness. However, the recognition of the vulnerabilities in transboundary peripheral areas is yet a struggle. This challenge is reflected in most South Asian countries' Disaster Management Plans. The components of integrated water resources management, reservoir management and basin level preparedness should demonstrate aspects of justice and equity for all the communities irrespective of their nationalities.

3.3.5 *Strengthening Dialogues over Shared Resources: Towards a Healthy Hydro-Diplomacy*

If anything is more harmonious in the symphony of transboundary co-operation, it is the idea of dialogue. It is the foundation on which communities, civil society and states together can potentially come together to devise plans for the governance of shared risks. In the Indo-Nepal border, the *Mahakali Samwad* (Mahakali dialogue)[12] has gained immense popularity among the civil society, local governments and the community of the river basin. It is a progressive platform started by Oxfam wherein communities sit together, map and understand common risks, and then chalk out plans for preparedness and mitigation. One of the crucial roles played is by a network of civil societies known as the *Indo-Nepal Joint Action Forum* which has been instrumental in bringing the stakeholders together for a dialogue on the issue of timely dissemination of early warning information about hazards.

Although the dialogue on co-operation is led by the civil society groups, they have been well-appreciated by the local government and policy makers. Through this transboundary dialogue, the *Samwad* also facilitates transformational change in the riparian societies through institutionalization. Sharda Nadi Nagrik Manch (SNNM)[13] in Uttar Pradesh which aimed to foster the values of transboundary co-operation among the community leaders, is an example. Mostly led by women, this network regularly meets with another formal network of water leaders known as *Women's Empowerment cell* located in various parts of the basin within the boundary

[12] Mahakali Samvad, A Civil Society Organization lead Dialogue in Mahakali basin, Published: December 2019. https://oi-files-cng-v2-prod.s3.eu-west-2.amazonaws.com/nepal.oxfam.org/s3fs-public/file_attachments/Mahakali%20Sambad%20English.pdf

[13] Rivers of Change, Oxfam India, 2019, https://oi-files-cng-v2-prod.s3.eu-west-2.amazonaws.com/nepal.oxfam.org/s3fs-public/file_attachments/Mahakali%20Sambad%20English.pdf

of Nepal. In 2021, women leaders on both sides sat together for another dialogue on risk co-operation wherein they volunteered to organize the *Samwad* through their own efforts and also took the lead to manage the early warning system through coordinated action under the banner of "*Water Governance Collective Action Network*".[14]

Transboundary Risk Co-operation is ensuring a scientific method of River Basin Management. It cannot be achieved unless co-operation is achieved at multiple levels. One of the most successful events in the Indo-Bhutan Border in the recent past. The civil society agreement to co-ordinate and manage water resources and an acknowledgement of that fact that effective cultural governance is the effective key to ensure justice for riparian communities. Through a consultative and deliberative processes, the civil society organizations from India and Bhutan came together to enter into a dialogue along with the government, academicians and the community in framing a transboundary policy. It was called the Kokrajhar Call for Action which advocates for 7 agenda points to secure transboundary co-operation for improved water governance: These are:

1. Strengthen transboundary community ties for ensuring effective riparian resources management.
2. Facilitate and promote community-led management and conservation practices, including cross-border basin-level interactions.
3. Acknowledging the role of gender in water governance and facilitating a space where gender voices take the lead in transformational changes.
4. Strengthen transboundary co-operation for preparedness to combat water-induced disasters, and lessen the stress generated by pollution and other forms of vulnerabilities impacting the general life and livelihood of river basin transboundary communities.
5. Continuous engagement with governments on both sides of the border for inclusive TBWG.
6. Media to highlight events of significance so that unheard voices may reach the concerned policy makers.
7. Dialogue and collaboration through research and knowledge partnerships.[15]

[14] TROSA Learning Brief ISSUE 4/MAY 2020 Photo: Kamal Rawal/ RUWUDC CITIZEN SCIENCE Engaging and empowering local communities, Available at https://oi-files-cng-prod.s3.amazonaws.com/asia.oxfam.org/s3fs-public/file_attachments/Citizen%20Science%20-%20Engaging%20and%20Empowering%20Local%20Communities%20-%20TROSA%20Learning%20Brief.pdf

[15] Bhutan-India CSO's Partnership for Inclusive Water Governance, Accessible at https://www.sentinelassam.com/news/001-bhutan-india-csos-partnership-for-inclusive-water-governance/.To mark the potential of such dialogues of generating co-operation between transboundary cultures and the immense impact it could produce, the words of the Former Prime Minister of Bhutan, KinzanDorje are noteworthy: "*I congratulate the visionary civil society leaders of both the countries for being able to arrive at an agreement, which will promote better ties for both the countries for a noble cause. This remain as a source of motivation for civil society leaders across the borders*". This in line with the vision that Oxfam India has set for ensuring transboundary co-operation. It believes that unless South Asian nations arrive at an agreement on the policies on Transboundary Rivers, ensuring justice to riparian communities and their ability to use their full potential, would not be possible.

The consultation emerged through an evolutionary process wherein the instrumental nature of culture paved the way for dialogue among representatives from the state and civil society to discuss and deliberate on existing and emerging shared risks. The pre-existing community ties which facilitated the governance of the traditional system of irrigation and agriculture, acted as a catalyst in boosting the efforts of civil society to improve water governance in the region. The civil society network—the Bhutan-India Friendship Association (BIFA) identified two major goals—one to foster socio-economic and cultural ties between the people of India and Bhutan and second, to promote community ties through "people to people network" and strive for peace in the region. As a matter of fact the transboundary communities at the Indo-Bhutan border areas have traditionally been maintaining a good relations and that is why such friendship networks could fructify. One such medium of co-operation is being manifested through the traditional *Jamfwi*[16] (an indigenous method of irrigation) maintained by the farmers in the Saralbhanga Basin and other adjoining areas, have been an intrinsic part of the indigenous culture since centuries. The Dong System which, although is traditional, embraces scientific method of water management which includes aspects of preparedness strategies and systematic waste management. The operational component of the entire Dong.[17] Management is through people's governance and cultural integrity where a canal is diverted to agricultural fields for irrigational purposes. From the small canal, other smaller units are made which reach the fields where crops are sown. While the practice has been traditional, it's more organized use in the Indo-Bhutan Border is seen since the early 1990s when informal agreement between the communities on both sides started to evolve into a more systematic form of water governance. The impact of such community-based governance is so immense that Baksa district of Assam sharing its borders with Bhutan has always remained impact-free from the tentacles of drought.

3.3.6 Protecting the Communities' "Right to the River"

In the Philippines, Thailand, Indonesia and parts of north east India, community managed irrigation system takes ownership of the common pooled resource through institutionalization, organization and operationalization where each member enjoys rights only in relation to their responsibilities which are not defined by the state but by a historical process embedded in community's traditions (Hirashima and Gooneratne 1996).Transboundary common property resources like rivers, mountains, lakes

[16] Refer to the article published by the third pole.net Yashwant (2018), accessible at https://www.thethirdpole.net/en/2018/08/27/villagers-in-bhutan-and-india-come-together-to-share-river/ This report is part of documentation by the author of Oxfam India's Transboundary Rivers of South Asia (TROSA) project in India. This five-year (2017–2021) regional programme, funded by the Government of Sweden, works with communities in the Ganges-Brahmaputra-Meghna (GBM) in Nepal, India and Bangladesh and the Salween in Myanmar.

[17] Refer to Rivers of Hope published by Oxfam India, accessible at https://www.oxfamindia.org/sites/default/files/Oxfam-Report Final.pdf

and others, constitute the subsets of the larger ecosystem which is not only shared by people but also governed by them through cultural governance practices through which traditional institutions protect the natural surroundings in a sustainable manner. In South Asia many transboundary villages have embedded conservation, preservation and preparedness in their way of life through traditional forms of institutionalization and co-operation with their neighbours who are otherwise administered by a different state.

The community's right to the river as a pragmatic solution can never be fulfilled unless the local inhabitants on both sides of the borders are given the right to maintain their local resources and the ownership of the risk governing apparatus. Common property resources are people's own resources and therefore their concerns should be at the core of the governance framework. For the community to be at the centre stage of governance, stakeholders of common property should also consider the normative forms to complement written constitutional laws. Some of these forms could also be seen more in practice rather than in any code. Such approaches are very functional when it comes to the governance of resources in border areas because populations living at the administrative peripheries often do not enjoy the privileges of freedom as easily as it happens in the mainland. Civic engagement over community resources refers to a freedom with ownership and the principle of ethical equilibrium. Here even though an individual or a group with vested interest tries to dominate the resources through planned hegemony, the ownership and solidarity principles prohibit them to take advantage of individual free-will to exploit resources for their own/private benefits. This could be best understood by observing the management of traditional canals and dams in many cultures as mentioned previously.

3.3.7 *Promoting Regional and Global Solidarity on Risks*

Transnational solidarity is the only ray of hope for an effective risk co-operation in transboundary areas. The global civil society is a powerful and progressive force which can shape ideologies and save communities from becoming the victims of neoliberal hegemonies and disaster capitalism (Klien 2007). Every risk has a cascading effect on regional and global economy and hence it is very important that the global community always has a strong voice on the rights of the community to a safe living environment. However it must be remembered that the global solidarity mechanism should be such that it appreciates community integrity and their right to a perspective. In the South Asian context, since the nations are sharing a common "water crisis" (Khalid et al. 2014), it is likely that disputes over water resources in the coming decades would aggravate further (Ranjan 2020). Therefore, it is important that states rise above the idea of sovereign ownership and endeavour to find ways for peaceful diplomacy. For instance, if two nations are sharing common hydro resources then the path of co-operation is only through water.

Absence of this fundamental understanding is, in fact, one of the major reasons why conflicts among nations are not yet resolved.

There is a need to revive the nature of diplomacy among nations because even in hydro-solidarity and hydro-diplomacy, water is just a medium. The very core of this is to strengthen the idea of peace among nations through economic, political and cultural co-operations. In 1996, scholars like Malin Falenmark during the IUCN World Conservation Congress pitched the idea of hydro-solidarity whose applications in the field of hydro ecology and integrated water resources management was further elaborated by Varady et.al (2014) for its multi-dimensional application. They emphasize that hydro-solidarity does not mean only co-operation on water resources. States, civil society and other actors should join hands to work together to reduce the impacts generated by structural inequalities leading to discrimination. Although climate activists globally have been demonstrating a sense of preparedness to the possible disasters due to climate change, there is a failure to acknowledge the real nature of risk lies not on the failure of diplomacy but on the very orientation of nation states.

3.4 Conclusion

The symphony of co-operation can only take place in a peaceful environment. Since disasters have become more frequent than before, it is an appropriate time for states to take the lead for preparing the governance apparatus through improvements in the current structure of diplomacy. Transboundary water co-operation also does not mean only fulfilment of international conventions and hydro ethics in policy documents but also signifies commitment through practical steps taken in consultation with the community on finding avenues to reduce the impact of risks. More importantly states should hold dialogues on reducing the structural inequalities leading to discrimination in South Asian societies. One principle which should be essentially inscribed in all diplomatic endeavours is that legal framework should first pass the litmus test of ethics. If people-centric risk apparatus are to be formed, then state-led development initiatives should imbibe the principle of "community's interest first". The local governments require more autonomy for decision making so that policies on "the right to the river" comes from the local context. Transboundary areas require special forms of poly-centric governance with a different set of rules and regulations which not only promise economic but also ecological sustainability. More importantly co-operation around shared risks may not necessarily be confined to hazards but other avenues of life which includes social interaction and livelihood exchanges. Unless the condition of the community during normal times is made better, any assurance for their well-being during disasters would ring hollow. Hence it is important that states and other stakeholders of risk governance acknowledge that mere production of white papers would only create a false hope of humanitarianism. It is high time that the world begins to understand the lived experiences of those whose daily survival is amidst risks.

References

Andharia J (2020) Blurred boundaries, shared practices: disaster studies and disaster management as an emerging discipline and a field of practice in Andharia (ed) disaster studies: exploring Intersectionalitiesin disaster discourse. Springer Nature, Singapore

Andharia J, Mathew T (2019) Common Property Resources and Environmental Justice in a World at Risk: What Kerala Has to Offer to Participatory Recovery and What It Can Learn, Social Science in Perspectives, Special Issue on Common Property and Citizens Rights: Issues of Reconstruction of Kerala, Vol.10, No.4: October–December 2018 and Vol.11, No.1:January—March. pg. 408–433. [ISSN: 0975–5497]. Published by C. Achutha Menon Study Centre & Library, Thiruvananthapuram

Baranyai G (2020) European water law and hydro politics: an inquiry into the resilience of transboundary water governance in the European Union. Springer Publication

Ebi K L (2010) Transboundary Risk Governance. Environ Health Perspective., Review of the Book 'Transboundary Risk Governance' by Rolf Lidskog, Linda Soneryd, and Ylva Uggla. Sterling, VA: Earthscan, 2010. 155 pp. ISBN: 978–1–84407-791-5

Filho M T F (2007) Westphalia: a paradigm? A dialogue between law, art and philosophy of Science. German Law Journal 8(10):955–975. https://doi.org/10.1017/S2071832200006118

Hirashima S, Goonerante W (1996) State and community in local resources management in Asia: an introductory discussion. In: Hirashima S, Goonerante W (eds) State and community in local resources management: the Asian experience. Haranand Publications, New-Delhi, India, pp 13–40

Khalid P I, Mukhtar A, Ahmed Z (2014) Water scarcity in South Asia: a potential conflict of future decades. 280, Journal of Political Studies, Vol. 21, Issue—1, 2014, 259:280

Klein N (2007) The shock doctrine: the rise of disaster capitalism. Metropolitan Books/Henry Holt and Company

Lucaciu M (2006) Territorial Flood defense: A Romanian Perspective, page 320–21 in the edited book by Marsalek, J., Stancalie, G., & Bliant, G., National Disaster Management Plan, 2019, National Disaster Management Authority, Ministry of Home Affairs, Government of India

NDMA, GoI (2019) National Disaster Management Plan. National Disaster Management Authority, Ministry of Home Affairs, Government of India

Nincic M, Weiss M (2016) *The Future of Transboundary Water Conflicts,* Political Science Quarterly, Vol. 131, No. 4 (Winter 2016–17), pp. 717–748, The Academy of Political Science

Olagunju A O, Blakley J A (2017) *Towards an Environmental Governance Agenda in Regional Environmental Assessment*, Journal of Environmental Assessment Policy and Management, June 2017, Vol. 19, No. 2 (June 2017), pp. 1–33, World Scientific Publishing Co., Inc.

Ranjan A (2020) Water issues in Himalayan South Asia: internal challenges. Disputes and Transboundary Tensions, Palgrave Macmillan, Singapore

Soini K, Dessain J (2016) Culture sustainability relation: towards conceptual framework. Sustainability 8(2):167. https://doi.org/10.3390/su8020167

Soliev I, Theesfeld I (2020) *Benefit Sharing for Solving Transboundary Commons Dilemma in Central Asia International Journal of the Commons*, International Journal of the Commons, Vol. 14, No. 1 (2020), pp. 61–77

Upadhyay M, Prakash A (2019) Understanding Cultural Sustainability in the Context of Transboundary Water Governance The Case of the Mahakali River Basin, Accountability is in Everyone's interest, Issue No: 184, December 2019, Southasiadisaster.net

UNECE (2009) United Nations economic commission for Europe convention on the protection and use of transboundary watercourses and International Lakes transboundary flood risk management: experiences from the UNECE region, United Nations New York and Geneva, page 1, para 3., Available at: https://unece.org/fileadmin/DAM/publications/oes/Transboundary_Flood_Risk_Management_Final.p df

Varady R G, Gerlak K A, McGovern D E (2014) Hydro solidarity and its place in international water diplomacy, Hydro Diplomacy*: Sharing Water across Borders* edited by Pangare, Published by

Academic Foundation, New Delhi, Financed by The Minister of Foreign Trade and Development Cooperation, The Netherlands and the International Conservation of Nature, 2014
Varis O, Tortajada C, Biswas A K (2008) *Management of transboundary lakes and rivers, water resources development and management*, series editors: Asit K. Biswas and Cecilia Tortajada, Springer Publication, Berlin
Yashwant S (2018) *Villagers in Bhutan and India come together to share river*. August 27, dialogue earth, London

Mukunda Upadhyay is a seasoned Disaster Management Professional based in India, with years of extensive experience in designing and implementing disaster risk reduction (DRR) policies and programs. He specializes in strengthening systems for effective emergency responses, developing technological solutions for project management and response governance, and designing recovery frameworks for long-term resilience. He has been credited with creating systems that enhances disaster preparedness, response, recovery, and resilience outcomes in different settings. His structural and non-structural models, designed and implemented to support vulnerable communities in disaster-prone regions, have received significant appreciation. He has also been instrumental in various transboundary water governance initiatives in the Mahakali basin, emphasizing citizen science, early warning systems, cultural sustainability, and promoting a sense of stewardship among various critical actors. In the past, he took the lead in organising global and regional dialogues with a wide range of stakeholders on water stewardship, riparian community rights, water footprints, and transboundary cooperation. His role in designing technology-embedded platforms for project management, disaster governance, and community-based transboundary early warning systems has effectively contributed towards achieving desired system outcomes. Apart from programmatic endeavours, he has also mentored scholars on critical disaster risk reduction research and has been an invited speaker at several state level, national and international forums. He is an alumnus of the prestigious Jawaharlal Nehru University (JNU), the International Institute for Population Sciences (IIPS), and Assam University (AU).

Chapter 4
Inclusive Collaborative Water Governance: Managing Risks by Drawing on Australian Experiences, Including Failures

Melita Grant, Cameron Holley, and Emma Carmody

4.1 Introduction

In December 2019, one thousand people travelled in a convoy of trucks to Australia's Parliament House to protest against the seven-year-old transboundary Murray-Darling Basin River Plan. Most of the people who attended the event were farmers, and held signs such as, "No water, no farm, no future" (ABC 2019a), and were calling for the end to the Murray-Darling Basin Plan. It was not the first time that community members and farmers had gathered to protest about the way in which Australia's most significant transboundary river was being managed. In October 2010, copies of a draft guide to the proposed Murray-Darling Basin Plan were burnt in piles (ABC 2016; Thompson 2011), and throughout the reform process, Indigenous people (Traditional Owners), community members, and environmental advocates had called for commitments to saving and sustaining the river system be honoured (Horne 2017; Holley and Sinclair 2018; Carmody 2018).

While these moments of protest tell only a small part of the Murray-Darling Basin transboundary water governance reform story, they are clear indications that during its evolution, the tensions felt by communities were significant and complex. They also point to larger lessons that could and should be shared with other jurisdictions

M. Grant (✉)
University of Technology Sydney, Sydney, Australia
e-mail: melita.grant@uts.edu.au

C. Holley
UNSW Law & Justice, Sydney, Australia
e-mail: c.holley@unsw.edu.au

E. Carmody
Director of Legal and Partnerships at Restore Blue, Sydney, Australia
e-mail: emma@emmacarmody.com

J. Andharia and P. Anand (eds.), *Shifting Lands, Flowing Waters*, Disaster Studies and Management, https://doi.org/10.1007/978-981-96-7691-0_4

considering transboundary water governance approaches. In this chapter we seek to identify and examine some of the key challenges Australia has faced with respect to governing and coordinating this significant river system, and share insights related to collaborative governance processes and opportunities.

This discussion supports the broader aim of supporting risk-informed approaches to transboundary water governance systems. In this chapter we use the definition of risk-informed approaches offered by Opitz-Stapleton et al.: "risk-informed development is a risk-based decision process that enables understanding of multiple concurrent threats and complex risks to and arising from development decisions and acting on that knowledge" (Opitz-Stapleton et al. 2019, p. 9). In order to achieve resilient and sustainable development, a broad range of threats need to be fully considered (beyond natural disasters and hazards) in terms of how they pose risks and opportunities to development (Mendoza et al. 2018; Opitz-Stapleton et al. 2019). As advocated by the Organisation for Economic Co-operation and Development (OECD) Water Governance initiative, "Further [water-related] conflicts will be inevitable without a serious transition from crisis management to risk management" (Akhmouch et al. 2018). In this chapter, we argue that risk management can't be achieved without more focus on inclusive and collaborative governance processes. Indeed, many of the challenges that Australia has faced in the Murray-Darling Basin may have been ameliorated with a more inclusive collaborative approach to policy making, implementation, and enforcement. Drawing on established collaborative governance and risk-informed frameworks, the overarching question we ask is: How can inclusive collaborative governance that draws on Australian experiences, including failures, support risk-informed planning in transboundary water resources management?

Collaborative governance involves new coalitions between government agencies and civil society organisations, supported by guidelines, plans and agreements (Holley 2015). Collaborative governance contributes to risk-informed approaches by: a) enabling a broad range of actors to contribute to the knowledge we need to assess risks; b) encouraging the involvement of a broad range of actors who can mitigate risks by promoting improved ownership and compliance, and c) demonstrating how to build shared ownership of the vision and values upon which transboundary water governance agreements are based.

This chapter delves into who needs to be part of collaborative governance approaches. We pay particular attention to the differing levels of power held by stakeholders, and we focus on marginalised groups such as Indigenous people, women and gender-discriminated peoples, and those with less financial and political power. A concept that is new to the water governance space, but becoming well-understood in water, sanitation and hygiene (WASH) programming, is that of intersectionality. This concept refers to multiple layers of marginalisation based on gender, socio-economic status, ethnicity, disability, and other experiences and identities. We also discuss how a wide range of stakeholders can be engaged in collaborative processes, drawing on the OECD Water Governance spectrum of participation (OECD 2015b) and other inclusion frameworks that consider intersectionality.

Water governance always involves compromises and trade-offs, especially if water sources are scarce and becoming scarcer as a result of overuse, drought, and climate change in particular. People are unlikely to sacrifice their own economic interests, which makes compromises challenging, and conflicting views inevitable. Access to water is access to money and livelihoods, culture, and wellbeing, and Australia, like all countries, has seen these sometimes-competing interests lead to conflict. Collaborative governance has been put forward as a way to manage compromises and conflicting views. However, it requires resourcing, it is lengthy, it is no "silver bullet", and the benefits have been challenging to monitor and measure (Tadaki et al. 2020). Nevertheless, collaborative approaches are widely recognised as necessary given the complexity and diversity of the interests involved in water management, and the long-term costs to policy and implementation efficiency and effectiveness that flow from non-transparent deal-making, corruption, mistrust, and adversarial positioning. We therefore argue that inclusive collaborative governance is a key pathway for transboundary water governance, even though the path is challenging and multifaceted, and even though it will inevitably reveal tensions.

Australia has been experiencing a painful period of transition in managing the Murray-Darling Basin over the last decade, one that is ongoing, slow and evolving, but it could have been improved with greater attention placed on not only identifying common values, but acknowledging that certain values (for example environmental and Indigenous knowledge) have been marginalised in order to maintain historic water sharing arrangements. Revealing this breadth of values has required educating certain communities, especially the majority centre, who have traditionally taken their privileges for granted. In international forums this story is sometimes not articulated as openly as the sophistication of Australia's water trading system, data management and sharing platforms, and the restoration of some environmental flows. Despite the billions of dollars committed to the water reform process, decades of work at the national, state and local levels, and political prioritisation at certain times, commitment to the established governance framework has been tested. As one inquiry into Australia's water strategy put it: "Governance arrangements…have been significantly eroded over recent years" (Productivity Commission 2021, p. 53). Another inquiry report assessing progress made towards implementing the Murray-Darling Basin Plan since 2018 recommended a range of actions needed to achieve full implementation of the Basin Plan and to strengthen and simplify the framework (Productivity Commission 2023).

What is possibly missing most is a value-based discussion to underpin the transition to a system which considers the sustainable management of the Basin's water resources, and supports people to move away from using it in a way that undermines its ecological basis. Experts warned in 2011 that, "If the conversation about the future of the Murray-Darling Basin continues to be framed in terms of numbers, it will inevitably end in political compromise and leave most stakeholders unsatisfied" (Wallis, quoted in Thompson 2011).

A "Values and vulnerability of the Murray-Darling Basin - Basin-wide rapid assessment" conducted by Wahid et al. in 2023 centred on the ways in which climate change will create diverse vulnerabilities across different communities and ecosystems within the MDB. They emphasised that economic impacts would hit agricultural regions hardest, while social vulnerabilities will be greatest in remote communities with limited access to services and reliable water supplies (Wahid et al. 2023). The study revealed that while communities possess various adaptive capacities—from economic diversity to strong social networks—it remains uncertain whether these mechanisms will be sufficient for long-term climate adaptation (Wahid et al. 2023). The authors found that environmental values face significant threats, with reduced flooding under warmer-drier conditions negatively affecting wetlands, native species, and important bird habitats, and stress the need for better data collection and understanding of First Nations' cultural values and vulnerabilities, noting the profound spiritual connection to water that current assessment methods struggle to capture (Wahid et al. 2023; Productivity Commission 2023).

A values-based discussion needs to be complimented with a government-funded and/or well-governed market-based structural adjustment approach to support changes in people's livelihoods (e.g. transitioning away from certain types of farming which overuse water to the detriment of the environment and other users). Such a discussion requires a compassionate and long-term perspective, and one that allows for change to happen for the greater good, with consideration of future generations and protection of the commons (Ostrom 1990; Yazd et al. 2020). There will be resistance from some actors, as has been seen in Australia, where a market-based system of adjustment was perceived by some to favour economic factors, for example the maintenance of particular industries or rural townships (for a discussion of other concerns see ACCC 2021). There will therefore be compromises to make between existing economic conditions and benefits and sustaining the commons, as in all natural resources management scenarios. Diffusion innovation theory (Rogers 2003) tells us that focusing on the small minority of "laggards" (those who resist and disrupt change) will not be the most effective strategy in terms of time and resources. Rather, educating and supporting the middle group, or the majority, is an important focus in the transition towards any change or reform. This is because the members of the middle group are likely to be supportive of changes when they have been involved through education and engagement, can see the benefits involved, when they have heard success stories, and when they have seen others adopt the changes (Rogers 2003). A recent review of the management of the Murray-Darling Basin found that education about the governance mechanisms was lacking, that misunderstandings were widespread, and that this has contributed to undermining the Plan and governance mechanisms that have been put in place (Keelty 2020).

The context in which transboundary water management takes place naturally informs and shapes possible governance pathways and mechanisms. The political economy defines how a water source is valued, managed, shared, or monopolised. While this chapter gives examples from Australia, the purpose is to share lessons learned, including failures, to inform other contexts, without advocating prescriptive solutions. We now turn to the theory and practice of collaborative transboundary water governance, and how it can support and inform risk-based approaches.

4.2 Collaborative Transboundary Water Governance—Theory and Practice

Collaborative governance involves new coalitions between government agencies and civil society organisations, supported by guidelines, plans, and nonbinding agreements (Holley 2015). Extensive literature has discussed multiple forms of collaboration in environmental and other contexts. However, it was in the 1980s that examples of collaboration in transboundary water governance first emerged in the form of Integrated Catchment Management and Integrated Water Resource Management (IWRM). Although there is literature on these early forms of collaboration, the evidence and practice are still evolving, and they are taking many different forms around the globe. In many countries in the Global South, multilateral organisations such as the World Bank have promoted collaboration in the form of IWRM (Lenton and Muller 2009), and more recently, the Sustainable Development Goals (SDGs) have an explicit goal for partnerships (Goal 17), and Target 6.5 calls for 'integrated water resources management at all levels, including through transboundary cooperation as appropriate' (UN General Assembly 2015). Additionally, Target 6.8 encourages participation of local communities in improving water and sanitation management (UN General Assembly 2015).

Collaborative governance requires skilful and meaningful discussions about the values of water resources, linked to people's real lives and visions for their communities' future (Daniell 2011). The emergence of collaborative governance in transboundary water management has occurred in response to the confounding nature of water governance challenges (Holley and Sinclair 2014; Holley 2015; Holley and Sinclair 2016a), and a recognition that top-down processes alone have lost political legitimacy, and that deliberative and bottom-up processes are needed (Akhmouch and Clavreul 2016). Another driver has been the need for leadership which reflects the values of the broader community and the common good given the shared and interconnected nature of water systems including rivers. Yet potentially conflicting values include environmental, cultural, human rights, economic, regional and rural growth, and intergenerational rights which make collaborative governance challenging, but likely more effective in the long-term. Differences in values may be expressed by some actors who will value the economic benefits of using water for agriculture and industry over Indigenous water rights and ecological values, and vice versa. We therefore do not suggest that collaborative governance will result in a homogenous view with respect to managing water resources. Rather, we argue that that by engaging in meaningful inclusive collaborative processes, management of a common resource will have a greater chance of success over the longer-term, and more buy-in from water users and communities dependent on water sources. This is particularly true if such processes support discussion around historic and ongoing power imbalances with respect to water management (for example corporate irrigators v Traditional Owners), and the real-world impact of such dynamics. As Benham et al. (2014) point out, the development of social capital, through stronger and more extensive stakeholder networks, can be drawn on to facilitate coordinated approaches to addressing catchment-scale problems. It appears that often in practice, there is no meaningful process for considering the values and aspirations which underpin the functions of a governance framework. As Jimenez et al. suggest, "A visionary approach would start by defining the desired outcomes" (Jimenez et al. 2020, p. 14).

Collaborative governance approaches can vary depending on the stages at which collaboration occurs in the policy cycle. Some collaborations may be one-off, short-term efforts (e.g. consultation with affected stakeholders about the broad parameters of a water plan), while others may involve significant long-term relationships and stakeholder commitment, including implementation and/or enforcement of agreed rules and responsibilities (Head 2005). Australia's approach to collaboration is perhaps most prominently embodied in the National Water Initiative, which sets out high-level planning principles and a vision of community involvement, but leaves it to each state and territory to determine how best to undertake planning activities (COAG 1994; NWI 2004). Regardless of the stage of the policy cycle, collaborative governance is increasingly identified as aspiring to involve inclusive participation, integration, knowledge generation, learning, and adaption (Holley 2015; de Burca et al. 2013). Collaborative approaches are capable of gathering data from alternative sources such as place-based knowledge and citizen science. These sources can be a valuable complement to scientific data (Holley et al. 2012).

Collaborative approaches may have the potential to better balance power differentials by diffusing decision-making and emphasising multiparty deliberation. Moreover, having a mix of decision-makers and capacities can create greater mutual accountability between affected actors and decision-makers, enabling conflicts of interest to be managed more effectively. In addition, collaborative decision-making can help to ensure that consideration is given to the interests of groups which may otherwise be overlooked (for example Traditional Owners, women and other marginalised groups). Farmers, communities, and environmental interest groups may also work together through collaborative governance mechanisms to manage the full spectrum of impacts, and this can in turn strengthen risk-informed approaches (Bice and Moffett 2014). More generally, given the resource-related limits of water regulation, collaborative approaches can mobilise the knowledge of non-government organisations (NGOs) for informing decisions; engaging communities; monitoring, evaluation, and learning, thus improving the effectiveness of water governance regimes.

The precise form that collaborative governance takes in practice remains an open question, and even when developed, it too will confront its own implementation challenges. Indeed, aiming for an ideal level or type of collaborative governance is likely to be costly to implement and time consuming to pursue. Nevertheless, glimpses of such a model at work have previously been identified in the 2000 EU Water Framework Directive, the California Bay Delta Plan, the Chesapeake Bay Plan, and other initiatives. In urban contexts, an examination of the application of a collaborative, risk-informed, triple bottom line, multi-criteria decision analysis planning framework for integrated urban water management in Fort Collins, Colorado[1] found that "more inclusive participation throughout the study and maintaining a cooperative approach allowed for the identification of obstacles to project success and the adaptation of a decision framework to address those obstacles" (Cole et al. 2018, p. 15). Contrastingly, a study of three collaborative water governance in three regions of New Zealand, did not find that community confidence in water management decisions was increased overall as a result of collaborative approaches being employed. The reasons for this finding, however, are not known, and the authors point to the need for qualitative information to accompany quantitative studies aiming to evaluate the effectiveness of collaborative water governance and some of the challenges that may occur in implementation (Tadaki et al. 2020).

Collaboration is not limited to visioning and planning, and can include enforcement and compliance mechanisms. Options to collaboratively enforce water governance agreements is an emerging field, and may include a range of options including traditional government regulators, stakeholders (e.g. collaborative water user groups) and third parties playing different co-regulatory roles. In contrast to situations where there is pure self-regulation (Ostrom 1992), under co-regulation a designated collaborative group can assume some of the compliance and enforcement functions of a government regulatory authority. In such models, the government regulator may retain some compliance and enforcement functions such as more formal and onerous sanctions, and may reserve the right to step in (by invitation or unilaterally) if and when it considers non-compliance has occurred in spite of collaborative enforcement efforts (Holley et al. 2012).

4.3 Collaborative Transboundary Water Governance as the Basis for Risk-Informed Approaches

The international standard on risk management (International Standards Organisation (ISO) 31,000) defines risk as "the effect of uncertainty upon objectives", and in a world of increasing uncertainty (climate change, pandemics), risk-informed approaches could not be more relevant. A thorough and robust risk assessment process will help to map the key risks to people, animals and environments within a river basin or transboundary water source context, drawing on experiences of that context, modelling of future scenarios, and information from other relevant contexts. Decisions about how to balance the identified risks are underpinned by people's values which are likely to be varied, informed by their worldviews (de Vries and Petersen 2009). Agreeing on shared values helps to identify which risks should be prioritised in the management framework (i.e. to what extent the consequences

[1] The study explored alternative strategies for dual water supply within a local government context (Cole et al. 2018)

matter to most): "Deciding on the 'acceptable' level of risk involves not only scientific and technological evaluations, but also ethical, economic cost–benefit analyses and political considerations such as acceptability to the public and the development community" (Opitz-Stapleton et al. 2019, p. 41). Traditionally, when using the ISO Standards process, a multidisciplinary group of experts identify hazards, assesses the likelihood of the hazard occurring, assesses the consequences, and then uses likelihood and consequence to rate risks as high, medium or low. This information can inform the identification of issues of common concern for inclusion in a situation assessment. In this chapter we call for risk assessment processes to be informed by experts and local knowledge holders, as well as the failures and lessons learned, both in their own jurisdictions and other relevant jurisdictions. We also recognise that risks are often multiple and concurrent, and that transdisciplinary approaches will be needed to consider risks that relate not only to water resources, but also energy, climate, and social justice, at a range of institutional scales:

> Development cannot be considered sustainable or resilient unless it accounts for multiple threats and associated risks. Actions to reduce those risks and avoid risk creation need to be integrated into multiple levels of planning and delivery, from NGO and community-based programmes to government plans and the policies of international donor agencies (Opitz-Stapleton et al. 2019, p. 33).

Risk-informed development, as described by Opitz-Stapleton et al. (2019), starts with knowledge generation, and leads to three key interrelated outcomes: 1) avoids creating complex risks, 2) reduces complex risks, and 3) promotes iterative and constant learning. The more collaborative risk-informed processes can be made transparent and ongoing, the greater the likelihood that there will be community involvement, and the greater the likelihood that reforms and implementation related to transboundary water governance will be accepted and implemented. Informed by these intended benefits, Fig. 4.1 considers how inclusive collaborative governance and risk-informed processes could be mutually supportive as outlined in this chapter.

Good practice in risk-informed development will be inclusive, transparent, iterative, flexible and reflective (Opitz-Stapleton et al. 2019, p. 34). Firstly, decisions will be inclusive and transparent. This is important because the assessment of the severity and importance of risks is a value judgement, and so needs to be informed by a wide range of perspectives and stakeholders. Secondly, risk-informed development needs to be phased and iterative—generating knowledge about risks, and putting in place mitigation strategies happens over time, and over a number of phases of a project lifecycle. Thirdly, flexible and adaptable processes are needed to respond to different contexts and the dynamics of water and social systems. Lastly, continuous reflection is needed to learn from past mistakes in order to avoid repeating them. From a transboundary water governance perspective, we examine these principles, in particular inclusive and transparent stakeholder engagement, and learning from other contexts in order to avoid making the same or similar mistakes (in this case, Australia). While other governance approaches also emphasise these principles (e.g. adaptive water governance and elements of IWRM), arguably one of the more prominent and more developed approaches is collaborative water governance.

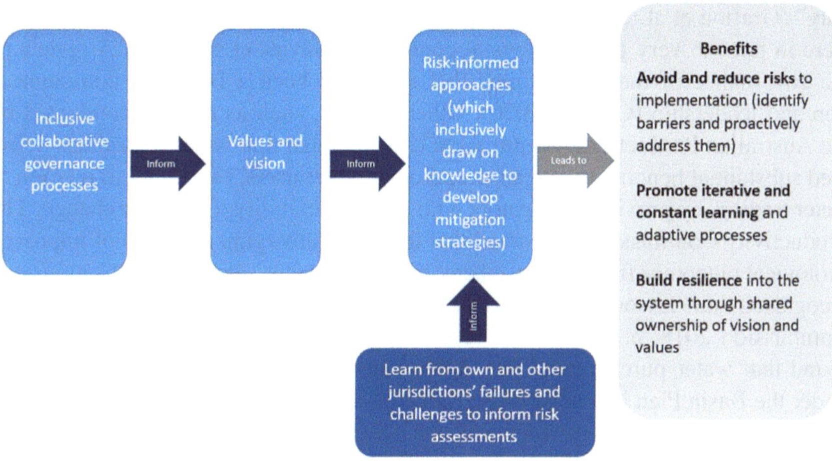

Fig. 4.1 Relationships between inclusive collaborative governance structures and processes, and risk-informed approaches for transboundary water management. Source: Authors, drawing on Opitz-Stapleton et al. (2019)

4.4 Background to the Murray-Darling Basin

Australia has relatively sophisticated and technologically enabled water governance systems, and it has adapted these systems as a result of crises, lessons learned, political interests, and community expectations and demands. The lessons learned are important to share globally, so that these can be considered by other jurisdictions in their risk-informed planning and options analysis.

The Murray-Darling Basin (MDB) is of great significance to Australia, due to its cultural and environmental importance, and because it produces one-third of the nation's food supply (Simons 2020). The MDB is home to more than two million people and 16 Ramsar-listed wetlands (Vidot and Jasper 2019). Furthermore, as explained by the Wentworth Group of Concerned Scientists, an eminent group of Australian water governance experts who convened in response to concerns about the lack of evidence-based policy making in relation to the MDB:

> The Murray Darling Basin is Australia's largest and most iconic river system and one of the biggest basins in the world by geographic extent. It covers over a million square kilometres in south-eastern Australia. An area larger than the combined size of France and Germany. It is Australia's most complex environmental asset and Australia's food bowl. The Basin is the traditional country of around 40 Indigenous nations. Healthy rivers, lakes and wetlands in the Basin are essential now and for future generations (Wentworth Group of Concerned Scientists 2019).

Despite the cultural, environmental, and economic value that the MDB provides to Australia, it has been described as being in a "poor state" (Carmody 2013; The Murray-Darling Declaration 2018; Australian Academy of Science 2019), and its governance has been described as a "failure of water reform, and a dereliction of duty" (Grafton et al. 2019). By 2012, 21 out of 23 major river valleys in the MDB were in poor to very poor ecological condition[2] (Davies et al. 2012). A review of the governance of the MDB found that there had been a failure in communication and leadership (Connell 2020). Other reviews, such as the one conducted by the Australian Productivity Commission (2018), noted that water reform has delivered substantial benefits to irrigators and other water users, especially in relation to water trading, by providing irrigators with greater flexibility to manage change. The Productivity Commission also noted that there was emerging evidence of improved ecological outcomes from increased environmental water, although the Commission recognised that "it would take time for the full benefits to be realised" (Productivity Commission 2018, p. 2). Another recent review by the Productivity Commission found that water purchased by the Australian Government to meet commitments under the Basin Plan had negative socio-economic impacts on some Basin communities. However, at a Basin-wide level, some aspects of economic performance have improved since the Basin Plan was put in place—for example, the gross value of irrigated agricultural production in the Basin increased by approximately 12% between 2013 and 2018, despite the volume of water used in irrigation declining by more than 16% over the same period (Productivity Commission 2023, p. 16).

Australia's water management reforms were developed in response to a "slowly evolving ecological catastrophe" (Carmody 2018, p. 38). Following a blue-green algae event on the Darling River in the early 1990s, and a water audit which revealed that water extraction had reduced the median annual flow of the Basin to 27% of the flows that would have occurred without development (Doolan 2016), a cap on extractions was introduced. This decision set up the drivers for water trading, water markets, and water accounting to measure compliance with the cap (Doolan 2016). In 2007, the Water Act was enacted and provided for Commonwealth oversight of water management in the MDB. Previously, water management had nearly exclusively been controlled by the state governments which struggled with transboundary contexts such as the MDB flowing through four states. A key ingredient in Australia's sophisticated and evolving governance system was setting a limit on the amount of water that can be taken out of the river system. Under this system, states are required by law to reinstate an "environmentally sustainable level of take", otherwise known as the sustainable diversion limit (SDL) which is the level of water extraction that can occur without compromising the health of the system. There is an SDL for the Basin as a whole, made up of SDLs for individual valleys and shared targets for connected systems. Fierce negotiations on what the SDL ought to be occurred in Australia, and under the finalised Basin Plan (which considers the management of the Basin as an interconnected system) SDLs were expected to deliver environmental outcomes, by setting a cap on extractions. However, the determined SDL set for the MDB was understood to fall short of being able to deliver on the intended environmental outcomes, and without providing scientific justification for the targets (Wentworth

[2] Only the Paroo Valley River system was rated in Good Ecosystem Health and only the Warrego Valley was rated in Moderate health. Fifteen valleys were rated in Poor health and six were rated in "Very Poor" health (Davies et al. 2012, p. viii)

Group of Concerned Scientists 2019). It has also been described as unlawful by a range of legal experts on the basis that it did not reflect the requirements of the Water Act, including because it failed to take into account climate change (Murray-Darling Basin Royal Commission 2019; Carmody 2018).

The AUD$13 billion Murray-Darling Basin Plan was signed into law in 2012, with water recovery for the environment expected to be fully operationalised by 2024. Unfortunately, Australia's transboundary water governance appears to have stagnated, and even retreated, due in part to two key issues: 1) lack of attention to implementation of law and governance measures, and 2) de-prioritisation of water governance mechanisms by federal and state governments (Holley and Sinclair 2018, p. 4). The creation of a new regulator in NSW resulted in some improvements to compliance and enforcement in terms of the water taken by individual licence holders. However, as further discussed below, legal issues remained in relation to the implementation of the Basin Plan itself—a Plan as described above, which does not reflect the requirements of the Water Act. This mismatch between law, policy and implementation has propelled a clash of interests. One of the principal architects of the Australian water trading system argued that while the concepts underlying the Plan are close to worlds' best practice, yet the implementation plan is flawed and the execution lacking (Mike Young, paraphrased in Simons 2020, p. 11). In response to implementation challenges, in 2023, the Productivity Commission recommended that "The Australian Government should – without delay – plan and implement a renewed water recovery approach, including voluntary water purchases. Purchases should be undertaken gradually, to avoid driving water market disruption and community adjustment pressures" (Productivity Commission 2023, p. 2).

It is important to note the challenges or failures that Australia has experienced and consider them in risk-informed and adaptive management processes so as to learn from other governance models. Some challenges we believe are important to share with other jurisdictions include: (A) political interference resulting in evidence-based policy not being implemented (B) water theft, and (C) a mistrust in and discrediting of the large-scale planning process and governance mechanisms due to failing to engage deeply and collaboratively with the community. By openly sharing and understanding the mistakes and challenges of other countries, risk-informed planning and programming, including collaborative approaches can be strengthened. In any risk-based assessment it is important to examine barriers to change, and potential risks at the start of the process. This is needed to ensure that efforts can be balanced with designing laws, which are often drawn up in haste. An understanding of existing and potential barriers is also needed to discern how laws can best be implemented and how to mitigate barriers to implementation. The following section explains how three key challenges, failures, or risks in the Australian context came to be, and what suggestions the authors have for mitigating them.

(a) Political Interference: Is it Inevitable, and if so, How to Manage it as a Risk?

Australia has had its fair share of political interference in the Murray-Darling Basin, with alleged backroom deals, special "rates for mates", public money for water buybacks being misused, and stacking decision-making boards with lobbyists

(Matthews 2017; NSW Ombudsman 2018). Political interference is not a high- or low-income challenge, but a risk that needs to be managed in all transboundary water governance contexts.

Given water is a precious commodity in agriculture and industry, decision-making about its management is vulnerable to influence by powerful lobby groups and politics. In Australia, the water trading system explicitly gives economic value to water resources, and decisions have been made in the interests of certain groups seeking to maximise economic gain from the system. This phenomenon is not unique to Australia, and the lessons learnt are also relevant to contexts without independent oversight of water governance mechanisms, and in contexts where corruption levels are high. Power and politics impact on water sharing arrangements all over the world, whether or not the resource has been commodified as in the Australian case.

Examples of political interference have included a NSW Minister for Water erroneously communicating that an embargo on water take had been lifted, and "not indicating any consideration for the needs of the environment but rather an approach that prioritised the needs of productive water users and an attempt to have the law implemented in the 'least worst' way for them" (NSW ICAC 2020, p. 10). The Minister in question in this case was in campaign-mode just a few days from a state election and was seeking to placate his constituents about the contentious embargo on the Barwon-Darling River (NSW ICAC 2020, p. 74). In another case, an agricultural water user received AUD$31 million in government grants (under the Healthy Headwaters program) and used them to complete earthworks rather than install efficient irrigation systems (Simons 2020, p. 54). Other allegations of misused funds have surfaced, demonstrating that without the correct checks and balances in the system, funding programs are likely to fall prey to misuse and corruption (ABC 2019b). Each context is different, so appropriate measures to manage non-compliance, rent seeking behaviour and corruption are needed.

Democracies are of course not immune to interference. Stigler (1971) put a finer point on this claim many years ago when he noted that state capture is a fundamental and intended feature of political processes, both totalitarian and democratic: "The state – the machinery and power of the state – is a potential resource or threat to every industry in the society" (Stigler 1971, p. 3). Acknowledging the inevitability of interference helps to include it as part of a risk assessment, and to manage rent seeking and short-term gains at the expense of broader values. Additional examples from Australia may be helpful in bringing these realities to life. For example, spending in politically sensitive electorates was also found to have occurred in response to a large-scale and high-profile fish kill in which a 60-kilometre stretch of the river system was littered with millions of dead fish in 2019 (Davies 2019; Australian Academy of Science 2019). One observer reported that:

> The fish kill became an issue in the NSW state election. In the lead-up to the March 2019 poll, the towns of the Lower Darling were suddenly a target of spending. For example, there was a commitment of $25 million in federal and state funding to seal the road which the community had been wanting for years to encourage tourism (Simons 2020, p. 39).

The risk of political interference in favour of any group or interest needs to be mitigated. This is true in relation to large farming industry interests, and in relation to consideration of the environment without measures in place to protect competing social interests. Any interference leads to ad hoc and ill-considered interventions, and an undermining of trust in institutions and the water sharing process overall. To mitigate the threat of political interference, it is important to conceive of water governance broadly so that it encompasses a strong civil society, investigative journalism and a free press, independent oversight of grants processes, and structural adjustment programs involving financial compensation. There also needs to be strong accountability and transparency measures, including publishing water use data, establishing independent oversight and compliance agencies (such as the Natural Resources Access Regulator discussed below; see also Productivity Commission 2021, p. 56 discussing the critical need for independent assessment of implementation). The boards of statutory authorities should represent different interest groups and be compelled to report publicly. What has been missing in Australia, is the addition of a system of risk-informed development that has achieved sufficient consensus and engagement, and that includes detailed rules on adapting long-term measures that will prevent or minimise political interference.

(b) Compliance: An Essential Though Under-Resourced Component of Implementation.

Delivering inclusive, transparent, and reflexive risk-informed development requires that transboundary water governance focuses not only on planning, but equally on ongoing resourcing and implementation. The latter has commonly been overlooked, particularly in Australia. While significant attention has been paid to establishing caps and markets, ensuring their effectiveness and efficiency through compliance and enforcement has largely remained an afterthought. Four key concepts, or pillars are important to consider with respect to compliance regimes in water management contexts, as explained by Holley et al. (2020) in relation to groundwater regimes. These four pillars are:

1. Understanding people's motivations for complying: people's motivations are related to knowledge, social norms, and economic factors.
2. The characteristics of water users: including their economic resources, knowledge/education, cognitive capabilities, technologies, management capacities, and worldviews.
3. Regulations and enforcement strategies: include risk-based regulation and resourced compliance pyramids.
4. Regulatory pluralism: involves seeking the context-specific optimal mix of stakeholders and developing structures which enable a range of actors (e.g. civil society, farmers, community) to generate norms which encourage others to comply with water governance strategies (Holley et al. 2020). These four pillars are helpful to consider when consulting on and planning compliance regimes suited to the context.

The example of Australia's Murray-Darling Basin shows how even in resource rich environments, compliance is a challenge, and relying purely on top-down enforcement strategies is unlikely to succeed unless it is extremely well resourced, and water metering is universal, well maintained and analysed in a timely manner. In 2017 it was alleged that large-scale water users tampered with water meters, and that some irrigators pumped water from the river system in periods when pumping was not permitted, and/or they pumped quantities greatly in excess of their entitlements (ABC 2017). Community concern about these compliance breaches was high, with a Basin-wide review of compliance finding that numerous stakeholders felt considerable frustration that the Murray-Darling Basin Association (MDBA) did not respond adequately to allegations of serious breaches (Commonwealth of Australia 2018). Independent investigations resulted in the creation of a new independent NSW regulator (called the Natural Resources Access Regulator—NRAR). Government directors/staff members involved in facilitating the acts of non-compliance were removed from the bureaucracy (NSW Department of Industry 2017) and the NSW Independent Commission Against Corruption (NSW ICAC) was tasked to investigate the allegations of political interference and corruption (2020, p. 8). Although the Commission was not satisfied that the evidence established any person as having engaged in corrupt conduct (as per the Independent Commission Against Corruption Act 1988), it was satisfied that those matters it found had "detrimental effect on the public's confidence in the ecologically sustainable, equitable, transparent and efficient management of the water sources of the state and in the integrity and good repute of public administration, more generally" (NSW ICAC 2020, p. 9). It further found that key elements of the state's water laws were not properly implemented in certain catchments so as to promote consumptive uses over the environment (NSW ICAC 2020, p. 20). Other independent investigations of past practices were also alarming: "Water compliance had been ineffectual and needed significant improvement. Allegations of water theft had not been acted on – sometimes for years" (Simons 2020, p. 52). In 2021, the Inspector-General of Water Compliance (IGWC) was established to provide oversight, monitoring, compliance, and enforcement of the Basin Plan and parts of the Water Act. Australia's experience with compliance challenges shows the importance of fostering buy-in from regulated actors such as farmers and other water users (Holley and Sinclair 2016b) to help counter illegal actions, as well as timely investigation of breaches. In addition, inclusive collaborative processes can inform and enhance opportunities for a wider range of actors to exert public pressure against maladapting compliance mechanisms. As Carmody concludes, "Constant vigilance – from civil society in particular – is required to ensure that these laws are properly implemented and enforced over time" (Carmody 2018, p. 52).

Research on compliance motivators has found that personal values (in relation to honesty and fairness) and social reputation are highly influential, and need to be better capitalised upon in order to design a robust, agile and adequately resourced compliance framework. It is recommended that regulators build their knowledge of, and partnerships with, water users in order to better understand compliance motivators, and to be close to what is happening on the ground (Holley et al. 2020). There is also a significant need in Australia for compliance systems to be adequately resourced, and includes mobilising third parties to widen monitoring networks and nurture desired changes in the community (e.g. training and policy guidance). Without out adequate resourcing, compliance and enforcement strategies will not meet policy expectations (Holley et al. 2020).

The above examples from Australia demonstrate that compliance challenges have been experienced at both the institutional level and at the individual user level, and had significant impacts societally, especially in terms of undermining trust and faith in the transboundary process. There has been failure to properly implement new water laws that introduce limits on extraction at the catchment scale, and there have been compliance and enforcement problems at the individual user level (Holley and Sinclair 2016a). These experiences demonstrate how important it is to identify early on what compliance factors might prevent implementation at the larger scale, and the role that evidence-based, plural, and adequately resourced compliance strategies that tap into stakeholder motivations will play.

(c) Failing to Engage Deeply and Inclusively with Community.

Another case from Australia illustrating governance failure involved the development and release of the draft Guide to the MDP plan in 2010. The draft was written without substantial community engagement, and when it was suddenly released, some farmers were so irate that they burned copies of the draft Guide to the Plan in a media event, and governments were not prepared for the backlash (ABC 2016). As a result, the proposed minimum 3000 GL/year to recover for the environment was met with hostility by some irrigation-dependent towns. This was despite the fact that 3000 GL/year was not an environmentally sustainable diversion limit, and fell well short of Australia's obligations under international agreements, such as the Ramsar Convention (Carmody 2018, p. 41). While state governments had experienced similar challenges almost a decade earlier (Miller 2003; Holley and Sinclair 2013; Tan et al. 2012), this experience reinforced to water management professionals and politicians that understanding stakeholder positions, interests, and values was imperative, as was sharing and co-analysing data to develop joint ownership of transboundary water management plans and governance options. Sadly, and possibly as a result of poor stakeholder engagement, securing water for the environment through mechanisms such as water buybacks[3] "have become the proverbial patsy for all manner of rural mischance, from unemployment to reduced school enrolments" (Carmody 2018, p. 45–46).

Poor attention to collaboration and consultation could be a key reason for the failure of the MDB Plan to achieve the outcomes intended, and for its failure to

[3] Water buybacks (or water purchases) are when the government purchases water rights for the environment through the water trading system. In 2020, 1230 gigalitres of water rights had been bought from farmers via buyback programs at a cost of about A$2.6 billion (Hughes et al. 2020). In 2023 the Australian Productivity Commission agreed with previous inquiries that: "purchasing water is the most cost-effective way for governments to obtain water for the environment—the average cost of water recovery via purchases is currently $2808/ML, compared to $8126/ML for water recovery via gap-bridging irrigation infrastructure (or almost three times more expensive per ML)." (Productivity Commission 2023, p. 15).

garner the necessary community and political support to enable its implementation. Indeed, the Productivity Commission (2021, p. 199), citing Sefton et al. (2020, p. 1), noted a range of concerns about the inclusion, extent and effectiveness of engagement processes, particularly within the MDB: "Across the [Murray–Darling] Basin trust in governments—particularly federal and state — to deliver good long-term policy and support rural and regional communities has been severely diminished. This fall in trust has resulted from a failure to adequately include people in conversations about government policy and their future, especially those who have not been on the upside of change".

Buying water back from irrigators for the environment has become a particularly vexed issue, involving a great deal of misinformation, misunderstanding, and populist political discourse. This is despite the fact that experts have argued that the socio-economic impacts of buybacks would be much less than the impacts of drought (Wittwer and Dixon 2011 in Carmody 2018, p. 46; Hughes et al. 2020) and that purchasing water is the most cost-effective way for governments to obtain water for the environment (Productivity Commission 2023). Owens notes that a "lack of responsiveness to economic, social and cultural values in the MDB has led to a general resistance from irrigation communities to water buybacks, leading to a shift to more costly and less effective (environmentally) and potentially less assured, recovery measures, like infrastructure upgrades" (Owens in Holley and Sinclair 2018, p. 17; see also Wheeler et al. 2020). Furthermore, there is considerable evidence that purported efficient irrigation infrastructure subsidies are not cost-effective, and are in fact leading to a net increase in extractions through a phenomenon known as the "rebound effect" (Wheeler et al. 2020; Hughes et al. 2020). By 2011, the proposed 3000 GL/year to be returned to the environment had been reduced to a target of 2750 GL (as a long-term annual average), even though experts were calling for 3200 GL for the environment as a minimum (Wentworth Group of Concerned Scientists, 2019). Other scientific and legal advice indicates that Australia needs to return at least 4000 GL/year to the environment to ensure the longer-term health and viability of this vital transboundary river system (Carmody 2018, p. 50), which demonstrates the disconnection between policy and the outcomes needed. Debates about how much water to return to the environment, and how to manage it in times of scarcity has split MDB communities, pitting farmers against farmers (depending on the type of crops they grow), and dividing upstream and downstream river communities. Unfortunately, it has also created an artificial divide between the concepts of environment and livelihoods, whereby Australia has witnessed a large-scale "tragedy of the commons".

Australia has employed a range of consultation modalities in the context of the Murray-Darling Basin. These have included: notice and comment (where something is announced and provided for review/comment), invited submissions, targeted consultation, and the establishment of consultation bodies such as the Basin Community Committee (Lindsay 2018, p. 175). Lindsay presents five key lessons about the aims and process of stakeholder engagement in the MDB context: 1) Consultation has been flexible and varied (ranging from perfunctory to deep engagement). 2) There has been an absence of overarching policy related to the meaning and exercise of public participation. 3) Policy approaches to participation need to mature. 4) There has been a gap between policy and implementation with respect to consultation with Traditional Owners. 5) More deliberative, robust, sustained modes of participation must be influenced and underpinned by legally contestable rights and interests (Lindsay 2018, p. 180–181).

The depth of engagement with Aboriginal and Torres Strait Islander peoples on water governance matters in Australia has been woeful and have in many cases served to continue the dispossession and exclusion of Traditional Owners (see generally Marshall 2017; Godden et al. 2020; Moggridge 2020, 2021). Inadequate engagement with Traditional Owners is an example of where policy makers and politicians fell short of inclusive collaborative governance as explained by Godden et al. (2020, p. 671): "The difficulties in the MDB of having to retrospectively purchase water entitlements for Aboriginal people, due to their exclusion in the initial 'sharing' arrangements, highlights the need to dedicate water reserves for Aboriginal purposes in advance of full marketisation of water". Indigenous Australians have lived in Australia for 50,000 to 120,000 years (O'Connell et al. 2018; Daley 2019), and despite their connection to country and immense knowledge of water bodies and water-dependent environments, consultation has been considered disingenuous by some: "At the same time as fighting the native-title claim, the Barkandji had been 'consulted' over the Murray-Darling Basin Plan–'Fake consultation' elder Badger Bates calls it, because nobody ever listened and nothing they said changed the result". (Simons 2020, p. 31–32). In contrast, in Canada there is an expectation that deep consultation will involve negotiation, supported by legal obligations grounded in specific constitutional conditions (Lindsay, 2018, p. 178–179). Moggridge (2020, 2021) calls our attention to groundwater as an integral interconnected component of transboundary river basins, and the knowledge and connection that Aboriginal people have had with ground and surface waters for thousands of years, but this is seldom recognised in governance arrangements. In 2023, the Productivity Commission advised with respect to Indigenous engagement that "Meaningful engagement is crucial to building trust and working in partnership. It is not enough to recognise First Nations values in Basin Plan implementation. Transparent, accountable mechanisms by which First Nations people can inform and share decision making are important (they are also a key element of the priority reforms)" (Productivity Commission 2023 p. 19).

A lack of deep consultation with Indigenous peoples coexists with "reform fatigue" experienced by some farmers and rural communities who have been part of the various phases and iterations of transboundary water governance reforms in Australia. These stakeholders have spent many hours, over many years, driving to meetings and consultations which they can't afford to miss (Simons 2020, p. 23). Tan et al. trialled a range of deliberative processes for water planning in Australia including interactive tools, and those that had high visual impact, and found that these approaches were consistently rated highly by all sectors of the community and also by water planners (Tan et al. 2012). The challenge is therefore, how to effectively and meaningfully engage with communities including Indigenous peoples and other marginalised groups, while at the same time being transparent about how stakeholders' inputs will influence decisions and avoiding consultation fatigue. As the Productivity Commission (2021, p. 200) observed "for both water resource management and water services provision going forward, all Australian governments should further build their capacity and capability to: undertake effective, thorough and well-informed community consultation and engagement". Collaborative governance is also dependent on a certain level of organisational capacity if institutional efforts to promote collaboration are to be successful (Ananda and Proctor 2013).

The experiences in the MDB remind us of the ongoing importance of ownership and uptake of collaborative processes, and as observed by Carmody (2018): "newly minted, best practice water laws are more of an armistice than an outright victory. The latter requires well-resourced government agencies, political champions committed to the success of the implementation process and an active civil society capable of holding governments to account over the long-term" (Carmody 2018, p. 51). It is the power and importance of active and inclusive civil society engagement that we now turn to.

4.5 Inclusive Collaborative Governance to Inform Risk-Based Approaches

> A part of my anger and disappointment comes from the billions of dollars spent, the laws written, the plans and policies made, leaving the river system still collapsing because it has been undermined by bigger, more powerful interests, and their political influence (Bill Johnson, park ranger with responsibility for Macquarie Marshes, quoted in Simons 2020, p. 97).

Collaborative water governance models are many and varied. Some of them combine 'hard' legal rules with "softer" collaborative approaches (Holley 2016). These combinations may be hybrids (in which both systems remain independent but ineract in mutually supporting ways) or they may be complementary (in which the two approaches are merged) (Holley 2015). A more nuanced understanding of "who is involved in various "hard" and "soft" collaborative governance approaches will now be discussed, with a focus on the need for explicit and sincere engagement with Indigenous communities, women and minorities. Inclusive approaches will enhance water governance regimes through building more equitable and resilient communities and the water resources they rely upon. In the following section, we explore the spectrum of the modes of engagement that can be used in collaborative transboundary water governance, and we consider some of the key features with respect to addressing asymmetries in access to information and power which impact on people's ability to participate in collaborative processes and subsequent decision-making.

4.6 Modes of Engagement

The 2015 Organisation for Economic Cooperation and Development (OECD) Water Governance Initiative's framework for stakeholder engagement draws on a spectrum of modes of consultation and participation, and is a useful tool for considering the depth and breadth of engagement modalities (OECD 2015a, b). The framework considers top-down and transactional approaches (one-way communication), through to collaborative and shared decision-making (co-decision, and co-production), as shown in Fig. 4.2 (see also Daniell 2011 for examples of types of engagement and the relative costs and benefits of each).

Principle 10 of the OECD Water Governance Principles highlights the need for stakeholder engagement to foster informed and outcome-oriented contributions to water policy design and implementation. The principles make specific reference to the need to pay: "special attention to under-represented categories (youth, the poor, women, Indigenous people, domestic water users) newcomers (property developers, institutional investors) and other water related stakeholders and institutions" (OECD 2015a). As shown in Fig. 4.2, it is important to design engagement processes that are clear about what level of influence the stakeholders will have, and how their inputs will be used.

The water, sanitation, and hygiene (WASH) sector has grappled with implementation challenges related to inclusive collaborative processes for decades, and has

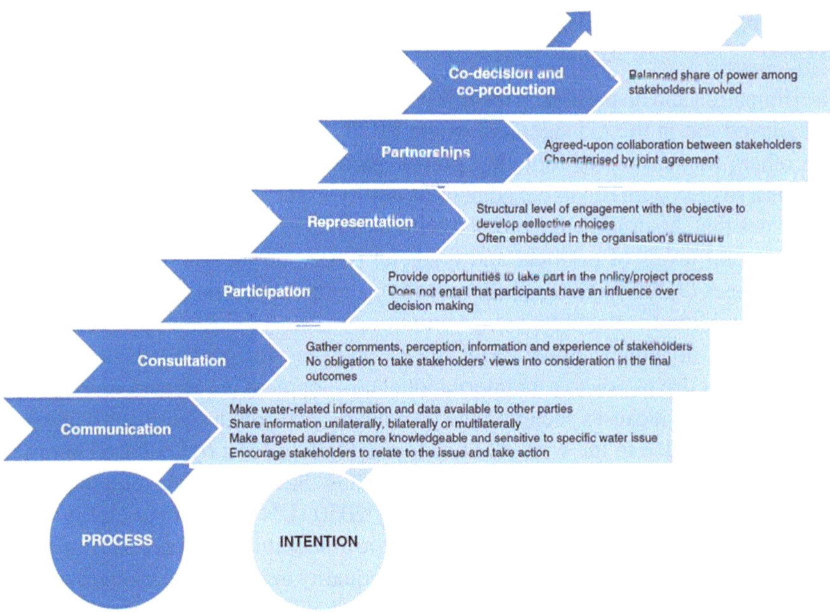

Fig. 4.2 Organisation for economic co-operation and development (OECD) typology of stakeholder engagement (OECD 2015b)

Fig. 4.3 Representation of intersectionality from the perspective of individual characteristics and societal power. (Soeters et al. 2019, p. 2)

found that without genuine and inclusive processes, designs will be flawed, systems will not be sustainable, and overarching economic and health related outcomes will not be delivered (Grant et al. 2019; Winfield et al. 2019; Water for Women 2022).

4.7 Inclusive Collaborative Governance

Discrimination and disparities in the levels of power that different stakeholder groups have played out in the management of the MDB: "Some have got away with more than others for reasons of lobbying power, money, networks and geography. The result is a river system run to the edge of its ability to survive". (Simons 2020, p. 68). The issues that Australia has grappled with in terms of managing various interests, with different levels of political and financial power, are not unique; they point to the challenges of interference, implementation, compliance, and mistrust that readily flow from minimal collaborative engagement.

Inequality in money, power and voice underpins the global phenomenon of inequality that drives poverty. In international development, awareness of inequality, and the harm that it causes societies and economies, has grown significantly in recent decades, in part due to the work of civil society calling for a rights-based approach to development, and initiatives such as the SDGs. Sustainable Development Goal 10 explicitly calls for reduced inequalities by 2030 (UN General Assembly 2015). Additionally, multilateral agencies such as the International Monetary Fund have identified that countries with high levels of inequality experience lower growth than countries that had less income inequality (Ostry et al. 2014). Inclusive collaborative governance is conscious of the need to reduce inequalities that arise from power imbalances, and seeks to address them in deliberative engagement processes. An intersectional approach to inclusion has been described as:

> Inclusiveness is recognising the rights of individuals and groups across different categories, needs and vulnerabilities, and without any kind of discrimination based on race, colour, age, gender, religious affiliation, ethnicity, language, disability, economic backgrounds or any other conditions of origin. It also concerns taking into consideration these diverse social, economic, and cultural aspects, along with taking special account of the vulnerable groups, minorities and indigenous people when performing different governance functions, contributing to balancing differences in power (Jiménez et al. 2020, p. 12).

As noted above, there are a number of ways in which discrimination manifests, and these are often "intersectional", in that people often experience marginalisation in a number of ways (Fig. 4.3). Consideration of intersecting forms of discrimination enables water governance practitioners to better understand the ways in which marginalisation may constrain or prevent people from engaging in water management decision-making processes. An intersectional approach also will help to prevent processes from being skewed towards elites, those with economic or other types of power (Brisbois and de Loë 2016).

Inclusive collaborative governance is conscious of power imbalances between stakeholders and it works to address them, given that not all community members have equal access to lobbying opportunities, engagement and decision-making forums. For example, in Australia, any attempt at inclusive collaborative governance would need to recognise the influence and power of the organised and well-resourced irrigation lobby and balance these interests with other voices and perspectives. In the case of the MDB, the power of the irrigation lobby, unchecked due to the absence of inclusive processes, was stronger than the power of environmental advocates and Indigenous people, despite the links between sustainable water extraction and long-term viability of Basin communities (Carmody 2018).

Global water management theory and practice and transboundary river politics have seen an increased focus on inclusion, though this focus remains at a policy rather than practice level (Grant 2017). Acknowledgement of women as holders of knowledge, water users, and managers was promoted by the Dublin Statement in 1992 as one of the four principles of IWRM. Yet still today, women are under represented in participation and decision-making of water governance processes and access to water management technologies (Zwarteveen 2008; Grant et al. 2019), the role of women as brokers of transformation within water decision-making spheres, including transboundary organisations and water bureaucracies, is not well recognised (de Silva et al. 2018; Sehring et al. 2022; Zwarteveen 2017). De Silva et al. call for the recognition of the roles that women play as agents of change, including in transboundary water conflicts (de Silva et al. 2018). Zwarteveen (2017) explains that the strong connection between masculinities and irrigation cultures and water infrastructure may provide an important explanation of why hydraulic bureaucracies are so resistant to change towards greater gender equality. To address this historical and cultural challenge, Zwarteveen and Boelens (2014) call for the water management sector to recognise the power and politics of water use, management, and governance. This includes understanding how injustices are produced, and how water rights are defined and understood (Zwarteveen and Boelens 2014). The types of mechanisms which can be used to address such injustices, and meaningfully include women and other marginalised voices in transboundary water management have been summarised by the Global Water Partnership across four areas of action (Grant 2017, p. 5):

- Action Area 1: Make gender equality and inclusion a core goal within transboundary water governance organisations and processes.
- Action Area 2: Conduct gender and inclusion analysis that drives change at all stages of the project cycle.
- Action Area 3: Adopt a "nothing about them without them" approach to meaningful and inclusive participation.
- Action Area 4: Create a level playing field between men and women with respect to access to and control of land and water resources.

These four areas are important factors to consider when pursuing meaningful and inclusive participation in decision-making and partnerships with marginalised peoples (such as women, Indigenous peoples, socio-economically disadvantaged, sexual and gender minorities, and people with disabilities). These factors include being conscious of, and not exploiting, the amount of unpaid work they are expected to do through being part of consultations, mobility issues related to getting to and accessing participation forums, events, and submission processes (high levels of literacy are often required), and cultural norms which may discourage women (for example) from speaking in front of men or openly in public forums. These four areas of action call for a redistribution of power and control over water resources, towards more equitable forms of governance drawing on diverse knowledges including traditional and cultural knowledge. Barriers to inclusion can all be managed with careful analysis and planning, and through being informed by local community organisations including disabled people's organisations, Traditional Owners, rights-based NGOs and gender equality and social inclusion experts (Grant et al. 2019; Earle and Bazilli 2013; Water for Women 2022). Additionally, the gendered nature of transboundary water governance organisations, forums and governance architecture needs to be addressed, in line with evolving international efforts to promote gender equality and more holistic approaches to natural resources management (Sehring 2022).

4.8 Conclusion

In this chapter we have shared a range of experiences pertinent to Australia's transboundary water governance history, particularly in relation to the MDB. We hope that by offering transparent accounts of the challenges faced and the lessons learned, we can support the risk-informed collaborative governance processes in other jurisdictions. We have argued that investment in deep and inclusive stakeholder engagement, in the form of inclusive collaborative governance, can enable a greater focus on shared values which can then be drawn on to inform water governance mechanisms and strategies, including shared and trusted compliance tools. We have revealed that while it is unlikely that perfect buy-in from all parties can be achieved anywhere, there are ways to work towards achieving balanced collaboration between governments, business and local communities. Communities are knowledgeable about the water resources they depend on, and therefore they need to be involved in governance decisions and regimes. Risk-informed approaches help public servants, researchers, politicians and communities to consider what the barriers to success might be, and they help them to deal with these barriers proactively rather than in a reactive manner. To support risk management processes, new models of participation are needed, with a focus on inclusion, an understanding of intersectionality (multiple layers of marginalisation), and thoughtful application of the spectrum of collaboration.

Transboundary water governance is evolving, iterative, and adaptive. The example of Australia shows that even in sophisticated governance contexts, ensuring buy-in, compliance, and implementation remains a challenge. We offer the following three key insights that could be useful to other jurisdictions, when planning models of inclusive collaborative water governance.

Firstly, it is important to put time and resources into developing the enabling conditions for water law and reform. Inclusive and collaborative processes are essential for developing these conditions, and for gaining acceptance from all those involved. Putting time and energy into processes from the outset will save time and political resources in the long run, through identifying and agreeing on values which can feed into risk-informed approaches.

Secondly, be mindful of the gap between law and implementation, and embed the necessary safeguards in governance frameworks to minimise issues related to potential challenges along these lines. Investing in governance mechanisms (such as laws) is important, but they are only a part of the broader picture. Investment in implementation will be needed to ensure that the laws have the desired effect.

Thirdly, civil society engagement is important as it ensures that the government is held to account and the gap between communities and public servants can be bridged. An understanding of intersectionality is needed to ensure that the power dynamics between different groups and individuals within society are taken into account, and to ensure that actions are taken to support more inclusive and equitable societies. Engagement processes must consider power differences between actors. Groups which may be vulnerable to being marginalised include small-scale farmers, Indigenous people, women, people considered lower caste, people with disabilities and sexual and gender minorities, as well as ethnic minorities. These people make up the majority of society, and yet do not have a say in many water management decisions which affect them. Community engagement processes need to consider power imbalances, and they need to be conscious of the dominant worldviews and values that may be put forward by groups and organisations with power. Culturally informed "do no harm" processes, involving people who are impacted by water management decisions is imperative to achieve outcomes for communities and environments dependent on water. Meaningfully and deeply listening to and considering a range of perspectives on values, water sharing, compliance and overall management options is central to inclusive collaborative transboundary water governance.

References

ABC. (2016). Young men burn copies of the guide to the Murray. ABC News. https://www.abc.net.au/news/2016-02-18/young-men-burn-copies-of-the-guide-to-the-murray/7180740?nw=0. Accessed on 1 June 2024

ABC. (2017). Pumped. ABC News. https://www.abc.net.au/4corners/pumped/8727826. Accessed on 1 June 2024

ABC. (2019a). Can the Plan Convoy Converges on Canberra. ABC News. https://www.abc.net.au/news/2019-12-02/can-the-plan-convoy-converges-on-canberra/11751398. Accessed on 1 June 2024

ABC. (2019b). Cash Splash. ABC News. https://www.abc.net.au/4corners/cash-splash/11289412. Accessed on 1 June 2024

Akhmouch A, Clavreul D (2016) Stakeholder engagement for inclusive water governance: practicing what we preach with the OECD water governance initiative. Water 8(5):204. https://doi.org/10.3390/w8050204

Akhmouch A, Clavreul D, Glas P (2018) Introducing the OECD principles on water governance. Water Int 43(1):5–12. https://doi.org/10.1080/02508060.2017.1407561

Ananda J, Proctor W (2013) Collaborative approaches to water management and planning: an institutional perspective. Ecol Econ 86:97–106. https://doi.org/10.1016/j.ecolecon.2012.10.018

Australian Academy of Science (2019) Investigation of the causes of mass fish kills in the Menindee Region NSW over the summer of 2018–2019. https://www.science.org.au/supporting-science/science-policy-and-sector-analysis/reports-and-publications/fish-kills-report. Accessed on 1 June 2024

Australian Competition and Consumer Commission (ACCC). (2021). Murray–Darling Basin water markets inquiry: Final report. https://www.accc.gov.au/system/files/Murray-Darling%20Basin%20water%20markets%20inquiry%20-%20final%20report%20-%20Overview.pdf. Accessed on 1 June 2024

Benham C, Hussey K, Beavis S (2014) Planning for success in a climate change future: collaborative water governance in the upper Murrumbidgee catchment, Southeastern Australia. Australian Journal of Water Resources 18(1):1–14. https://doi.org/10.7158/W13-027.2014.18.1

Bice S, Moffat K (2014) Social licence to operate and impact assessment. Impact Assess Proj Apprais 32(4):257–262. https://doi.org/10.1080/14615517.2014.950122

Brisbois MC, de Loë RC (2016) Power in collaborative approaches to governance for water: a systematic review. Soc Nat Resour 29(7):775–790. https://doi.org/10.1080/08941920.2015.1080339

Carmody E (2013) The silence of the plan: will the convention on biological diversity and the Ramsar convention be implemented in the Murray-Darling basin? Environmental and Planning Law Journal 30(1):56–73

Carmody E (2018) The unwinding of water reform in the Murray-Darling basin: a cautionary tale for transboundary river systems. In: Holley C, Sinclair D (eds) Reforming water law and governance. Springer Nature, pp 35–55

Cole J, Sharvelle S, Grigg N, Pivo G, Haukaas J (2018) Collaborative, risk-informed, triple bottom line, multi-criteria decision analysis planning framework for integrated urban water management. Water 10(12):1722

Commonwealth of Australia (2018) The integrity of the water market in the Murray-Darling Basin. https://www.aph.gov.au/Parliamentary_Business/Committees/Senate/Rural_and_Regional_Affairs_and_Transport/MurrayDarlingPlan/Report. Accessed on 9 June 2025

Connell D. (2020). No water no leadership. The Conversation. https://theconversation.com/no-water-no-leadership-new-murray-darling-basin-report-reveals-states-climate-gamble-136514. Accessed on 1 June 2024

Council of Australian Governments (COAG). (1994) Water Reform Framework. http://www.ielrc.org/content/e9401.pdf. Accessed on 1 June 2024

Daley P (2019) A big jump: People might have lived in Australia twice as long as we thought. The Guardian. https://www.theguardian.com/australia-news/postcolonial-blog/2019/mar/11/a-big-jump-people-might-have-lived-in-australia-twice-as-long-as-we-thought. Accessed on 1 June 2024

Daniell KA (2011) Enhancing collaborative management in the basin. In: Connell D, Grafton RQ (eds) Basin futures: water reform in the Murray-Darling basin. ANU E-Press, pp 413–438

Davies A (2019) Hundreds of thousands of native fish dead in second Murray-Darling incident. The Guardian. https://www.theguardian.com/australia-news/2019/jan/07/hundreds-of-thousands-of-native-fish-dead-in-second-murray-darling-incident. Accessed on 1 June 2024

Davies P E, Stewardson M J, Hillman T J, Roberts J R, Thoms M C (2012) Sustainable Rivers Audit 2: The ecological health of rivers in the Murray–Darling Basin at the end of the Millennium Drought (2008–2010) (Vol. 1). https://www.mdba.gov.au/publications-and-data/publications/sustainable-rivers-audit-2-sra-2. Accessed on 1 June 2024

de Búrca G, Keohane R O, Sabel A (2013) New modes of pluralist global governance. IILJ Working Paper 2013/1 (Global Administrative Law Series). https://www.iilj.org/publications/new-modes-of-pluralist-global-governance/. Accessed on 1 June 2024

de Silva L, Veilleux JC, Neal MJ (2018) The role of women in transboundary water dispute resolution. In: Fröhlich C, Gioli G, Cremades R, Myrttinen H (eds) Water security across the gender divide. Springer, pp 129–146

de Vries HJM, Petersen AC (2009) Conceptualizing sustainable development: an assessment methodology connecting values, knowledge, worldviews and scenarios. Ecol Econ 68:1006–1019. https://doi.org/10.1016/j.ecolecon.2008.11.015

Doolan J (2016) The Australian water reform journey: An overview of three decades of policy, management and institutional transformation. Australian Water Partnership. https://waterpartnership.org.au/publications/aust-water-reform-journey/. Accessed on 1 June 2024

Earle A, Bazilli S (2013) A gendered critique of transboundary water management. Fem Rev 103:99–119

Godden L, Jackson S, O'Bryan K (2020) Indigenous water rights and water law reforms in Australia. Environmental and Planning Law Journal 37(6):655–678

Grafton Q, Carmody E, Colloff M., Williams J (2019) Toxic: What is rotten in the Murray-Darling Basin? Its not just fish. Asia and Pacific Policy Society. https://www.policyforum.net/toxic-what-is-rotten-in-the-murray-darling-basin/. Accessed on 1 June 2024

Grant M (2017) Gender equality and inclusion in water resources management–Action piece. Global Water Partnership. https://www.gwp.org/globalassets/global/about-gwp/publications/gender/gender-action-piece.pdf. Accessed on 1 June 2024

Grant M, Willetts J, Huggett C (2019). Gender equality and Goal 6 The critical connection: An Australian perspective. Australian Water Partnership. https://waterpartnership.org.au/publications/gender-equality-goal-6/. Accessed on 1 June 2024

Head BW (2005) Governance. In: Saunders P, Walter J (eds) Ideas and influence: social science and public policy in Australia. UNSW Press, pp 44–57

Holley C (2012) Compliance and enforcement of water licences in NSW: limitations in law, policy and institutions. The Australasian Journal of Natural Resources Law and Policy 15:149–189

Holley C (2015) Crafting collaborative governance: water resources, Californias Delta plan and audited self-management in New Zealand. Environmental Law Reporter: News and Analysis 45:10324–10337

Holley C (2016) Linking law and new governance: Examining gaps, hybrids, and integration in water policy. Law and Policy. 38: 24–53. https://onlinelibrary.wiley.com/doi/abs/10.1111/lapo.12048

Holley C, Gunningham N, Shearing C (2012) The new environmental governance. Routledge, London

Holley C, Sinclair D (2013) Deliberative participation, environmental law and collaborative governance: insights from surface and groundwater studies. Environmental and Planning Law Journal 30(1):32–55

Holley C, Sinclair D (2014) A new water policy option for Australia?: collaborative water governance, compliance and enforcement and audited self-management. The Australasian Journal of Natural Resources Law and Policy 17:189–216

Holley C, Sinclair D (2016a) Governing water markets: achievements, limitations and the need for regulatory reform. Environmental and Planning Law Journal 33:301–324

Holley C, Sinclair D (2016b) Regulation, technology, and water: "buy-in" as a precondition for effective real-time advanced monitoring, compliance, and enforcement. George Washington Journal of Energy and Environmental Law 7:52–66

Holley C, Sinclair D (2018) Reforming water law and governance: From stagnation to innovation in Australia. Springer Nature Singapore. doi:https://doi.org/10.1007/978-981-10-8977-0

Holley C, Mutongwizo T, Pucci S, Rho JC, Sinclair D (2020) Groundwater regulation, compliance and enforcement: insights on regulators, regulated actors and frameworks in New South Wales, Australia. In: Holley C, Sinclair D (eds) Sustainable groundwater management: a comparative analysis of French and Australian policies and implications to other countries. Springer, pp 51–69

Horne J (2017) The politics of water reform and environmental sustainability in the Murray–Darling basin. Water Int 42(8):1000–1021. https://doi.org/10.1080/02508060.2017.1412201

Hughes N, Galeano D, Hatfield-Dodds S (2020 September 1) Recovering water for the environment in the Murray-Darling: Farm upgrades increase water prices more than buybacks. The Conversation. https://theconversation.com/recovering-water-for-the-environment-in-the-murray-darling-farm-upgrades-increase-water-prices-more-than-buybacks-145289. Accessed on 1 June 2024

International Standards Organisation (2021) Risk management–Guidelines (ISO 31000). https://www.iso.org/iso-31000-risk-management.html. Accessed on 1 June 2024

Jiménez A, Saikia P, Giné R, Avello P, Leten R, Lymer BL, Schneider K, Ward R (2020) Unpacking water governance: A framework for practitioners. Water 12(3):827. https://doi.org/10.3390/w12030827

Keelty M J (2020). Interim Inspector-General of Murray–Darling Basin Water Resources March 2020, Impact of lower inflows on state shares under the Murray–Darling Basin Agreement. Canberra. https://www.igwc.gov.au/sites/default/files/2020-09/iig_final_report.pdf. Accessed on 1 June 2024

Lenton R, Muller M (2009) Integrated water resources management in practice: Better water management for development. Earthscan

Lindsay B (2018) Public participation in water resources management in Australia: procedure and possibilities. In: Holley C, Sinclair D (eds) Reforming water law and governance. Springer Nature, pp 147–164

Marshall V (2017) Overturning aqua nullius: Securing Aboriginal water rights. Aboriginal Studies Press. https://aiatsis.gov.au/publication/35022. Accessed on 1 June 2024

Matthews K (2017) Independent investigation into NSW water management and compliance: Advice on implementation, final report. NSW Department of Industry. https://www.industry.nsw.gov.au/__data/assets/pdf_file/0019/131905/Matthews-final-report-NSW-water-management-and-compliance.pdf. Accessed on 1 June 2024

Mendoza G, Jeuken A, Matthews J, Stakhiv E, Kucharski K (2018) Climate risk informed decision analysis (CRIDA). United Nations Educational, Scientific and Cultural Organization (UNESCO). https://agwaguide.org/docs/CRIDA_Sept_2019.pdf. Accessed on 1 June 2024

Miller I (2003). Legal aspects of environmental allocations in water management planning. Paper presented at the 2003 NELA Annual Conference, Broken Hill, 23–25 October 2003. http://www.austlii.edu.au/au/journals/NatEnvLawRw/2003/40.pdfAccessed on 1 June 2024

Moggridge BJ (2020) Aboriginal people and groundwater. Proceedings of the Royal Society of Queensland 126:11–27. https://doi.org/10.5962/p.357835

Moggridge BJ (2021) Indigenous water knowledge and values in an Australasian context: special issue of the Australasian journal of water resources (AJWR). Australian Journal of Water Resources 25(1):1–3. https://doi.org/10.1080/13241583.2021.1935919

Murray-Darling Basin Authority (MDBA) (2017) The Murray–Darling Basin water compliance review. https://www.mdba.gov.au/publications/mdba-reports/murray-darling-basin-water-compliance-review. Accessed on 1 June 2024
Murray-Darling Basin Royal Commission (2019) Murray-Darling Basin Royal Commission report, 31 Jan 2019. https://cdn.environment.sa.gov.au/environment/docs/murray-darling-basin-royal-commission-report.pdf. Accessed on 1 June 2024
National Water Initiative (NWI) (2004). Intergovernmental agreement on a national water initiative. https://www.agriculture.gov.au/sites/default/files/sitecollectiondocuments/water/Intergovernmental-Agreement-on-a-national-water-initiative.pdf. Accessed on 1 June 2024
NSW Department of Industry (DoI) (2017) Independent investigation into NSW water management and compliance—final report. https://www.industry.nsw.gov.au/__data/assets/pdf_file/0019/131905/Matthews-final-report-NSW-water-management-and-compliance.pdf. Accessed on 1 June 2024
NSW Independent Commission Against Corruption (NSW ICAC) (2020) Investigation into complaints of corruption in the management of water in NSW and systemic non-compliance with the Water Management Act 2000. Independent Commission Against Corruption. https://www.icac.nsw.gov.au/investigations/past-investigations/2020/nsw-government. Accessed on 9 June 2025
NSW Ombudsman. (2018). Investigation into water compliance and enforcement 2007–17: A special report to Parliament under section 31 of the Ombudsman Act 1974. 17 August 2018. https://www.ombo.nsw.gov.au/__data/assets/pdf_file/0011/132977/Water-compliance-and-enforcement-a-special-report.pdf. Accessed on 1 June 2024
O'Connell JF, Allen J, Williams MAJ, Williams AN, Turney CSM, Spooner NA, Kamminga J, Brown G, Cooper A (2018) When did Homo sapiens first reach Southeast Asia and Sahul? Proc Natl Acad Sci 115(34):8482–8490. https://doi.org/10.1073/pnas.1808385115
Opitz-Stapleton S, Nadin R, Kellett J, Calderone M, Quevedo A, Peters K, Mayhew L (2019). Risk-informed development: From crisis to resilience. United Nations Development Programme (UNDP). https://odi.org/en/publications/risk-informed-development-from-crisis-to-resilience/Accessed on 1 June 2024
Organisation for Economic Co-operation and Development (OECD) (2015a) OECD principles on water governance. http://www.oecd.org/governance/oecd-principles-on-water-governance.htm. Accessed on 1 June 2024
Organisation for Economic Co operation and Development (OECD) (2015b) Stakeholder engagement for effective water governance. OECD Publishing. https://www.oecd.org/gov/stakeholder-engagement-for-inclusive-water-governance-9789264231122-en.htm. Accessed on 1 June 2024
Ostrom E (1990) Governing the commons: the evolution of institutions for collective action. Cambridge University Press
Ostrom E (1992) Crafting institutions for self-governing irrigation systems. ICS Press
Ostry JD, Berg A, Tsangarides C G (2014) Redistribution, inequality, and growth. International Monetary Fund Research Department. https://www.imf.org/external/pubs/ft/sdn/2014/sdn1402.pdf. Accessed on 1 June 2024
Productivity Commission (2018) National water reform inquiry report (Submitted to Government 2017, released for the public in 2018). https://www.pc.gov.au/__data/assets/pdf_file/0007/228175/water-reform.pdf. Accessed on 1 June 2024
Productivity Commission (2021) Assessment of National Water Initiative implementation progress (2017–2020) draft report. https://www.pc.gov.au/inquiries/current/water-reform-2020/draft/water-reform-2020-draft-assessment.pdf. Accessed on 1 June 2024
Productivity Commission (2023) Murray–Darling Basin Plan: Implementation review 2023 (Inquiry report no. 103). https://www.pc.gov.au/inquiries/completed/basin-plan-2023/report/basin-plan-2023-overview.pdf. Accessed on 1 June 2024
Rogers EM (2003) Diffusion of innovations, 5th Edition (5th ed) edn. Free Press
Sefton R Peterson D, Woods R, Kassebaum A, McKenzie D, Simpson B, Ramsay M (2020) Panel report: Independent assessment of social and economic conditions in the Basin-Draft

report. Melbourne. https://www.mdba.gov.au/publications-and-data/publications/independent-assessment-social-and-economic-conditions-basin. Accessed on 1 June 2024

Sehring J (2022) Gendered river basin institutions. In: Zwarteveen M, Sehring J, ter Horst R (eds) Gender dynamics in transboundary water governance, 1st edn. Routledge

Sehring J, Zwarteveen M, ter Horst R (eds) (2022) Gender dynamics in transboundary water governance. Taylor & Francis

Simons M (2020) Cry me a river: The tragedy of the Murray-Darling Basin. Quarterly Essay, 77. Black Inc. https://www.quarterlyessay.com.au/essay/2020/03/cry-me-a-river. Accessed on 1 June 2024

Soeters S, Grant M, Carrard N, Willetts J (2019) Intersectionality: Ask the other question. Water for Women: Gender in WASH-Conversational article 2. Institute for Sustainable Futures, University of Technology Sydney. https://www.waterforwomenfund.org/en/resources/en/resourcesGeneral/news/1903ISF-UTS_2019_IntersectionalityArticle2_GenderinWASH_WaterforWomen.pdf. Accessed on 1 June 2025

Stigler GJ (1971) The theory of economic regulation. The Bell Journal of Economics and Management Science 2(1):3–21. https://doi.org/10.2307/3003160

Tadaki M, Sinner J, Stahlmann-Brown P, Greenhalgh S (2020) Does collaborative governance increase public confidence in water management? Survey evidence from Aotearoa New Zealand. Water Alternatives 13(2):302–323

Tan P-L, Bowmer K-J, Mackenzie J (2012) Deliberative tools for meeting the challenges of water planning in Australia. J Hydrol 474:2–10

The Murray-Darling Declaration. (2018). https://murraydeclaration.org/. Accessed on 1 June 2024

Thompson M (2011) Murray-Darling Basin Plan draft released: Expert reactions. The Conversation. https://theconversation.com/murray-darling-basin-plan-draft-released-expert-reactions-4478. Accessed on 1 June 2024

UN General Assembly (2015) Transforming our world: The 2030 agenda for sustainable development.. https://www.refworld.org/docid/57b6e3e44.html. Accessed on 1 June 2024

Vidot A, Jasper C (2019). What is the Murray-Darling Basin Plan and why are we still talking about it? ABC News. https://www.abc.net.au/news/2016-11-21/what-is-the-murray-darling-basin-plan/8043180. Accessed on 1 June 2024

Wahid S, Nolan M, Dunlop M, Watto MA, Robertson DE, Jalilov S, Pirzl R, Lyons P, Bettini Y and Harper R (2023) Values and vulnerability of the Murray-Darling Basin–Basin-wide rapid assessment. CSIRO, Australia. https://www.mdba.gov.au/sites/default/files/publications/values-and-vulnerabilities-murray-darling-basin-rapid-assessment.pdf. Accessed on 9 June 2025

Water for Women. (2022). Partnerships for transformation: Guidance for WASH and rights holder organisations. https://www.waterforwomenfund.org/en/news/partnerships-for-transformation-guidance-for-wash-and-rights-holder-organisations.aspx. Accessed on 1 June 2024

Wentworth Group of Concerned Scientists. (2019). Significance of the Murray-Darling Basin. https://wentworthgroup.org/wp-content/uploads/2019/02/Wentworth-Group-of-Concerned-Scientists-MDB-snapshot.pdf. Accessed on 1 June 2024

Wheeler SA, Carmody E, Grafton RQ, Kingsford RT, Zuo A (2020) The rebound effect on water extraction from subsidising irrigation infrastructure in Australia. Resour Conserv Recycl 159:104755. https://doi.org/10.1016/j.resconrec.2020.104755

Winfield D, Harriss D, Grant M, Weber T (2019). River basin planning: An Indian guide. Australian Water Partnership. https://waterpartnership.org.au/publications/river-basin-planning-an-indian-guide/. Accessed on 1 June 2024

Yazd D, Wheeler S, Zuo A (2020) Understanding the impacts of water scarcity and socio-economic demographics on farmer mental health in the Murray-Darling basin. Ecol Econ 169:106564

Zwarteveen M (2008) Men, masculinities and water powers in irrigation. Water Alternatives 1(1):111–130

Zwarteveen M (2017) Hydrocracies, engineers and power: questioning masculinities in water. Eng Stud 9(2):78–94. https://doi.org/10.1080/19378629.2017.1358730

Zwarteveen MZ, Boelens R (2014) Defining, researching and struggling for water justice: some conceptual building blocks for research and action. Water Int 39(2):143–158. https://doi.org/10.1080/02508060.2014.891168

Melita Grant Melita Grant is a Research Director at the Institute for Sustainable Futures, University of Technology Sydney. She specialises in water, sanitation, and hygiene (WASH) and integrated water resources management (IWRM) in international development, with a focus on gender equality and inclusion. Melita has worked in local and state government, civil society, and university sectors. With an academic background in environmental management and international relations, Melita has expertise in a range of sustainable development issues, and has applied this in Indo-Pacific and Australian contexts. Melita is the Co-Chair of the Australian WASH Reference Group, an Asialink leader, member of the Expert Review Panel for the Australian Water Partnership, and a Board Member of the Australian Network for Arts and Technology (ANAT).

Cameron Holley Professor Cameron Holley is a Professor in the School of Law, Society and Criminology at UNSW Law & Justice. He is a former Head of School and is a current co-chair of the International Network of Environmental Compliance and Enforcement (INECE) Academic Committee, a member of the Taskforce on Earth System Law, a member of the Steering Committee for the Global Water Institute and a former Director of the Connected Waters Initiative, UNSW Sydney. He was also a member of The Australian Panel of Experts on Environmental Law (APEEL). Professor Holley publishes widely in the areas of environmental law, natural resources law, energy law and water law, with a focus on regulation and governance. His contribution in this book was funded by an Australian Research Council Discovery Grant (DP190101584) and Linkage Grant (LP220200350).

Dr. Emma Carmody Emma is a Co-Founder and the Director of Legal and Partnerships at Restore Blue. Emma has almost 20 years' experience working in Australia and abroad as an environment and planning lawyer and policy advisor, including 11 years at the Environmental Defenders Office in Australia and 5 years as the Legal Advisor to the Ramsar Convention on Wetlands in Switzerland. She is particularly well known for her expertise in the fields of water law and international environmental law, as well as her deep understanding of the complex legal and financial frameworks governing nature restoration projects.

Chapter 5
India-China Water Issues: Hydro-Politics, Disputes and Cooperative Arrangements

Amit Ranjan

5.1 Introduction

After a slump due to the pandemic in 2019–20, the Chinese economy has managed to revive relatively faster than many other economies of the world. In 2021, China recorded a growth of 8.1% in 2021 (China Briefing 2022). Such recovery reflected the release of "pent-up demand" and a "quicker than expected" resumption of the production process and exports of products (Global Economic Prospects 2021). Growth in 2021 was encouraging after a COVID-19 linked contraction of 2.6% in 2020 (The World Bank 2021). However, amid rift with the United States, in 2022, China's growth was plunged to 3% and expected to rebound to only 5% in 2023 (The World Bank 2023).

To maintain and sustain economic growth, availability of sufficient amount of water is key. Water is needed to support industrial production, agricultural activities and for human consumption. However, due to growing population and increase in consumption of waters for economic activities, per capita amount of available water resources in China is contracting, and the demand-supply gap is widening. In 2020, water resources per capita data was 2239.756 cubic meters which reached at 2098.488 cubic meters in 2021 (Ministry of Water Resources, People's Republic of China, n.d.a).

It is estimated that China has 2963.82 billion cubic meters (BCM) of water (China Water Resources Bulletin 2021,1). China shares about 80 major transboundary rivers with 14 immediate neighbouring states, and three additional riparian countries (Yangtso, 2017; Wouters et al., 2024; Zhang and Li 2018a: 24). The mean annual outflow from the transboundary rivers in China is about 728 BCM, around 34 times more than the water inflow in the country. Only 0.96 per cent of total water flows into

A. Ranjan (✉)
National University of Singapore, Singapore, Singapore

J. Andharia and P. Anand (eds.), *Shifting Lands, Flowing Waters*, Disaster Studies and Management, https://doi.org/10.1007/978-981-96-7691-0_5

China from outside. Other South and Southeast Asian countries' dependence on transboundary rivers are very high such as Bangladesh (91.44%), Cambodia (74.67%), Vietnam (59.35%), India (30.52%), Thailand (48.81%), Vietnam (59.35%), and Myanmar (14.13%) (Cited in Yangtso 2017, 51). In terms of fresh water availability, China is at number six in the world next to Brazil, Russia, Canada, the USA, and Indonesia (Ministry of Water Resources, Peoples Republic of China a). It has around 19 per cent of the world's population (Zhang and Mingjiang Li 2018: 3).

The available water resources are unevenly distributed in China: northern part of China (north to the Yangtze River Basin) has 63.5% of the national territory with about 19% of the country's total water resources (Ministry of Water Resources, People's Republic of China a, 3) while Southern China with 37% territory possesses 80% of the country's waters. Consumption wise, according to data published by the Chinese government in 2015, about 13% of the total available water is used for domestic consumption, 21.9% is used for industrial purposes, 63.1% is used for agricultural activities and 2% flows freely to maintain environmental sustainability (Ministry of Water Resources, Peoples Republic of China).

By 2030 China's population is likely to be more than 1.4 billion (Reuters 2019) at that time the demands for water will increase to—50% for agriculture, 32% for industry, and 18% for domestic use (Osborn 2010). This sector wise increase will widen the demand-supply gap to 24% (ibid). Some of the signs of water-related stress are tangible: About 65 per cent of the Chinese cities suffer from domestic water shortage (ibid); and a few of the provinces, mainly in North of the country, are already facing water scarce[1] situation (See China Statistical Year Book 2017). Analysing the water situation in country, the United Nations World Water Development Report of 2018 observes that if it remains business as usual, parts of China (like many other parts of the world) may face severe water scarcity by 2050 (United Nations World Water Development Report 2018, 12).

Other than widening demand-supply gap of water, China also faces water-related challenges due to climate change. It is known that climate change causes unseasonal rainfall and beyond the managed precipitation during the rainy season which causes floods. This will affect the Chinese economy and its people. For example, the 2017 flood affected about 11 million people and China incurred a loss of about US $ 3.7 billion. The most affected provinces were Hunan, Sichuan, and Guangxi (South China Morning Post 2017). Paradoxically, in 2022, parts of China faced severe drought. The Chinese province of Jiangxi declared "red alert" as water levels in Poyang Lake, a flood outlet for the Yangtze River, had fallen. Water level in ten of the reservoirs in Anhui province fallen below the "dead pool" level, where they were unable to discharge water downstream. Drought affected the middle and lower reaches of the Yangtze River (Al Jazeera 2022).

As China experiences water shortage in the northern provinces and face damages during the rainy season, it is being accused of diverting waters from the transboundary

[1] Availability of water of 1700 cubic meters per person in country makes it water stressed. Below it, availability of 1000 cubic meters of water per person represents a state of "water scarcity"—and below 500 cubic meters is "absolute scarcity". (Human Development Report 2006:135).

rivers in lean seasons and releasing more to the lower riparian States in monsoon (See Chellaney 2011). This diversion and release of more than the managed waters damages the economy and affects people from the lower riparian States. One of the lower riparian States is India. It is estimated that around 354 BCM of waters flow from Tibet into India (MPIDSA Task Force Report 2010, 44).

This chapter is divided into three sections. The first part discusses India-China transboundary water issues in Eastern and Western sectors. The second section examines China's water transfer projects and their impact on India. The third section analyses cooperative arrangements between India and China on their transboundary rivers. The chapter argues that growing demand for water in India and China will escalate their tensions over transboundary rivers. Secondly, it argues that Sino-India political and military tensions have impact on their cooperative arrangements on transboundary river water.

5.2 India-China Water Issues

India and China share 3488 square kilometres (kms) of border of which, the Union Territories of Jammu & Kashmir and Ladakh share 1597 kms, Himachal Pradesh 200 kms, Uttarakhand 345 kms, Sikkim 220 kms, and Arunachal Pradesh 1126 kms. There is a total of 176 Border Out Posts set up by the Indo-Tibetan Border Police Force which guards the Indian border with China (Ministry of Home Affairs, Government of India n.d.a). India considers the 3488 square kilometers as Line of Actual Control (LAC) while for China only 2000 square kilometers of land is LAC (Singh 2020). The LAC was introduced by the then Chinese Prime Minister Zhou Enlai in a press conference in Delhi. Earlier he mentioned it in 1959 in letter to the then Indian Prime Minister Jawaharlal Nehru. Zhou proposed six points to solve the border issue (Apri 2023). Of the six, the first two points were: There exist disputes with regard to the boundary between the two sides; and, there exists between the two countries a line of actual control (LAC) up to which each side exercises administrative jurisdiction (ibid). After months of talks between Indian and Chinese officials following Zhou's visit to Delhi in 1960, India stated (cited in ibid),

> The India-China boundary starts from the tri-junction of the boundaries of India, China and Afghanistan and runs eastward through the Kilik Pass, Mintaka Pass, Kharchanai Pass, Parpik Pass, and the Khunjerab Pass. These passes lie on the watershed between the Hunza river flowing into the Indus system in India and the Qara Chukar river flowing into the Yarkand system in Sinkiang. From the Khunjerab Pass the boundary lies along a spur down to the north-western bend of the Shaksgam or Muztagh river which it crosses at that point and ascends the crest line of the Aghil mountains. It then runs along the crest of the Aghil watershed through the Aghil Pass the Marpo Pass and the Shaksgam Pass to the Karakoram Pass.

As India considers Gilgit-Baltistan, administered by Pakistan, its own part "illegally occupied" by Islamabad, it is argued that then the total International Border between India and China would be around 4056 square kilometers (ibid). In 1963,

Pakistan and China exchanged land to settle their boundary-related differences bordering Gilgit-Baltistan.

Existing disputes over their border in the Eastern and Western sectors have occasionally created military stand-off between India and China. In 1988, India and China decided to set up joint groups to negotiate their border-related issues and since 2001 the talks have been taking place between the Special Representatives from the two countries. In September 1993, India and China signed an "Agreement on the Maintenance of Peace and Tranquility along the Line of Actual Control in the India-China Border Areas" (Ministry of External Affairs, Government of India 2005). In 1996 the two countries signed "Agreement on Confidence Building Measures in the Military Field along the Line of Actual Control in the India-China Border Areas" (ibid), in 2003 they signed "the Declaration on Principles for Relations and Comprehensive Cooperation between India and China", and in 2005 India and China agreed on "Political Parameters and Guiding Principles for the Settlement of the India-China Boundary Question" (ibid). In 2012, India and China established Working Mechanism for Consultation and Coordination (WMCC) on India-China Borders. Its aim is to study how to strengthen cooperation between military personnel; explore the possibility of cooperation in the border areas; and, address the situation and matters that may affect the maintenance of peace and tranquility in the border areas and work towards maintaining the friendly atmosphere. Emergency consultations can be convened after mutual agreement. The WMCC can meet once or twice every year (Ministry of External Affairs 2012). In 2013, India and China signed a Border Defense Cooperation Agreement. Under this Agreement, it was agreed to further enhance cooperation by exchanging military information, assist each other in locating personnel, livestock, means of transport, and aerial vehicles that may have crossed or are possibly in the process of crossing the LAC, not follow or tail patrols of the other side in areas where there is no common understanding of the LAC and maintain maximum restraint if the two sides come face-to-face situation in areas where there is no mutual understanding (Press Information Bureau 2013). In 2015, during the Indian Prime Minister Narendra Modi's visit to Beijing, India-China signed on measures to boost cooperation on the border-related matters.

Nevertheless, India-China borders are not entirely peaceful; military standoffs between the two armies occur at regular interval. The latest one occurred in the western sector of Ladakh where in June 2020 Indian and Chinese military personnel engaged in a fight in which about 20 Indian and four Chinese soldiers lost their lives.

Besides physical, political, and historic factors, one of the chief reasons why China eyes Indian territories and lays claim over them is because of water. The Indian regions in the Himalayas that borders China are rich in water resources. For example, by laying its physical claim on the Indian State of Arunachal Pradesh, China also claims almost about 200 million cusecs of waters in the State (MPIDSA Task Force Report 2010 44). This is also the case in the western sector of the India-China border.

5.2.1 India-China Transboundary Rivers in the Eastern Sector

In the Eastern sector, India and China share source to the Brahmaputra River or the Yarlung Zangbo River as it is called in the upper stream. Brahmaputra originates in Angsi glacier in south western part of Tibet Autonomous Region (TAR) in China. Its total length is 2906 kilometers. Of this, 1625 kilometers is in TAR, 918 kilometers in India (278 kilometers in Arunachal Pradesh and 640 kilometers in Assam) and 363 kilometers of the river flows in Bangladesh before its outfall in the Bay of Bengal. Brahmaputra drains about 5,80,000 square kilometers of area. Out of that, 2,93,000 square kilometers is in TAR, 240,000 square kilometers in India and 47,000 square kilometers is in Bangladesh (Ministry of Jal Shakti, River Development & Ganga Rejuvenation, Government of India a 2017). On an average, the mean annual flow of the Brahmaputra that comes to India from Tibet is about 131 BCM (MPIDSA Task Force Report 2010 44 & Chellany 2011, 143). From Tibet, it cuts through Gyala Peri and Namcha Barwa and takes about 180 degrees turn southward to enter in Arunachal Pradesh at a place called Gelling in upper Siang district where it is called Dihang or Siang. The river is further joined by two mountain streams Dibang and Lohit near Sadiya in Tinsukhia District of Assam and lands in Assam Valley. Below that confluence, it gets its name Brahmaputra.

To secure its water interests and generate hydropower China has been constructing hydropower on the Brahmaputra's upper stream that falls in the TAR. It is estimated that the number of such constructions was 198 by 2016 (Deepak 2016). One of the major projects on the Brahmaputra in the TAR is Zangmu hydropower project that is located in the middle reaches of the river in Gyaca in the TAR. In 2014 the first unit of the US $1.5 billion Zangmu project became operational (Ramachandran 2015) and in 2015 it was fully operationalised. Zangmu is considered to be the world's highest-altitude hydropower station, located at 3370 feet (India Today 2011). It has a capacity of producing 2.5 billion kilowatt hours of electricity annually. Zangmu has been built by China Huaneng Group which operates the Zangmu plant that comprises six units (VOA 2015).

Unfolding energy plan for the 12th Five-Year Plan (2011–2015) Chinese State Council listed two more hydropower projects-Dagu and Jiacha- besides Jixu on the upper stream of Brahmaputra (Krishnan 2013). Among the three projects planned, Dagu (640 MW) is bigger than Zangmu and lies 18 kilometers upstream of Zangmu. Jiacha is 320 (MW) is located in the middle reaches of Brahmaputra's upper stream. Jixu (560 MW) is 11 kilometers upstream of Zangmu (ibid). Referring to these three planned dams on the upper stream of Brahmaputra in a press conference on 4 February 2013, the then China's foreign ministry spokesperson, Hua Chunying said (Office of the Commissioner of the Ministry of Foreign Affairs of the People's Republic of China in the Macau Special Administrative Region):

> China has always been responsible on the development and utilisation of cross-border rivers. We carry out the policy of placing equal importance on development and protection and take account of the influence on the downstream countries. The planned dams will not affect flood

> control, disaster mitigation and ecological environment in the downstream areas. China and India have maintained communication and cooperation on the issue of cross-border rivers.

Despite such assurances, all such dams in the upper stream of the Brahmaputra are looked upon as a means to control and regulate water flow into India's northeast States (See Chellaney 2011). Such structures and their regulations are mainly blamed for regular occurrence of floods in Assam and other northeast States. It is also believed that with such potent weapon, China can inflict severe damage to India (ibid).

Contrary to such a widely perceived opinion, there is an argument that the hydro structures in the upper stream of the Brahmaputra do not have serious impact on its waters flowing beyond Arunachal Pradesh. The river receives more water from its tributaries, who join it in plains, and from rainfall in the region. Citing data published by the Chinese researcher L Jiang and team, Ghosh shows that the total annual outflow of upper stream of Brahmaputra from China is estimated to be about 131 BCM while by the time it reaches Bhadurabad, the gauging station near the end of the rivers' sub-basin in Bangladesh, its outflow is about 606 BCM (Ghosh 2018). Referring to the works of Datta & Singh (2004), Ghosh writes that Brahmaputra's peak flow during the monsoon at Nuxia and Tsela Dzong, water measuring stations in Tibet, is about 5000 and 10,000 cubic meters per second respectively while this is around 40,000 cubic metres per second in Guwahati and 50,000 cubic meters per second at Bhadurabad (Ghosh 2018). During the lean season, the flow at Nuxia is about 300–500 cubic metres per second while it is about 4000 cubic metres per second at Guwahati and 5000 cubic meters per second at Bhadurabad (ibid). This means, as Ghosh observes, Brahmaputra becomes fatter only after entering India where it is joined by tributaries such as Dibang, Lohit, Subansiri, Manas etc. (ibid).

Due to physiography of the region, often landslides occur in the Upper Stream of the Brahmaputra. On 9 April 2000 there was a landslide in Yigong Tsangpo, a major tributary of Yarlung Tsnagpo/Zangbo in the TAR. The landslide was due to two earthquakes. It blocked the floodplain river and created a dam area of about 2.5 square kilometers and 90 meters deep (Sheth & Anirban Datta-Roy 2017). The natural dams were breached on 1 June 2000, leading to a massive flood in Arunachal Pradesh and Assam (Ibid). In that flood a large swathe of forest was destroyed and 50,000 people from five districts of Arunachal Pradesh were left homeless (Rediff.com 2000). Villages and people in Assam were also affected.

As in 2000, in the year 2017 fears were expressed by India when waters from the Brahmaputra turned dark due to debris flowing into it because of landslides in Upper Stream (Mazoomdaar 2017). The news reports also talked about the Chinese purported plan to construct about 1000 kilometers tunnel to divert waters from Yarlung Tsangpo/Zangbo to deserted Taklamakan (Chen 2017 & Saikia 2017). China refuted such reports that the flows of debris in the Siang River are due to its attempts to build tunnel to divert waters in Xinjiang region (Varma 2017). The Public Health Department report of the Arunachal Pradesh blamed the increase in the nephelometric turbidity unit level in Siang's Waters by several times as a reason for the waters turning black (Saikia 2017).

In October 2018 also, flash floods occurred in Arunachal Pradesh because of a lake caused by the landslide in the upper stream of the Siang River. However, the timely information provided by China helped the administration to move the people living near the Siang in Arunachal Pradesh to safe places. On this issue spokesperson from China's Ministry of Foreign Affairs said that (Ministry of Foreign Affairs, People's Republic of China 2018):

> After that happened, the Chinese departments for water resources immediately notified the Indian side of the relevant situation, activated the emergency information reporting mechanism and warned the Indian side to get prepared to deal with it. Starting from October 17, the Chinese side offered the temporary hydrological data on an hourly basis to the Indian side. As of October 22, we have provided seven bits of hydrological information and 110 bits of hydrological data to India. On October 19, China emergently notified the Indian side of the situation regarding the natural flow of the barrier lake. According to Chinese competent authorities, as of 12 o'clock, October 20, the section of Yarlung Tsangpo where the barrier lake is located has basically resumed normal natural flow and the Mo Tuo part in the lower reaches has also basically come back to normal. The Chinese side is still closely monitoring the landslide and will maintain communication and cooperation with the Indian side through the existing channels.

Chinese information in 2018 prevented damage in Arunachal Pradesh but concerns remain.

In its fourteenth five-year plan (2021–2025) China has planned to develop 60 Gigawatt of hydropower on the upper stream of stream of Brahmaputra in Tibet. Chairman of Chinese state-owned Power Construction Corporation of China, Yan Zhiyong called the damming of the river a "historic opportunity" and would meet the country's clean energy plans and also strengthen water supply security (Reuters 2020). The Chinese Communist Party said that the party would "implement the development of hydropower resources on the lower reaches of the Yarlung Tsangbo river" which are mentioned in a list of 2021–2025 "five-year plan" policy recommendations (ibid). This is likely to affect the water flows in downstream region of India and also in Tibet (ibid). The dam China is planning in Tibet's Medog County is expected to dwarf the Three Gorges Dam on the River Yangtze in central China and is expected to produce about 300 billion kilowatts of electricity each year. The 14th five-year plan was unveiled in March 2021. Earlier, in October 2020, Tibet government signed a "strategic cooperation agreement" with Power Construction Corporation of China to produce hydroelectricity (CNA 2021). Once it comes into operation this "super dam" may affect the downstream basin. There is also ecological risk in the region where the dam is going to be built as the region is known for seismic activity and contains unique biodiversity (ibid).

As a sovereign country, India which shares waters of Brahmaputra from China, is entitled to have freedom from any harm by any diversion of the river in the upper stream. China is obligated to demonstrate that the hydro structure it is developing in the upper stream is legal, built scientifically and will not damage the water interests of India. China is required to conduct transboundary environmental impact assessment (TEIA) which is a part customary international law obligation (Sadat 2021). International jurisprudence recognises the obligation to conduct in the Pulp Mills case on River Uruguay between Argentina and Uruguay in 2010, and the Advisory

Opinion on the Responsibilities and Obligations in the Area and Kishanganga Arbitration case between India and Pakistan in 2013 (ibid). China claims that it adheres to the TEIA (ibid).

5.2.2 *Transboundary Rivers in Western Sector*

Two of the major rivers of the Indus Rivers System (IRS)-Indus and Sutlej- have their source of origin in Tibet. Waters of these two rivers are governed between India and Pakistan by the Indus Waters Treaty of 1960. Under the treaty major portion of river Indus was allocated to Pakistan while larger amounts of waters of river Sutlej are used by India.

Due to past water behaviours and activities of China, India has concerns over the flow of the Sutlej River (called Langqen Zangbo in China) into India from Tibet. In the year 2000, as timely information was not provided to India, flash floods occurred that caused huge damage in the Indian state of Himachal Pradesh. In that mishap, more than hundred people died, 120 kilometers of old Hindustan-Tibet highway was washed away, and 98 bridges destroyed (Gupta 2001). Four years after, in 2004 once again similar "natural processes" occurred in upper basin of the river. An artificial lake was formed near river Pareechu in Tibet due to seasonal landslides. This lake was sixty meters deep and covered an area of about 230 hectares (Gautam 2004). To meet the situation a red alert was issued by the Himachal Pradesh government and armed and paramilitary forces were deployed and put on high alert. Nathpa-Jhakri project on river Sutlej had to be shut down (Water Power and Dam Construction 2004). Nothing like 2000 happened in 2004, though many people left their houses in anticipation of repetition of 2000 like disaster. However, in June 2005, floods occurred in Himachal Pradesh due to release of millions of cusecs of waters from upstream overflows of Sutlej. Though there were no loss of lives, the economic damage caused by dam was about INR 700 crore (US $ 9.3 billion) (Vinayak 2005).

There are also reports about hydro structure constructions by China on Sutlej (Tibet Support Association Sambhala Tibet Center 2010). One of the dam-related issues on the Chinese side of Sutlej was first raised by the Indian media in March 2006. Reacting to the reports the then Foreign Ministry spokesperson said that "In order to satisfy the electricity needs of the local population, the Chinese side has built a small-scale hydro-electric station on the Sutlej River at Zada County recently" (DNA 2006). However, addressing the fears expressed by the Indian media about the dam, the spokesperson added that "In the process of development of trans-border rivers, the Chinese side fully considers the impact on its lower reaches" (ibid).

5.3 China's Water Transfer/Diversion Projects and India

To meet its water needs in water scarce north, China is engaged in developing south-north water diversion projects. The first phase of the South-to-North Water Diversion Project/South to North Water Transfer Project (SNWDP/SNWTP) or *Nanshuibeidiao* started operation in December 2013. The objective of this project is to move about 44.8 billion cubic meters of water across the country every year (Kuo 2014). Two lines of this project—The Eastern Line and the Central Line[2]—are only concerned with China's domestic water issues. But the third line in this project is the Great Western Line which is a concern for India, if it fully implemented. This line is sub-divided into two—The Western and the Greater Western Routes (Wirsing et al., 2013, pp. 115).

The Greater Western Route was first conceived in 1950s and again in 1990s it was proposed by Guo Kai, vice director and secretary-general of the Shuotian Canal preparatory committee, who said that, "So much water of the Yarlung Zangbo runs out of China, it's a huge waste" (Global Times 2017a). Kai proposed to divert around 200 billion cubic meters of water every year to North and Northwest parts of China through tunnels and pipelines. It found support from the military officers, including 118 major general and above ranked officers (ibid). In 1999, Kai also came out with a positive report on the feasibility of the project. However, many other hydrologists and Wang Sucheng, the then Minister of Water Resources defined the plan "unnecessary, unscientific and unfeasible" (ibid). Later in 2006, supporting its vitality, Li Ling wrote a book titled *Tibet Water Will Save China.* This plan envisages diversions of waters from six rivers such as Yarlung Zangbo, Nu, Lancang, Jinshu, Yalong, and Dadu (Deepak 2016, 44) through canals and tunnels across Qinghai Tibetan plateau. Again, the opponents of this project questioned its significance as they found that only 17 billion cubic meters of waters could be diverted, at the cost of around 390 billion yuan, and it would take about 40 years to be completed (ibid).

In 2011, China's Vice-Minister of Water Resources, Jiao Yong had publicly stated his government's position against this project. The vice-minister categorically stated that there are technical difficulties, and it would affect China's relationships with its co-riparian neighbours (Wirsing et al., 2013, 120; Deepak 2016 & Zhang 2016). Despite the official position of China on the Greater Western Line project, the debate in support of it continued. In July 2017, around 20 scholars met outside Urumqi in Northwest China's Xinjiang Uyghur Autonomous Region and discussed the feasibility of diverting waters from Qinghai-Tibet Plateau to Xinjiang's lowland plains. Supporting this project Professor Ren Qunluo said that "Water from rivers such as Yarlung Zangbo River can help turn the vast deserts and arid lands into oasis and farmlands, alleviate population pressure in the east, as well as reduce flood risks in the countries through which the river travels downstream" (Global Times 2017b).

[2] Eastern Line is also known as Beijing-Hangzhou line. In this water from Yangtze is transferred through a canal. The Central Line has an objective is to transfer about 13 billion cubic waters from the river Han to Beijing and Tianjin from Danjiangkou reservoir.

5.4 Cooperative Water Arrangements between India and China

States in international system do cooperate when there are some economic or political payoffs from such cooperation. Lower riparian states do not have many options other than to cooperate with the upper riparian state for their water security. Unlike them, the advantaged state opens to an agreement with the lower riparian state only if it leads to some form of political or economic payoffs (Haftendorn 2000).

Cooperative arrangements between India and China over water issues were first signed in 2002 and 2004 respectively when China agreed to provide hydrological information of major transboundary rivers to India in flood season (Ministry of Water Resources, Peoples Republic of China b). One of the prime reasons for signing such cooperative arrangements was much improved bilateral ties with economic payoff to the upper riparian. For instance, in 2001 India-China trade was about US $3.6 billion in which China enjoyed a trade surplus of about US $ 196 million (The Times of India 2002). Politically, the two countries agreed to appoint Special Representatives to discuss their border disputes.

In 2002, India and China signed cooperation on provision of Hydrological Information of the Brahmaputra/Yarlung Zangbo in Flood Season by China to India.[3] The MoU was for five years. The MoU was renewed in 2008, 2013 and 2018. It expired on 5 June 2023 and is under the process of renewal (Ministry of Jal Shakti, Government of India n.d.a). In 2004, a MoU was signed on Sutlej/Langqen Zangbo whose main purpose was flood control and disaster mitigation in downstream areas. In April 2005, a MoU was signed to share hydrological information about Sutlej River during flood season from Tsada station. The MoU was renewed in 2010, and 2015 but now under the process of renewal. However, China supplied the monsoon data on the Sutlej between 1 June and 15 October 2021. There is an institutional Expert Level Mechanism set up between India and China. The 14th meeting of ELM was held in New Delhi on 20–21 June 2023 (Ministry of Jal Shakti, Government of India n.d.a).

However, data is not free. Between 2002 and 2007 China shared the hydrological information without any price; afterwards, India pays RMB 850,000 or around USD 134,000 every year in April for maintaining three hydrological datacenters on the Chinese side (Zhang and Li 2018a: 15). Also, this hydrological data information sharing and MoU between India and China often fails to sustain the pressure of India-China military stand-off. This was proved during 73 days of stand-off between the Indian and the Chinese soldiers in Doklam in 2017. During that stand-off hydrological information sharing process was interrupted (Gupta 2018) and resumed only in March 2018 (India TV 2018). Resuming the data sharing the Chinese Foreign Ministry spokesman Lu Kang stated, "The two sides have agreed to continue with

[3] On condition of anonymity an officer from Central Water Commission, Ministry of Jal Shakti, Department of Water Resources and River Development and Ganga Rejuvenation, Government of India told the author that many times the data India gets does not match with the real flow in the river. Also, he said that it is not known how much water is available in the upper region of the transboundary rivers.

such cooperation. On the grounds of humanitarian principles and the basis of bilateral relations China will continue with this cooperation with the Indian side on the provision of hydrological information and emergency response cooperation" (The Pioneer 2018). During Doklam, the Chinese government said that the interruption was due to then ongoing some technical upgrading works (Gupta 2018) in the water measuring stations and had nothing to do with the Doklam crisis. The then spokesperson from the Chinese Foreign Ministry, Geng Shuang said, "Last year [in 2016], due to the needs for reconstruction after being damaged by the flood and out of such technological reasons as upgrading and renovation, the relevant hydrological stations in China do not have the conditions to collect relevant hydrological data now [in 2017]" (Khadka 2017). However, during the same period, China shared hydrological data with Bangladesh. Mofazzal Hossain, a member of the joint rivers commission of Bangladesh, told *BBC* that "We [Bangladesh] have been receiving such data from three hydrological stations in Tibet since 2002 and they have continued to share the figure with us even during this [2017] monsoon season] (ibid).

5.5 Conclusion

Increasing economic activities and growing population in India and China will raise their water demands which is likely to escalate tensions between the two politically rival riparian countries. As mentioned above, India accuses China for building hydro structure with an objective to divert waters from transboundary river. This accusation is often inflated by Indian media and civil society. However, as pointed out above, Chinese projects do not have serious water impact beyond Arunachal Pradesh. Nevertheless, India's water concerns cannot be entirely overlooked.

Although under renewal process, there are cooperative arrangements between India and China to share water-related data. However, as mentioned above, implementation of such cooperative arrangements comes under question at a time when political or military tensions built up between India and China. In the post-covid years, Sino-India relationship has plummeted. India along with the USA, Japan, and Australia has re-energised QUAD (Quadrilateral Security Dialogue) to check Chinese assertion in Asia and near its border. In such situation, cooperative arrangements cannot remain unaffected.

References

China Statistical Year Book 2017 Compiled by National Bureau of Statistics of China. http://www.stats.gov.cn/tjsj/ndsj/2017/indexeh.htm Accessed on 1 August 2018

Al Jazeera (2022 23 September) Red alert in China as drought dries up countrys biggest lake https://www.aljazeera.com/news/2022/9/23/red-alert-in-china-as-drought-dries-up-countrys-biggest-lake Accessed on 25 December 2022

Apri Claude (2023 15 February) LAC vs IB: Why Indias assessment of the length of border along China is misleading Firstpost. Accessible at https://www.firstpost.com/opinion/lac-vs-ib-why-indias-assessment-of-the-length-of-border-along-china-is-misleading-12157882.html?fbclid=IwAR0ffCHV4gIlJxK8pt1p9E5ePLnKBZd5GQQvCxCRGYmEQFWiVpw-cHAPTvI. Accessed on 12 July 2023
Chellaney B (2011) Water: Asia's new Battlegroun. Harper-Collins, Noida
Chen Stephen (2017, 29 October) Chinese engineers plan 1,000km tunnel to make Xinjiang desert bloom South China Morning Post. https://www.scmp.com/news/china/society/article/2116750/chinese-engineers-plan-1000km-tunnel-make-xinjiang-desert-bloom. Accessed on 21 March 2020
China Briefing (2022, 7 February) Chinas Most Productive Provinces and Cities as per 2021 GDP Statistics Available at https://www.china-briefing.com/news/chinas-2021-gdp-performance-a-look-at-major-provinces-and-cities/#:~:text=Chinas%20GDP%20in%202021%20reached,Bureau%20of%20Statistics%20(NBS). Accessed on 20 June 2023
China Water Resources Bulletin 2021. https://www.yicode.org.cn/en/chinas-2021-water-resources-bulletin-released/. Accessed on 17 May 2023
CNA (2021, 11 April). Chinas plans for Himalayan super dam stoke fears in India https://www.channelnewsasia.com/news/world/china-plans-himalayan-tibet-super-dam-stoke-fears-india-14600280. Accessed on 13 April 2021
Datta B, Singh VP (2004) Hydrology. In: Singh VP, Sharma N, Ojha CSP (eds) The Brahmaputra Basin water resources. Springer, Dordercht, pp 139–195
Deepak BR (2016) India and China: foreign policy approaches. Viz Books, New Delhi
DNA. (2006, 7 July). China Defends Dam on Sutlej https://www.dnaindia.com/world/report-china-defends-dam-on-sutlej-1040220 Accessed on 19 July 2017
Embassy of India, Bejing, Government of India (2018). Implementation Plan: Provision Of Hydrological Information of the Yarlung Zangbu/ Brahmaputra River in Flood Season by China to India http://indianembassybeijing.in/implementation-plan.php Accessed on 24 September 2018
Gautam P.K (2004, 30 November) The Parechu Lake Incident: A Preliminary Analysis IPCS http://www.ipcs.org/comm_select.php?articleNo=1569. Accessed on 12 March 2021
Ghosh Nilanjan (2018, 9 November) China Cannot Rob us from Brahmaputra The Hindu Businessline Available at: https://www.thehindubusinessline.com/opinion/china-cannot-rob-us-of-brahmaputra/article9974000.ece Accessed on 25 September 2018
Global Economic Prospects, East Asia and Pacific, 5 January 2021. https://pubdocs.worldbank.org/en/693711599838723416/Global-Economic-Prospects-January-2021-Regional-Overview-EAP.pdf. Accessed on 8 March 2021
Global Times (2017a, 8 July). Scholars mull project to divert water from Tibet to arid Xinjiang' http://www.globaltimes.cn/content/1060047.shtml. Accessed on 12 September 2018
Global Times (2017b, 21 June) Biggest water transfer project ever benefits 100 million in China Global Times. http://www.globaltimes.cn/content/1052749.shtml. Accessed on 23 April 2018
Global Water Partnership (2015). Chinas Water Resources Management Challenges: The Three Red Lines. https://www.gwp.org/globalassets/global/toolbox/publications/technical-focus-papers/tfpchina_2015.pdf. Accessed on 12 June 2017
Grieco Joseph (1988). Anarchy and the Limits of Cooperation: A Realist Critique of Newest Liberal Institutionalism. International Organization 42(3) Summer 427–60. Published in Baumann, Rainer, Peter Mayer and Bernhard Zangal (ed) (2011) International Relations: The Great Debates Volume I Cheltenham: An Elgar Research Collection, p 332–365
Gupta Joydeep (2018) China, India, and water across the border. https://www.thethirdpole.net/en/2018/05/01/china-india-and-water-across-the-border/. Accessed on 12 October 2018
Gupta Sishir (2001, 25 June) Floods ravage Himachal and Arunachal Pradesh, satellite pictures suggest Chinas hand India Today. https://www.indiatoday.in/magazine/neighbours/story/20010625-floods-ravage-himachal-and-arunachal-pradesh-satellite-pictures-suggest-china-hand-774997-2001-06-25. Accessed on 13 March 2021
Haftendorn H (2000) Water and international conflict. Third World Q 21(1):51–68

Hardiman Richard (2018) Chinas Water Crisis Revisited China Policy Analysis Institute. https://cpianalysis.org/2018/05/28/chinas-water-crisis-revisited/. Accessed on 10 June 2018

Human Development Report (2006). Beyond Scarcity: Power, poverty and the Global Water Crisis, United Nations Development Programme, New York. http://hdr.undp.org/sites/default/files/reports/267/hdr06-complete.pdf p 135. Accessed on 20 April 2018

India Today (2011, 20 June). China not reducing height of dam on Brahmaputra, says intelligence report https://www.indiatoday.in/world/neighbours/story/china-not-reducing-height-of-dam-on-brahmaputra-says-intelligence-report-135952-2011-06-20. Accessed on 24 March 2017

India TV (2018, 28 March). China to resume sharing hydrological data with India on Brahmaputra river https://www.indiatvnews.com/news/world-china-to-resume-sharing-hydrological-data-with-india-on-brahmaputra-river-434818. Accessed on 12 October 2018

Khadka Navin Singh (2017, 18 September) China and India water dispute after border stand-off BBC. https://www.bbc.com/news/world-asia-41303082. Accessed on 12 April 2020

Krishnan Ananth (2013, 30 January). China gives go-ahead for three new Brahmaputra dams The Hindu. https://www.thehindu.com/news/international/China-gives-go-ahead-for-three-new-Brahmaputra-dams/article12323702.ece. Accessed on 19 July 2020

Kuo, Lily (2014, 7 March). China Has Launched the Largest Water-Pipeline Project in History The Atlantic 7 March. https://www.theatlantic.com/international/archive/2014/03/china-has-launched-the-largest-water-pipeline-project-in-history/284300/. Accessed on 21 April 2018

Mazoomdaar Jay (2017, 22 December) Whats darkening Brahmaputra: Landslide, not Chinese machines. The Indian Express http://indianexpress.com/article/india/whats-darkening-brahmaputra-landslide-not-chinese-machines-4993783/ Accessed on 23 April 2018

Ministry of External Affairs Government of India. (2014) Joint Statement between the Republic of India and the People's Republic of China on Building a Closer Developmental Partnership. http://www.mea.gov.in/bilateral-documents.htm?dtl/24022/Joint_Statement_between_the_Republic_of_India_and_the_Peoples_Republic_of_China_on_Building_a_Closer_Developmental_Partnership. Accessed 12 July 2015

Ministry of External Affairs, Government of India 2017 Press Statement on Doklam disengagement understanding. http://www.mea.gov.in/press-releases.htm?dtl/28893/Press_Statement_on_Doklam_disengagement_understanding. Accessed on 21 May 2018

Ministry of External Affairs, Government of India (2018) India-China Bilateral Agreements signed in Qingdao, China 2018. https://mea.gov.in/bilateral-documents.htm?dtl/29966/IndiaChina+Bilateral+Agreements+signed+in+Qingdao+China. Accessed on 15 July 2019

Ministry of External Affairs, Government of India 2012 Agreement between The Government of the Republic of India and The Government of the People's Republic of China on the Establishment of a Working Mechanism for Consultation and Coordination on India-China Border Affairs 2012. https://www.mea.gov.in/Portal/LegalTreatiesDoc/CH12B0194.pdf. Accessed on 15 June 2015

Ministry of External Affairs, Government of India 2005 Political Parameters and Guiding Principles for the Settlement of the India-China Boundary Question https://www.mea.gov.in/bilateral-documents.htm?dtl/6534/Agreement+between+the+Government+of+the+Republic+of+India+and+the+Government+of+the+Peoples+Republic+of+China+on+the+Political+Parameters+and+Guiding+Principles+for+the+Settlement+of+the+IndiaChina+Boundary+Question. Accessed on 12 January 2020

Ministry of Foreign Affairs, Peoples Republic of China 2018 Spokesperson Hua Chunyings Regular Press Conference https://www.fmprc.gov.cn/mfa_eng/xwfw_665399/s2510_665401/t1606198.shtml. Accessed on 15 June 2019

Ministry of Home Affairs, Government of India (2022) Management of Indo-China Border. chrome-extension://efaidnbmnnnibpcajpcglclefindmkajhttps://www.mha.gov.in/sites/default/files/INDO_China_16022018.pdf

Ministry of Jal Shakti, Department of Water Resources River Development and Ganga Rejuvenation (2022) Government of India India-China Cooperation https://jalshakti-dowr.gov.in/india-china-cooperation/

Ministry of Jal Shakti 2018 Department of Water Resources, Rivers Development & Ganga Rejuvenation b, Government of India 'Annual Report 2016–17-Brahmaputra Board'http://mowr.gov.in/sites/default/files/AR_BB_2016-17-English.pdf. Accessed on 21 April 2018

Ministry of Jal Shakti 2018a Department of Water Resources, Rivers Development & Ganga Rejuvenation, Government of India a, INDIA-CHINA COOPERATION http://mowr.gov.in/international-cooperation/bilateral-cooperation-with-neighbouring-countries/india-china-cooperation Accessed on 12 April 2018

Ministry of Water Resources, People's Republic of China (2015a) Statistic Bulletin On China Water Activities. http://www.mwr.gov.cn/english/publs/201702/P020170207383603707034.pdf. Accessed on 12 April 2018

Ministry of Water Resources, People's Republic of China (2015b) International Cooperation on Trans-boundary Rivers between China and its Neighbouring Countries http://www.mwr.gov.cn/english/mainsubjects/201604/P020160406513798903048.pdf. Accessed on 12 April 2018

Ministry of Water Resources, Peoples Republic of China 2015 Water Security in Chinas Modernization Process' http://www.mwr.gov.cn/english/speechesandarticles/jiaoyong/201511/t20151106_725196.html. Accessed on 10 March 2018

Ministry of Water Resources, Peoples Republic of China, 2022 China Water Resources Per Capita https://www.ceicdata.com/en/china/water-resource/cn-water-resource-per-capita#:~:text=China%20Water%20Resource%20Per%20Capita%20data%20was%20reported%20at%202%2C098.488,to%202021%2C%20with%2023%20observations

MPIDSA (Manohar Parrikar Institute for Defence Studies and Analysis) (2010) Water Security for India: The External Dynamics https://idsa.in/book/WaterSecurityforIndia. Accessed on 21 March 2017

Office of the Commissioner of the Ministry of Foreign Affairs of the Peoples Republic of China in the Macau Special Administrative Region 2020 Foreign Ministry Spokesperson Hua Chunyings Regular Press Conference on February 4, 2013. http://www.fmcoprc.gov.mo/eng/gsxwfb/lxjzzfh/t1011493.htm. Accessed on 18 July 2020

Osborn, Jeremy. 2010. China faces tough choices on water' The Guardian, 22 December. https://www.theguardian.com/sustainable-business/china-water-choices Accessed on 25 April 2018

Press Information Bureau, Ministry of Defence, Government of India 2013 Indo-China Border Defence Cooperation Areement. 09 December 2013. https://pib.gov.in/newsite/PrintRelease.aspx?relid=101179. Accessed 12 May 2020

Ramachandran, Sudha. 2015 Water Wars: China, India and Great Dam Rush https://thediplomat.com/2015/04/water-wars-china-india-and-the-great-dam-rush/ Accessed on 12 September 2018

Rediff.com (2000, 10 July). Chinese dam breach caused northeast floods: AFP. http://www.rediff.com/news/2000/jul/10china.htm. Accessed on 16 March 2019

Reuters (2019, 5 January) Chinas population set to peak at 1.44 billion in 2029: government report https://www.reuters.com/article/us-china-population-idUSKCN1OZ08A. Accessed on 12 July 2020

Reuters (2030, 20 November) China eyes 60 GW of hydropower on Tibets Brahmaputra river-state media". https://www.reuters.com/article/china-hydropower-idINKBN28A11S?edition-redirect=in. Accessed on 12 March 2021

Sadat Anwar (2021, 6 April) Is Chinas hydrological design in Brahmaputra legitimate?" The Financial Express. https://www.financialexpress.com/defence/is-chinas-hydrological-design-in-brahmaputra-legitimate/2227774/. Accessed on 8 April 2021

Saikia Arunabh (2017,7 December) As the Siang river in Arunachal turns black, China is the usual suspect. But is it really to blame? Scroll.in. https://scroll.in/article/860304/as-the-siang-river-in-arunachal-turns-black-china-is-the-usual-suspect-but-is-it-really-to-blame. Accessed on 21 March 2020

Samanta Pranab Dhal. (2009, 15 October) China begins building dam on its side of Brahmaputra The Indian Express. http://archive.indianexpress.com/news/china-begins-building-dam-on-its-side-of-the-brahmaputra/529244/ Accessed on 23 April 2018

Scott D (2012) Sino-India territorial issue: the razors edge. In: Pant HV (ed) The rise of China: implications for India. Cambridge University Press, New Delhi, pp 195–217

Sheth Chintan and Anirban Datta-Roy (2017, 22 December) Earthquakes in Tibet Are Behind the Mysterious Darkening of Arunachal's Siang River The Wire. https://thewire.in/environment/arunachal-pradesh-siang-river. Accessed on 12 July 2020

Singh Sushant (2020, 1 June) Line of Actual Control (LAC): Where it is located, and where India and China differ The Indian Express. https://indianexpress.com/article/explained/line-of-actual-control-where-it-is-located-and-where-india-and-china-differ-6436436/. Accessed on 15 July 2022

South China Morning Post (2017,4 July) Over 50 dead, hundreds of thousands evacuated as floods ravage southern China. http://www.scmp.com/news/china/policies-politics/article/2101179/over-30-dead-hundreds-thousands-evacuated-floods-ravage. Accessed on 22 May 2018

The Pioneer (2018, 29 March). China, India to share hydrological data https://www.dailypioneer.com/2018/world/china-india-to-share-hydrological-data.html Accessed on 18 July 2018

The Times of India (2002, March 6) India-China Trade sets new record in 2001. https://timesofindia.indiatimes.com/business/international-business/India-China-trade-sets-new-record-in-2001/articleshow/2946205.cms. Accessed on 20 October 2018

The World Bank (2023, 20 April) The World Bank in China. https://www.worldbank.org/en/country/china/overview. Accessed on 12 July 2023

The World Bank 2021 Global Economy to Expand by 4% in 2021; Vaccine Deployment and Investment Key to Sustaining the Recovery 5 January 2021. https://www.worldbank.org/en/news/press-release/2021/01/05/global-economy-to-expand-by-4-percent-in-2021-vaccine-deployment-and-investment-key-to-sustaining-the-recovery. Accessed on 9 March 2021

Tibet Support Association Sambhala Tibet Center. (2010) Damming Tibets Yarlung Tsangpo-Brahmaputra and other south Asian Rivers http://www.tibet.hu/tibetpressen/szabalyozni-akarjak-a-tibeti-jarlung-cangpo-avagy-brahmaputra-illetve-mas-del-azsiai-folyok-vizet-. Accessed on 12 June 2017

United Nations World Water Development Report 2018. http://unesdoc.unesco.org/images/0026/002614/261424e.pdf. Accessed on 12 May 2018

Varma K.J. M. (2017) China to maintain communication with India on artificial lakes Live Mint. https://www.livemint.com/Politics/Nxmanja68r4b5sqIrj8GhP/China-to-maintain-communication-with-India-on-artificial-lak.html. Accessed on 25 February 2018

Vinayak Ramesh (2005,11 July) Sutlej in Himachal swells as Pareechu lake bursts in Tibet https://www.indiatoday.in/magazine/indiascope/story/20050711 sutlej-river-in-himachal-pradesh-swelled-up-after-pareechu-lake-burst-in-tibet-787274-2005-07-11. Accessed on 12 March 2021

VOA 2015 Report: China Completes Tibets Biggest Hydropower Plant 13 October 2015. https://www.voanews.com/east-asia/report-china-completes-tibets-biggest-hydropower-plant. Accessed on 18 June 2020

Water Power and Dam Construction (2004, 25 August) Nathpa Jhakri shut down for a second time https://www.waterpowermagazine.com/news/newsnathpa-jhakri-shut-down-for-a-second-time. Accessed on 12 March 2021

Wines Michael (2011, 19 May). China admits problems with Three Gorges Dam The New York Times. https://www.nytimes.com/2011/05/20/world/asia/20gorges.html. Accessed on 21 April 2018

Wirsing RG, Stoll DC, Jasparro C (2013) Transnational conflict over water resources in Himalyan Asia. Palgrave Macmillan, New York

Wouters P, Daza-Clark AM, Devlaeminck DJ (2024) Chinas transboundary hydropower development at home and abroad: exploring the regulatory interface between international water law and international economic law. Front Clim 5:1302103. https://doi.org/10.3389/fclim.2023.1302103

Yangtso L. (2017). Chinas River Politics on the Tibetan Plateau: Comparative Study of Brahmaputra and Mekong. The Tibet Journal, Vol. 42, No. 2, Special Issue: Third Symposium in Memory of Prof. Dawa Norbu. Pp. 49–58

Zhang Hongzhou (2016) Sino-India Water Disputes: the coming water wars? Volume 3, March/April Wires Water 155–166 http://www.indiawaterportal.org/sites/indiawaterportal.org/files/sino_indian_water_disputes_wires_water_2016.pdf. Accessed on 18 October 2018)

Zhang H, Li M (2018a) Thirsty China and its transboundary waters. In: Zhang H, Li M (eds) China and transboundary water politics in Asia. Routledge, Oxon, pp 3–24

Zhang H, Li M (2018b) China and global water governance: new developments. In: Zhang H, Li M (eds) China and transboundary water politics in Asia. Routledge, Oxon, pp 219–236

Amit Ranjan is a Research Fellow at the Institute of South Asian Studies (ISAS) National University of Singapore. His areas of interest are: Water Politics, South Asian politics Indias regional policy and Indias internal security. His latest book is The Breakup of India and Palestine: The causes and legacies of partition (edited with Victor Kattan Manchester University Press Manchester 2023). He is also author of Federalism and Inter-State River Water Disputes in India (Routledge, London and New Delhi, 2023), Contested Waters: Indias Transboundary River Water Disputes in South Asia (Routledge Oxon and New Delhi, 2021) and India- Bangladesh Border Disputes: History and Post-LBA Dynamics (Springer, Singapore, 2018). Amit has also edited Migration, Regional Autonomy, and Conflicts in Eastern South Asia (with Diotima Chattoraj, Palgrave, Switzerland, 2023), Environment, Climate Change and Migration in South Asia (with Rajesh Kharat and Pallavi Deka, Routledge, London, 2023), Urban Development and Environmental History in Modern South Asia (with Ian Talbot, Routledge, Oxon, 2023), India in South Asia Challenges and Management (Springer, Singapore, 2019), Partition of India: Postcolonial Legacies (Routledge, London and New Delhi, 2019) and Water Issues in Himalayan South Asia: Internal Challenges, Disputes and Transboundary Tensions (Palgrave, Singapore, 2019).His paper review essay and book review have been widely published in journals, including Asian Survey Asian Affairs Asian Ethnicity Asian Journal of Comparative Politics, Economic and Political Weekly, India Review, Indian Journal of Public Administration India Quarterly Journal of Migration Affairs Journal of the Indian Ocean Region The Roundtable: The Commonwealth Journal of International Affairs, Social Change, Studies in Indian Politics, Society and Culture in South Asia South Asia Research, Social Identities, Journal of Asian Security and International Affairs Water History and World Water Policy among others. Further details of publications can be found at: https://www.isas.nus.edu.sg/research_team/amit-ranjan/

Chapter 6
Approaching Transboundary Perspectives on Water in South Asia from a Security Perspective

Nitin Sathe

6.1 Introduction

As every person would attest to, water is one of the most important resources required for the survival of the human race. Despite three-fourths of the world being covered in water, usable freshwater supplies are limited mostly to rivers and lakes across the world. This makes the issue of water a major concern in densely populated areas such as South Asia, where there are increasing apprehensions regarding water security, both due to climate change and noncooperation over available resources. This paper deals with the water resources of South Asia which are significant but diminishing. Disputes over trans boundary water sharing remain between South Asian nations due to fluctuating bilateral relations which do not provide opportunities for settling disputes. This paper attempts to analyse the issue of water relations and water security in South Asia within the larger question of better bilateral relations and closer security cooperation. One major reason for this approach is the increasing perception of water resources within the ambit of national security. South Asian states are developing and require significant water resources, leading to more demands made by states. Such changes can pose threats to national security through water wars, refugee crises, and internal disquiet. Bilateral relations generally shape perceptions in international affairs and issues like transnational water sharing and security cooperation, which form a part of bilateral relations, significantly depend on the relationships between states. Also, this paper attempts to analyse the role played by China in South Asian bilateral relations and how this factor has influenced security perceptions in India and her neighbours and what India may do to rebuild her influence in the region while negating the Chinese.

N. Sathe (✉)
Air Commodore Nitin Sathe (Retd), Pune, Pune, India

J. Andharia and P. Anand (eds.), *Shifting Lands, Flowing Waters*, Disaster Studies and Management, https://doi.org/10.1007/978-981-96-7691-0_6

6.2 South Asia: An Introduction to the Geopolitics of the Region

South Asia, as a geographical region, has access to some of the largest and longest perennial rivers in the world, with the most important being the Indus river system, the Gangetic river system and the Brahmaputra river system, all of which are transboundary rivers. Despite this, the region suffers from a serious shortage of water due to multiple factors. Climate change remains one of them; the other factor is the significantly large and growing population with a very high consumption of water. This is because of the largely agrarian economies requiring a lot of water, apart from the domestic demands. River waters also serve as the basis for hydroelectric power. This puts South Asia in danger of significant water stress given the pressure on the region's river systems. Many areas in South Asia currently suffer from frequent shortages of water. All these factors have led to states trying to ensure as much water as possible for themselves.

The Indus has a total basin area of 1.14 million square kilometers, covering the countries of Pakistan, China, India, Afghanistan, and Nepal. The Ganges-Brahmaputra-Meghna basin covers an area of 1.63 million square kilometers spread over Bangladesh, India, China, Nepal, Bhutan, and Myanmar (Bala, 2018). As a result, water sharing is a major cause for concern among South Asian states, especially the lower riparian states in the region as their supplies are subject to their relations with the upper riparian states. This makes water security an important part of the lower riparian's national security concerns.

The region is marked by fluctuating bilateral relations, making it an extremely complex region for both water and national security. The relations between India and her neighbours fluctuate at times, with tensions arising primarily over the issue of border security, often leading to a degree of breakdown in bilateral relations. Border disputes exist between India and most of her neighbours. Despite these, India has relations ranging from cordial to friendly with most of its neighbours. However, tensions arise frequently over concerns of Indian high-handedness concerning her smaller neighbours. Pakistan, Nepal, and Bangladesh consider some of the existing water sharing agreements to be biased towards India with less consideration towards their requirements. Thus, bilateral tensions shape perceptions about water sharing and concerns regarding water security.

Along with this, one major factor for the volatility of South Asia is the presence of the Chinese factor. It is an important player in the shaping of bilateral relations and security dynamics in the region. The Chinese occupation of Tibet (1950) has provided it with an extremely long land border with South Asia, which it shares with Pakistan, India, Nepal, and Bhutan. Sino-Indian relations have ranged from strained to cordial due to border disagreements. Also, due to their status as emerging powers, both India and China remain in a state of rivalry with each other.

Chinese involvement in South Asia is linked to its global ambitions and its need to project power. Attempting to dominate South Asia, strategically placed at the head of the IOR, not only allows it to make inroads into India's sphere of dominance but also a way to get access to West Asia. The latter is to ensure its West Asian oil trade and growing Chinese clout in the region. Dominating the IOR also allows it to safeguard its trade routes in the region.

To facilitate this domination of Asian geopolitics, the Belt and Road Initiative (BRI) has been initiated at a rapid pace. BRI has had a big impact on South Asia, especially in India from a security point of view as it is an attempt to surround the sub-continent within a Chinese circle of influence and reduce Indian influence in the region. Pakistan, Sri Lanka, Nepal, Bangladesh, and Myanmar have all been roped in as part of the BRI with significant Chinese economic and security investments in these states. This would force Indian strategic thinkers to re-evaluate Indian strategy with regard to China, as the latter would not just have access to India's borders and the headwaters of the Brahmaputra under its control, but could also regulate India's export-import markets in the IOR. This threatens India's position as the dominant power in the region and also provides her neighbours with an alternative to counter India's economic and political predominance in the region.

This has been made very visible in attempts by the Chinese to develop cordial relations with Pakistan, Bangladesh, and Nepal through defence and economic infrastructure developments. As relations of these states with China get closer, it negates India's negotiating strength on bilateral issues. China's presence in the region, therefore, requires careful consideration for India's interests.

With this as the background to the region, it is essential now to focus on the bilateral water sharing and water security issues India has with other South Asian states.

Case Studies: The following case studies would look at the bilateral water sharing agreements concluded between India and other states in South Asia, with whom India shares trans boundary rivers, namely Nepal, Bangladesh, and Pakistan. It would also look at India-China water relations within the ambit of Indo-China bilateral relations and how Chinese influence has impacted India's bilateral relations, and whether existing water sharing agreements have been affected by this. The agreements under consideration are the Indus Water Treaty with Pakistan (1960), the Ganges Water Sharing Treaty (1996) with Bangladesh, the Indo-Nepal Koshi agreement (1954), and the Mahakali Treaty (1996).

6.3 India-Nepal Agreements and Bilateral Relations

India and Nepal bilateral relations have not always been the steadiest. Relations were initially close in the 1950s culminating in the treaty of friendship and peace. This was due to India being Nepal's primary trading partner in that period. Nepal had no established relations with China till 1955 when the Chinese occupation of Tibet brought them to the Nepalese doorstep.

The year 1954 also saw the Koshi river agreement signed between the two states (Tabassum, 2019). The agreement allowed India to dam the river for hydroelectricity generation. Aside from the Koshi agreement, the dispute over the Tanakpur barrage also developed around that time. These incidents led to the souring of bilateral relations; and the Nepalese government considered that India had taken advantage of Nepal in the provisions of the agreement (Tabassum, 2019) while the Indian government saw it as a breach of trust by the Nepalese. Although diplomatic efforts between the two countries led to a revision of the terms of the treaty, it entrenched the Nepalese mind that India was a dominating neighbour who would put Nepal at a disadvantage to further its interests whenever it liked.

The Koshi barrage agreement was one of the few water sharing agreements between India and Nepal in the second half of the twentieth century, the others being the Gandak agreement of 1959, the Tanakpur agreement of 1991, and the Mahakali water treaty of 1996 (Adhikari 2014). Water agreements were initially proposed and financed by India due to the fact that most headwaters of the Gangetic river system lay within the territory of Nepal. Not including consumption, water sharing was primarily along the lines of distribution of water during the wet and dry season for irrigation, flood control measures and hydroelectric power production. The power generation aspect has immense potential on almost all rivers flowing into India from Nepal, thereby being of interest to India as a long term, sustainable and renewable source of electricity for the Indian states bordering Nepal.

However, these agreements have caused grievances in Nepal leading to accusations of India acting as a patronising big brother and unfairly taking advantage of Nepal's goodwill; and, in case of the initial agreements, of the mountainous nation's inexperience in diplomacy. This has led to resentment in Nepal and suspicions regarding India's intentions, which further affected water sharing agreements. Such accusations probably stem from a realisation of the economic potential of Nepal's water resources and how it could be beneficial to Nepal. Thus, since the Koshi barrage agreement, Nepal-India relations regarding water sharing have not been very cordial, despite fairly good relations overall.

There is a distinctive nationalist tendency in Nepal concerning India, which traces its roots to India's insensitive handling of Nepal's water resources and the limited benefits that Nepal received from the projects, such as fixed quantities of electricity generated from hydropower and no Nepalese control over the waters near the projects (Adhikari, 2014).

Only with the signing of the Mahakali water sharing treaty in 1996, there was a balanced approach to water sharing terms. Under this treaty, both states were equal partners in the development of the project and the hydropower was supposed to be integrated in nature. Although it marked a significant improvement to previous agreements, there have not been major developments in the area after the Mahakali treaty, although there were discussions on the development of the Pancheshwar project in the aftermath of PM Narendra Modi's visit to Nepal in 2014.

Overall, India-Nepal relations have seen their share of ups and downs over the decades, which have been reflected in their water sharing agreements and practices. This has happened despite the Indian government's current "Neighbourhood First Policy". Understandably, Nepal's lack of diplomatic experience became a bane for it during the early water sharing agreements. Souring of the bilateral relationship was also caused due to India's rash decisions with respect to Nepal's internal and foreign affairs, such as the imposition of the two economic blockades in 1989 and 2015, which allowed China to fill India's void (Budhathoki, 2018; Stone, 2019).

The first blockade was due to Nepal's purchase of Chinese arms while the second, soon after the 2015 earthquake, was over the Madhesi Crisis. These have led to the development of feelings of hostility towards India. Also, India's responses to growing Sino-Nepal relations have, therefore, not always been diplomatic, with India overtly showing its displeasure towards Nepal at times. For example, India was displeased with Nepal not allying with India during the 1962 war (Budhathoki, 2018). Before the 1960s, India was the sole trading partner of Nepal. Nepal-China relations started developing from 1955 onwards. One of the reasons was probably to reduce Indian influence in their country by developing relations with other neighbouring powers.

The other reason has more to do with Nepal's strategic thinking. Nepal occupies a strategic location in South Asia, nestled in the heights of the Himalayas and lying between the Indian and Chinese behemoths. Being open to improving relations with both its larger and stronger neighbours allows Nepal to not choose sides in the rivalry of these emerging powers, thereby ensuring its security. It can therefore engage with both states over matters of infrastructure development.

In the case of India, improvement in infrastructure in Nepal is worrisome since the land provides a clear buffer zone between Chinese Tibet and the heavily populated and industrialised areas of northern India. Nepal has traditionally been a close partner of India, even in matters of security. However, with Nepal's increasing cooperation with China, there are fears in the Indian security establishment that this could be used by China to undermine Indian influence in Nepal and increase Indian vulnerability along the northern border.

For China, Nepal's importance lies in its proximity to Tibet and Tibetan refugees using Nepal as a base for undermining the Chinese (Jaiswal, 2014). Hence, for the Chinese, having Nepal in their sphere of influence is essential. Under the BRI, there is already extensive Chinese investment in the development of Nepal's economic infrastructure, apart from investment in Nepal's armed forces. Nepal has been offered access to Chinese ports for its trade and Kathmandu is supposed to be linked by a railway to Tibet in the near future.

Therefore, one can see that China has filled in the void left by India in Nepal, from supplying oil in 2015 to providing support to the government of Nepal in its battle against Maoist insurgency, at a time when India refused to support it (Jaiswal, 2014). Nepal also conducted military exercises with China after refusing to join the BIMSTEC (Budhathoki, 2018). While bilateral relations between India and Nepal are generally good, closer ties in the future not only depend on India's ability to aid in the development of the country without interfering in Nepal's internal affairs, but also in attempting to solve the existing water disputes.

Being an upper riparian, Nepal's view of its water resources is mostly economic while for lower riparian India, it is a matter of both economy and security. Also, losing Nepal to the Chinese sphere of influence would put most of India's water resources under threat in the long term. Future agreements with Nepal must include discussions on the concerns of both states regarding water and then attempt to find an amicable solution to the dispute. This can be done by removing Nepal's insecurity over being arm twisted by India into tilted agreements and negotiating existing and future treaties based on the idea of equal sovereign states with similar interests in the region.

6.4 India-Bangladesh Agreements and Bilateral Relations

Compared to other states of South Asia, India-Bangladesh bilateral relations are comparatively recent, due to the creation of Bangladesh only in 1971 in which India had a major part to play. Indian armed forces played a part in pushing Pakistan out of present day Bangladesh. The relations have generally been friendly, but, with the recent ouster of the President Sheikh Hasina, relations have seen a downswing, with the Bangladesh government in power openly siding with Pakistan and China and also trying to change the historical facts. Indian involvement in the freedom struggle is now being questioned and there is an attempt to obliterate historical facts. Additionally, the present leadership in Bangladesh is upset about the fact that their president in exile has been given asylum in India.

India had been grappling with occasional issues arising over a porous border and water resources sharing with Bangladesh ever since its birth in 1971; but largely due to good diplomacy and a soft attitude that India has had towards the country, things had moved positively in the past.

The water sharing and border disputes form significant areas of discussions between the states in bilateral talks. Bangladesh and India share 54 rivers (Islam, 2012), with India being the upper riparian state and Bangladesh the lower. The major sources of conflict are the three major river systems—the Brahmaputra, the Ganges, and the Meghna-Barak systems. Islam (2012) argues that conflicts over water sharing existed even before 1971, but since the creation of Bangladesh, it became a bilateral issue with the newly formed state. Although an Indo-Bangladesh Joint River Commission has been in existence since 1972, Bangladesh has believed that India being a larger and stronger state and its land upstream, certain unilateral decisions have been made regarding joint usage of waters of the river systems mentioned above, without considering Bangladesh's concerns.

Water resources are extremely important to both states having extremely large populations and agriculture being a major sector of both economies. South Asian rainfall is dependent on the monsoons, and in the dry period water supply comes from these large perennial rivers. Thus, river water is of utmost importance to Bangladesh to sustain itself.

Bangladesh's concerns about India's high-handedness in water sharing stemmed from the development of the Farakka barrage in 1975. It was an Indian initiative to divert Ganges water to the River Hugli to flush silt from the latter, increasing the usability of the Kolkata port. Despite assurances from India of the barrage not impeding water flow to Bangladesh, issues on water supply have been raised by the latter. There was a shortage of water after the monsoon, with the minimum water level recorded at 5.5 ft. This caused immense problems for agriculture in Bangladesh and affected fishing, navigation, and the ecology of the area (Khalid, 2010). Such negative views regarding India did not solely emerge from the Farakka barrage by itself, but also by the change in political leadership in Bangladesh in 1975. We are yet to see the current responses from the new regime on this aspect.

This context and the related issues led to the Ganges water agreement of 1977 whereby India recognised the international character of the waters of the Ganges River and terms were laid out regarding water management at Farakka (Khalid, 2010). A major development came in 1996 with the Ganges Water Treaty (1996), which decreed that Ganges water below Farakka would not be reduced further and fixed the water India would release to Bangladesh for the next 30 years.

The other issues have been over the sharing of the Teesta River waters. This has caused problems since the government of West Bengal has also been involved in the issue. The Teesta is an important river both in Bangladesh and in India; in the latter the river provides sustenance to the hilly districts of north Bengal and is one of the primary sources of water there. The issue has not been resolved as the government of West Bengal has refused to provide more than 25% of Teesta river water to Bangladesh (Islam, 2012), stalling any possible agreement on the issue during the 2011 Manmohan Singh and Sheikh Hasina summit. Another issue is regarding the Tipaimukh dam project on the Barak River, which has caused unrest in Bangladesh. A multi-purpose hydroelectric dam has been planned here by India and there has been assurances from the Indian side about the usage of water solely for power generation and not for agriculture (Khalid, 2010). But concerns exist in Bangladesh that it would become a Farakka 2.0, with Bangladesh suffering from India's development.

As in the case of Nepal, a lot has been made of Indian high-handedness in matters of water sharing due to its dominance in the region. While it is certainly true to some extent that India had tried to extract more benefits in the agreements, it is part and parcel of international diplomacy to do so. However, it is in both India and Bangladesh's interests to ensure an amicable solution to the water dispute for several reasons. For one, the dispute arises from both countries' concerns over water security in the future. This can be solved through discussions given that bilateral relations between the two states are generally good. With regard to the Brahmaputra water, however, it must be remembered that the "tap" of this river, i.e. its source is under the control of China. This makes both countries vulnerable to Chinese interference in their water supply.

From a security perspective, India and Bangladesh should sort their water issues mutually and then present a united front to China, if in the future there is an attempt to control or divert the waters of the Brahmaputra, which can adversely impact both India and Bangladesh. At this point, it would be reasonable to mention that the Indo-Bangladesh border dispute (over the issue of illegal migration) needs to be resolved alongside the water dispute. There needs to be better security of the border. In case of a future water conflict with China, a significant portion of Bangladesh's water starved population would possibly attempt to enter India under humanitarian grounds. A solution needs to be in place in case such a scenario does arise.

Until recently, Indo-Bangladesh relations have been friendly largely due to Indian attempts to cultivate a stronger relationship with her eastern neighbour. However,

Bangladesh is also cultivating a closer relationship with China, being a part of the BRI. Under the present arrangement, the country is enjoying the benefits of both Indian and Chinese investments and experiencing rapid development. However, it must be wary of China's growing global ambitions.

Under such circumstances, India should attempt to address Bangladesh's water security concerns and ensure that its concerns are also heard in Bangladesh. Such clarity would lead to a buildup of trust between the neighbours. As with Nepal, there also needs to be significant Indian involvement in the infrastructure development in Bangladesh to counter China's efforts and retain Bangladesh in the Indian sphere of influence. In 2016 alone, Chinese projects announced in Bangladesh approximately amounted to a sum of 38 billion dollars (Anwar, 2019). In contrast, India pledged to invest 5 billion dollars in Bangladesh. While this sum is the largest amount India has invested in Bangladesh at any single time, it is much lesser than the investment being made by China. Furthermore, China is the biggest arms exporter to Bangladesh; and as of 2017, there has been a significant transfer of military hardware from China to Bangladesh (Anwar, 2019).

It is in India's best interests to stall the development of a China-Bangladesh nexus by proactively engaging itself with Bangladesh and aid its development. It should also engage with the Bangladesh defence forces and improve security cooperation and interoperability in South Asia. From a security point of view, there are two major reasons for doing so. First, an improvement in the economic condition of Bangladesh would likely lead to a possible reduction of illegal migrants from Bangladesh. This would in some way solve one aspect of the Indo-Bangladesh border problem. Secondly, some influence of China might be negated with such an approach and Indian influence in Bangladesh would possibly be revived.

6.5 Indo-Pakistan Bilateral Relations and the IWT

India-Pakistan bilateral relations have been rocky since independence. In the case of India, there has traditionally been much more focus towards Pakistan from a bilateral point of view. This has been due to the security aspects of the western border with multiple conflicts having been fought at the western border since 1947; and the area as of today remains volatile. The Jammu and Kashmir issue has a lot to do in this regard with both India and Pakistan laying claim to the state; both having control over portions of it. While bilateral relations between the two states have been unstable, the same may not be said of the water sharing agreement between them till recently. The Indus Water Treaty (IWT), under the aegis of the World Bank has been one of the stable points of Indo-Pakistan relations, albeit with a third party involved.

The IWT came into being in 1960, after years of political wrangling between Indian and Pakistani authorities over the sharing of water from the Indus river system. Partition had caused significant insecurity in Pakistan about Indian interference in the steady supply of water in the Indus and its tributaries. Pakistan's primary freshwater source was, and remains, the Indus river system. The river system drains the province of Punjab which is the most fertile area in Pakistan. With the headwaters of the Indus lying in India—in the state of Jammu and Kashmir—independent Pakistan's water supply depended on the goodwill of India.

However, the two states fought a war within a year of independence and therefore, there was a marked absence of trust between them. Pervez Musharraf reportedly said that the Kashmir dispute arose over the sharing of Indus waters and a solution of either issue would solve both for good (Musharraf, 2006). Water sharing negotiations before the World Bank intervention saw no progress with India attempting to protect its rights to the Indus basin as the upper riparian and Pakistan, being the lower riparian, demanding almost complete access to it as its very survival depended on it. Being a state with limited water resources, Pakistan understood early the importance of water in national security.

The present IWT is based on a World Bank proposal tabled in 1954. According to the treaty, the Indus basin consisting of six rivers—the Indus, Chenab, Jhelum, Beas, Ravi and Sutlej—was divided between the two states. The three eastern tributaries—the Beas, Ravi and Sutlejwere given to India and the three western tributaries—Indus, Chenab, and Jhelum to Pakistan (Salik, 2015). India was allowed access to the western rivers for the generation of power and commercial irrigation purposes. Both sides agreed to the treaty and a "Permanent Indus Commission" was set up to deal with conflict resolution (Salik, 2015).

India's access to the western rivers has been a cause of insecurity for Pakistan. It has feared the headwaters of these rivers being extensively dammed by India, which would severely restrict its water supply. Such projects as the Wullar barrage, Baglihar, and Kishenganga dams have become bones of contention between the two states (Salik, 2015). Pakistani authorities believe that the terms of the IWT were relevant for when it was signed and with changing conditions, especially in terms of populations and climate change, the terms no longer remain valid. Salik (2015) suggests that they also feel that India benefited more from the treaty and from the international court of arbitration, which Pakistan had approached.

India however, has consistently denied that the projects on the western rivers would interfere with water supply to Pakistan. Indian authorities have claimed that due to an increasing population, the projects are essential for increased power generation and all such projects have been informed to the Pakistani authorities (Salik, 2015). Disagreements over the IWT continue till today. In 2019, there was an uproar in Pakistan in the face of Indian comments about restricting water flow to Pakistan. Pakistan, therefore, considers water to be a national security issue that has the potential to wreck it from the inside. Being a lower riparian like Bangladesh, it fears Indian interference over its supply of water during peace or, more importantly, during war, which may render the state helpless.

In April 2025, the IWT faced an unprecedented challenge. Following a militant attack in Pahalgam, Jammu and Kashmir, which resulted in the deaths of 26 individuals, India accused Pakistan of supporting cross-border terrorism: and in response, the Indian government suspended the IWT on April 23, 2025. This suspension included halting the flow of water from the Baglihar Dam on the Chenab River and conducting reservoir flushing at Salal and Baglihar projects without prior notification to Pakistan, actions that contravened treaty obligations.

Prime Minister Narendra Modi reinforced this stance by stating, "Terror and talks cannot go together; water and blood cannot flow together," signaling a firm position against reversing the treaty's suspension.

Pakistan viewed India's actions as a severe breach of the treaty, warning that any attempts to disrupt water flow could be considered acts of aggression. Reports emerged of unannounced water releases from the Uri Dam into the Jhelum River, causing flooding in Pakistan-administered regions, and significant reductions in the Chenab River's flow near Sialkot. These developments escalated tensions, with Pakistan threatening severe retaliatory measures.

The suspension of the IWT occurred amidst heightened military tensions, including cross-border skirmishes and missile strikes. International actors, notably the USA and Gulf countries, intervened to mediate, leading to a ceasefire. However, key agreements like the IWT and visa services remain suspended, indicating persistent underlying frictions. With the "ceasefire" in place at the moment, it remains to be seen if India will relent to Pakistan's requests for a status quo on the treaty.

Concurrently, environmental concerns have intensified. The Hindu Kush Himalayan region has experienced below-average snow levels for three consecutive years, threatening water availability for nearly two billion people dependent on meltwater. Such climatic shifts underscore the urgency for adaptive water governance and cooperative management of shared resources.

The suspension of the Indus Waters Treaty marks a critical juncture in India-Pakistan relations, intertwining issues of national security, environmental sustainability, and regional stability. As both nations navigate this complex landscape, the imperative for renewed dialogue and cooperative frameworks becomes increasingly evident to prevent further escalation and ensure the equitable management of shared water resources.

Indo-Pakistan relations are the most difficult to improve, with the shared history of the region and the sufferings of partition along with extreme distrust between each other constantly proving to be insurmountable odds in the path to improved bilateral relations. For every single attempt made towards better relations, unfriendly actions of the Pakistani side push back relations by a few years, especially due to Pakistani support to terrorist elements. Pakistan's close participation in the BRI and Chinese involvement in the Gwadar port has further damaged Indo-Pakistan relations. Pakistan is China's closest ally in South Asia which allows it to violate the peace of the region with impunity. This latter factor has been a significant problem for India in recent years.

The Chinese presence and influence in the entire sub-continent has given a new dimension to these disputes. Almost all of India's neighbours are trade partners of China; and consider Chinese influence to be a way of negating or at least neutralising India's dominance in South Asia. Although most South Asian states neighbouring India have generally balanced relations with both China and India fairly well till now, it remains to be seen how much of it is possible in the future.

6.6 Sino-Indian Bilateral Relations and Water Disputes

China and India are among the two most dominant powers in Asia and, based on the trajectory of their growth, have been predicted to become two foremost powers in the world in a few decades. The two states have had historic ties stretching back centuries. Buddhism reached China from India and trade and cultural ties existed between the two from those times. These ties continued even after Indian independence with the principles of Panchsheel forming an important part of Sino-Indian relations reiterating India and China's long relations and a peaceful, co-operative approach to development.

India is generally considered to dominate the Indian sub-continent and South Asia overall; Chinese presence traditionally has been in East and Southeast Asia. However, with the occupation of Tibet by Chinese military forces in the 1950's, China and India have come to share a border at the Himalayas leading to a conflict of interests over Tibet and also conflict over Chinese influence in India's immediate neighbours. Chinese claims over sections of the Himalayan range, primarily in the western Himalayas (Aksai Chin region) and in the eastern Himalayas (Arunachal Pradesh) have led to souring of relations between the two states, also affecting economic cooperation.

China and India fought a war over the eastern Himalayas in 1962, beginning the overt hostilities between the two states. Their most recent standoff was in 2017 over the Doklam plateau, located at the junction of the Chinese, Indian, and Bhutanese border. Such standoffs along the border are frequent but they have not proved to be non-negotiable through diplomatic channels. There are historical areas of cooperation between the two states and some of them stand even today, especially in the realm of trade, with China being India's largest trade partner. Currently, diplomatic relations with China are not so stable, with the Indian establishment calling the Chinese as bigger enemies than Pakistan.

In a blow hot blow cold situation, it is difficult to predict how and when relations will degenerate. With both India and China attempting to emerge as superpowers, it is more than likely that conflicts will continue to emerge in the long term. The cracks are already visible. The BRI, as and when it fructifies, is sure to increase the Chinese influence in the region. More importantly, the new silk route being planned is not restricted to the land routes alone; it traverses through central Asia and West Asia by land and takes the sea route into Africa—with Southeast Asia in the middle. This has aptly been called the String of Pearls policy—a policy that threatens to strangle India from all sides.

Given its scope, Indian authorities have consistently raised concerns about how the BRI and the String of Pearls would completely encircle the country and intrude into Indian spheres of influence leading to compromise of Indian territorial integrity. BRI projects are being undertaken in Pakistan, Bangladesh, and Sri Lanka, with China increasing investments in these countries for infrastructure development.

Bangladesh has received significant investments to develop its economic infrastructure and defence capabilities. Meanwhile in Pakistan, China's closest South Asian ally, and in Sri Lanka, China has significant stakes in the ports of Gwadar and Hambantota respectively. These bases are more than likely to be used for military use by the Chinese Navy, (PLAN) providing it unhindered access to the IOR. China has also offered Nepal the use of some ports and agreed to build a railway linking Tibet to Kathmandu. Apart from these, in 2017, the Chinese Premier Xi pledged 500 million dollars as financial aid to build infrastructure in the south-east Asian region.

These burgeoning relationships have been considered with caution by the Indian security establishment as it puts India's sphere of influence in jeopardy. Apart from the above, overt Chinese military help to Pakistan and tacit support to Pakistani hand in destabilising India causes strain on Sino-Indian relations.

The major dispute over water between the two nations is over water usage of the river Brahmaputra. This stems primarily from India's side, since India is downstream of the Brahmaputra which originates in the Tibetan highlands, (where it is known as the Yarlung Tsangpo) travelling westwards through Tibet before turning eastwards into India and Bangladesh, draining into the Bay of Bengal. The river drains both India and Bangladesh and forms the most important river system in eastern South Asia.

Since Tibet is under Chinese control, India's control over the Brahmaputra is under threat. Indian concerns lie in the fact that China could severely restrict the water supply to the Brahmaputra causing immense damage to India and possibly leading to internal discontent within India. India's water and energy requirements are largely due to a large population and a growing economy.

Furthermore, in the year 2000, it was feared that the floods in India's North East were caused by sudden release of water from the dams across the Brahmaputra by the Chinese. (Holsag, 2011). This was followed by rumours of Chinese plans to dam and divert waters from the Tsangpo to feed the Yellow River system. Reports also indicate dams on the Tsangpo along with ambitious plans for a south-north water diversion project (Holsag, 2011). This has led to numerous studies on both sides about the "weaponised" use of water systems and media has painted scenarios of what would happen in such a case.

With the Chinese authorities providing a delayed acknowledgement about the Zangmu project, (Holsag, 2011) Indian fears regarding the damming of the Brahmaputra have proven true. The Zangmu dam is less than 200 kilometres from the Indian border and any sudden water release by China may not give adequate time to India to evacuate those who live and sustain themselves downstream. Indian media has also written about certain "secret" projects of water diversion by the Chinese which do not appear to be mainly for supporting agriculture/local population—but as a tool to threaten Indian security.

From an Indian standpoint, this is a cause for great concern as each of such projects may be used later to regulate the water supply of the Brahmaputra, leading to severe consequences for the north-eastern states of India. The shortage of water could further lead to disturbance in this region, which was beset by insurgency and has only recently been brought under control. Water stress in the Brahmaputra would inevitably lead to deterioration of Indo-Bangladesh relations as both states would try to maximise their gains from the limited supply.

Chinese authorities have, however, stressed that the Zangmu dam has been built only for hydroelectric purposes and have denied plans for diversionary projects on the Tsangpo. However, there is evidence of Chinese studies which aim to check the viability of such diversion projects with an aim to fuel the water supply in the Chinese mainland north of Tibet. While some experts have rubbished such ideas stating that they are too huge in magnitude and may not be economically and physically viable, there are a few who say that the Chinese are capable of surprising the world. What remains to be seen is if these would have environmental consequences, if brought to fruition.

6.7 Conclusion

As can be seen from the discussion, South Asian bilateral relations are often complicated which makes dispute resolution difficult. Some bilateral disputes exist among the states and they have often continued to remain unresolved for decades, water being one among them. A dearth of trust between the neighbouring states is one of the many factors which plague South Asian bilateral relations. India has often been branded as a big brother with "not so friendly" intentions-especially relating to sharing of water.

Water security is a sensitive matter with the developing states of the sub-continent requiring significant quantities of water. This has led to all agreements being scrutinised solely through a prism of national development and security. Most of India's neighbours sharing transboundary water with India have multiple grievances regarding India's dominant position in the matter. While it is true to some extent, it must also be considered that with water security as a prime concern, all the states are attempting to ensure the availability of water for the present and future.

South Asian states appear to be quite conscious of the security aspect of water in terms of national security. However, the understanding of the security implications of scarce water resources at the regional level is stunted. Given the increasing impact of climate change, water crises are highly likely in the future and could lead to the internal collapse of states. Thus, having narrow water security goals at the national level would not be enough.

Furthermore, with China's growing influence, South Asian states should consider resolving old disputes and developing stronger bilateral relations in an emerging world where Asia is likely to be a key player and regional cooperation required to stop the slow but steady Chinese juggernaut. This should begin by initiatives to rebuild trust among neighbours. Mechanisms must be established to look into existing bilateral disputes over water security especially with regards to trans boundary water. The SAARC can become a significant platform in this regard depending on the agreement of all stakeholders.

It can generally be agreed that water is an extremely precious resource for all South Asian states and therefore needs to be shared and managed on an equitable basis to ensure long-term water security for all stakeholders and stability in the region. At a bilateral level, the concerns of both countries regarding water need to be clearly expressed; and subsequently, efforts must be made to reach an unbiased solution which is acceptable and beneficial to both parties. Such agreements must also include water management mechanisms. This would not only ensure water security at the national level but also the regional level.

It must be understood by states that are developing close relations with China, i.e. Pakistan, Nepal and Bangladesh (Anwar, 2019)—that Chinese goals are aimed at increasing their global power and ensuring their stability and long-term success. The BRI is one such initiative. While partners in it may currently derive significant benefits from it, in the long run, their national policies and security would get subordinated to the policies of the BRI, which will directly be under Chinese control.

Should China face a serious water shortage in the future, it is obvious that it would divert water from major rivers originating in its territory. The present state of divisiveness in the Indian sub-continent has developed historically and has been encouraged by China, especially in matters concerning India, in order to fulfil its own geopolitical goals. South Asian states must understand this and try to keep Chinese influence to a minimum without unnecessarily antagonising it.

Therefore, there is a need for a regional cooperative mechanism in South Asia to protect the region's water security and overall regional security in the long term. India, as the biggest and most powerful nation must ensure the build-up of trust and cooperation in the region leading to closer bilateral ties, not only for its own sake but to ensure the security of all of South Asia. It must initiate the process of building a closer and stronger South Asia, not just in terms of water relations but also with closer security ties to counter the impact of China in the long run.

Such steps must, however, be taken with the greatest caution and with respect towards the sensitivities of its neighbours, in order to achieve desired goals. For states on which persuasion may not work, it may become imperative to inform them of India's active rejection of Chinese influence in South Asia to pressurise them into agreement. Pakistan, for example, may be arm-twisted with a threat of possible flooding of the Punjab plains by excessive water release from Indian dams as has been seen in the current Indo-Pak standoff. However, such a step requires great discretion and must only be taken in the worst-case scenario to ensure India's respect and credibility. In the final analysis, India can, and must rebuild and increase its influence in the region and ensure South Asian water security, while negating Chinese influence in the region.

References

Adhikari, K.N. (2013) https://www.researchgate.net/publication/330512925 January 2019

Adhikari KN (2014) Nepal-India water relations: time for change in approach. Pakistan Horizon 67:114

Anwar A (2019) How Bangladesh is benefitting from the India-China rivalry, The Diplomat, July 12, 2019 https://thediplomat.com/2019/07/how-bangladesh-is-benefiting-from-the-china-india-rivalry/

Bala A (2018) South Asian Water Conflict. South Asian Journal, May 11. https://southasiajournal.net/south-asian-water-conflict/
Budhathoki A (2018) How China and India assert military dominance through Nepal, TRT World, 30 November 2018 https://www.trtworld.com/opinion/how-china-and-india-assert-military-dominance-through-nepal-22100
Holsag J (2011) Assessing the Sino-Indian Water Dispute, *Journal of International Affairs,* Vol. 64, No. 2, Spring/Summer 2011, pp. 22–23
Islam MS (2012) Bangladesh-India water sharing disputes: possible policy responses. Journal of Bangladesh Studies 14(1):38–39
Jaiswal P (2014) Caught in the India-China Rivalry: Policy options for Nepal, Institute of Peace and Conflict Studies, Issue Brief, 249, March 1, 2014, http://www.ipcs.org/issue_select.php?recNo=556
Khalid I (2010) Bangladesh Water Concern, *South Asian Studies,* Vol. 25, No. 1, 2010, pp. 80–84
Musharraf P (2006) In the Line of Fire: A Memoir, Simon and Schuster, New York, p 352
Salik MAN (2015) Water security: challenges of transboundary water issues between Pakistan and India. Strategic Studies 35(4):86–92
Stone R (2019) Chinas Silent Foray into Nepal, TRT World, 21 October 2019 https://www.trtworld.com/opinion/china-s-silent-foray-into-nepal-30727
Tabassum S (2019) Assertion or Abjuration: India-Nepal Water Relations, Revisiting Nepals Foreign Policy in Contemporary Global Power Structure, G.B. Books, New Delhi, pp. 104–113, July. https://www.researchgate.net/publication/334737482

Air Commodore Nitin Sathe (Retd) is a distinguished veteran of the Indian Air Force, with dedicated 36 years to the armed forces of India before retiring in 2020. In his career, he amassed approximately 6,000 hours of flying experience across various helicopter platforms, participating in numerous critical missions both domestically and internationally. His operational tenure includes commanding a frontline base in Jammu and Kashmir, contributing to UN peacekeeping efforts in Congo, and managing the rescue and relief operations post the tsunami of 2004. Beyond his military accomplishments, Air Commodore Sathe has significantly influenced the training and selection processes of officer cadres within the Indian Defence Forces. Post-retirement, he has emerged as a prolific writer and columnist, with contributions appearing in leading electronic and print media. His literary works, including Born to Fly, A Few Good Men and the Angry Sea, Up in the Sky, and 1971. 1999. War Stories and The Siachen Soldier offer insightful narratives on military life and experiences. Residing in Pune, he continues to inspire as a cyclist, adventure enthusiast, and motivational speaker, particularly encouraging youth to consider careers in the armed forces. His diverse interests also encompass graphology and culinary arts, reflecting a multifaceted personality committed to lifelong learning and service.

Chapter 7
Democratising Early Warning Systems: Oxfam's Experience with Community-Based Early Warning Dissemination Systems in India

Animesh Prakash and Pankaj Anand

7.1 Introduction

A hazard event or emergency crisis becomes a 'disaster' relative to the human, ecological and economic costs that it accrues (Andharia 2020). Many disaster deaths are preventable, but most actions that are needed to save lives during disasters must be taken before they strike. Effective early action, however, is dependent on a multitude of factors such as community awareness of preparedness measures and protocols, evacuation capacities and ready access to shelters, as well as early warnings. This was demonstrated by the State Government of Odisha, India, during cyclone *Phailin* in 2013, where it set a target of zero casualties, and nearly achieved it, containing the number to 21. Phailin's severity was comparable to the Odisha super cyclone of 1999, which took a toll of nearly ten thousand lives. The State Government then adopting a zero-casualty approach in preparedness of another such super cyclone demonstrated a significant milestone during *Phailin*. It must be noted that for Odisha is one of the most disaster-prone States in India. It also reflected confidence in the state's capacity for early action built up since the early 2000s.

The steady integration of disaster risk reduction (DRR) measures in Odisha have been a result of both internal capacity building measures as well as paradigm shifts taking place in disaster risk management at the national and international levels. In India, the High Powered Committee on Disaster Management through its recommendations in 2001, gave an impetus to Asia's largest community-based disaster risk

A. Prakash (✉)
Mumbai, India
e-mail: animeshprakashog@gmail.com

P. Anand
Dili, Timor-Leste
e-mail: anandpankaj@yahoo.com

J. Andharia and P. Anand (eds.), *Shifting Lands, Flowing Waters*, Disaster Studies and Management, https://doi.org/10.1007/978-981-96-7691-0_7

management programme from 2002–2009. The passing of the Disaster Management Act by Parliament in 2005; and raising of the world's largest dedicated force for disaster management—The National Disaster Response Force (the NDRF) in 2006 as well as its accurate forecast and warning systems—all constituted pioneering reforms and contributed towards building a strong DRR agenda at the state level. The State Government of Odisha's zero-casualty approach was possible due to the capacity building efforts which allowed for the evacuation of nearly one million people before the cyclone made landfall; the sufficient number of safe shelters to accommodate the one million; the community preparedness and resilience-building measures taken; and most importantly, an accurate forecasting system developed that informed their early action at the community levels. This paper is based on learnings from Oxfam India's community-based early warning initiatives in Odisha and elsewhere. It also presents some of the gaps in early warning systems at the last mile, and opportunities that informal community-based networks can offer in strengthening the systems.

Oxfam India, UNDP and other organizations had partnered with the State Government on DRR projects at the regional, national and international levels which contributed to Odisha's successes in evacuation before the cyclones Phailin (2013), *Fani* (2019) and *Amphan* (2020). The main focus areas of these projects were: promoting stronger housing technologies to build more resilient and thereby safer homes along its highly vulnerable coastline[1]; strengthening community preparedness and engaged participation in the disaster management activities, as well as effective cooperation with the local and national authorities; securing safe water and sanitation and establishing an early warning system for cyclones and tsunamis for the people living along this 480-km long coast—becoming the first Indian State to do so by April 2018.[2]

7.2 Climate Change Impacts on Disaster Risks—Indian Realities

In the near 10 years since Cyclone Phailin, climate-change impacts and inequitable development have exacerbated Odisha's proneness to disaster risks and its people's socio-economic vulnerabilities, even amidst the State's actions and investments towards integrating DRR into development planning. Podempeta in Ganjam district

[1] For example, UNDP partnered with the Government of Odisha (2000–2007) on a project which constructed more than 60 buildings serving as Technology Demonstration Units for low-cost disaster resilient building technologies combined with indigenous building materials, across 4 districts, contributing to over 4000 houses and other public infrastructure in these vulnerable areas across the State. See UNDP Press Release (2015).

[2] The EWDS addresses the existing gap in disseminating disaster warnings, strengthening emergency operation centres in the State with over 1000 villages from 22 blocks of coastal districts being covered, and over 100 watchtowers set up within 1.5 km of the coastline for cyclone or tsunami warnings using sirens and mass messaging. See Mohanty (2021).

is a fishing village which was completely devastated in 2013 and received relief and assistance through a variety of humanitarian actors. A decade later, coastal erosion has left its people on the verge of permanent displacement, as with other settlements in the districts of Jagatsinghpur, Puri and Ganjam, joining already displaced villages like Satabhaya in Kendrapada. As the risk landscape changes, the development gains made are often lost. This highlights the ever-rising exposure to hazards, and the fact that system improvements that do not factor such in changes, would not sustain given the vulnerabilities that would develop in the face of future risks. Therefore, it is imperative for all provinces and nations to constantly evolve and improve to prepare not only for current risks but also for future ones.

India's geodiversity exposes it to myriad climate risks, further aggravated by the socio-economic vulnerabilities across various demographics that make up its 1.4 billion population, the largest in the world. It is currently ranked 7th in the Global Climate Risk Index (Eckstein et al. 2021), but it is home to the largest number of people exposed to the disaster risks and the compounding effects of climate change.

Between 1901 and 2020, India's average annual mean temperature rose by 0.62°C/100 years, with the warming trend peaking in the pre-monsoon season by 0.88°C/100 years, maximum temperature rising by 0.99°C/100 years, and minimum temperatures by 0.24°C/100 years (MoEs 2020). Fluctuations in the temperature extremes in the past 30 years are observed in northern, central, eastern and north-eastern India (MoEF 2021), corroborated by the Berkeley Earth Dataset, observing the strongest warming in India in the northern and north-eastern region (Carbon Brief 2018), notably closer to the Himalayan range which feed the glacial rivers that both nurture and devastate the floodplains in this region.

Precipitation trends have seen significant fluctuations in recent years. The El Niño Southern Oscillation (ENSO) impacts precipitation as it increases the sea surface temperature and reduces monsoon rainfall in India. Although data on inter-annual variability and total precipitation in India have remained stable in last 120 years, there have been more frequent dry spells and more intense wet spells during the summer monsoons. Dry spells in India have been 27% higher between 1981 and 2001, compared to 1951–1980. Rainfall intensities have also increased, especially in central India, where the daily precipitation extremes of rainfall exceeding 150 mm per day has increased by 75% during 1950–2015 period (Krishnan et al. 2020). These trends on climate variability in India suggest a significant impact on the economy, with agriculture being most affected. With nearly 43% of the population dependent on agriculture as their main employment source, it is imperative to look at the climate resilience of the sector in a changing climate context. The 6th IPCC report already indicated a 1.20°C rise in temperature and steadily rising to 1.50°C with a projection of reaching it in less than 20 years, if unprecedented measures are not taken. It is the unpredictability of a disasters, combined with a lack of emergency preparedness that leads to the loss of lives and assets. An effective early warning system can ensure that 'the at-risk community' and local actors have sufficient lead time to prepare and respond to disasters. An early warning system can provide more than tenfold returns-on-investment; just 24-h-notice of an impending hazardous event can cut the ensuing damage by 30% (Global Commission on Adaptation 2019). It is well acknowledged that an effective end-to-end early warning system is essential for pursuing a zero-casualty goal in highly populous countries like India.

7.3 Developments in Early Warning Systems (EWS) in the Global Context: The Last Mile Challenge

The 1990s saw paradigm shifts in approaches to disaster risk management and DRR, at least in policy documents, which emphasised the significance of early warning systems. This has evolved through prior efforts such as the Famine Early Warning Systems Network (FEWS NET), developed in the US after the 1984–1985 famines in Sudan and Ethiopia which caused an estimated one million deaths. FEWS NET is one of the leading providers of early warning and analysis of acute food insecurity around the world. Further the 1994 World Conference on Natural Disaster Risk Reduction highlighted the importance of early warnings in prevention, preparedness and mitigation; in 1995, the UNISDR advocated for the examination of new science-based methods to improve the accuracy and timeliness of short-term forecasting.

The Indian Ocean Tsunami of 2004 which took more than 225,000 lives, gave further momentum to developing a global EWS framework for ocean-related hazards, through a resolution adopted by the Intergovernmental Oceanographic Commission (IOC). Key cooperative measures through SAARC were also established in the process, and a special session of the SAARC environment ministers in 2005 led to the establishment of an expert group to formulate a comprehensive framework on early warning, disaster management and disaster prevention.

The Hyogo Framework of Action (HFA) adopted in 2005 also included early warning as one of its five priority actions. The following year (2006) saw the formation of the Inter-governmental Coordination Group for the Indian Ocean Tsunami Warning and Mitigation System, as well as the establishment of the Third International Conference on Early Warning in Bonn, Germany which led to establishment of the Indian Ocean Consortium (for strengthening national capacities for early warning and response in the Indian Ocean). The UNISDR also developed a Checklist, to help governments and communities set up effective early warning systems.

A prime example of effective early warnings and early action at a national level is the Japanese Meteorological Agency whose scientific systems are run by the National Research Institute for Earth Science and Disaster Prevention and issues earthquake warning alerts directly to factories, schools, TV networks, radio stations and mobile phones in an at-risk or affected location through the 'Area Mail Disaster Information Service'. These timely warnings saved several lives during the 2011 earthquake and tsunami. Post HFA (2005), the Sendai Framework for DRR (SFDRR) for 2015–2030 has also included improved access to Multi-Hazard, Impact-Based Early Warning Systems (MH-EWS) as one of its global targets. Most recently, the UN Global Early Warning Initiative for the Implementation of Climate Adaptation was launched in March 2022, aiming to achieve universal access to early warning systems by the end of 2027 through a plan worth 3.1 billion USD which was unveiled in November, at COP 27.

Even though the investments, developments and collective capacities in early warning structures have improved across the board, critical early warning information is still not accessible to all populations exposed to hazards. However, last mile connectivity and outreach are still a significant concern in most countries, with those living in the remotest areas often being the most vulnerable to hazard risks. Beyond improving access at the last mile—albeit a significant first step in the direction towards universal access to an effective people-centred multi-hazard impact-based early warning system—other socio-political and economic as well as technological considerations for a well-functioning system need identification and careful integration. The following sections elucidate gaps, as well as ways in which community action integrated with governance helps foster more resilient networks that translate early warnings into early actions that help save lives and minimise damage to property.

7.4 Oxfam India's Gap Assessments on EWS in Odisha

For a country as populous and geographically widespread as India, the exposure to a variety of hazard risks and socio-economic vulnerabilities makes operationalising an EWS at different levels, a herculean task for governance at different levels. Adequate information dissemination, ensuring people's access to this information, covering the most vulnerable and clarity of actions to be taken in response to hazard warning-all require considerable thought, planning and effort. In vibrant democracies, non-governmental organisations (NGOs) and civil society organisations (CSOs)—whether local, regional, national or international—all need to play the essential role of supporting communities as well as the government for effective governance and welfare outcomes. Oxfam India generally plays this role working with local NGOs and the community and has worked systematically to strengthen social networks and communication channels. It has also endeavoured to build socio-economic resilience through interventions in sectors such as WASH (Water Sanitation and Hygiene) and livelihoods to best better prepare households and individuals to weather the vagaries of nature and effectively bounce back.

In 2015, Oxfam India conducted an early warning gap assessment to analyse the existing structure of EWSs in Odisha, an eastern state of India. Oxfam was already working in several disaster-prone coastal districts of this state which is exposed to multiple disasters (such as tsunamis, cyclones, storm surges, floods, heat waves, and droughts). With a coastline of 482 kms, Odisha is more prone to cyclonic storms. An analysis of the frequency of cyclones in the country between 1891 and 2000 showed that 86.5% of cyclones were in the eastern coast of India as against 13.5% on the western coast. Of all the cyclones of the eastern coast of India since 1891—the highest (31.8%) occurrences were in Odisha (Ramesh 2018). This made Odisha an ideal site for this study. The focus was to identify the factors that contribute to a delay in dissemination of warnings and inability of people to take appropriate actions, during impending hazards such as cyclones and floods. The assessment considered four aspects of the early warning framework including:

i. Risk assessment,
ii. Monitoring and detection,
iii. Communication and dissemination and
iv. Community Action.

The major focus of the study in Odisha was on communication and dissemination, as well as community action. The assessment methods included structured and non-structured interviews with coastal communities from different occupational groups including traditional small-scale sea fisherfolks, marine fisherfolks, farmers and others. Those interviewed during the study were: local officials and community leaders like the Community Mobilizer of the OSDMA (Odisha State Disaster Management Authority) and Shelter Committee Members; District officials including the District Emergency Officer, Additional District Magistrate and Emergency Operation Staff; and State level Government officials including the Director IMD, Special Relief Commissioner, and IAG[3] chairperson.

The detailed gap analysis is illustrated in Fig. 7.1

The gaps in Early Warning System (EWS) in Odisha that emerged from the assessment are:

1. Despite state efforts, from the District Emergency Operation Centre (EOC) to the community, warning communication at the last mile is dependent largely on a manual system, resulting in late dissemination, exclusion and limited outreach. Respondents across all levels agreed that the existing communication and dissemination system may not be effective in disasters like tsunamis, where the lead time could be less than 4 h.
2. Dissemination during odd hours is difficult (For instance people working in the early hours before dawn groups such as deep-sea fisher-folk would be left-out of the loop).
3. Communities lack trust in the warning messages issued by the government.
4. Communities' risk perceptions and priority to safeguard assets, lead to delays in evacuation.
5. The existing relief entitlements are based on ownership of assets; and a large dependent workforce is denied compensation.
6. Due to the lack of privacy, women and girls avoid going to shelters unless extremely necessary, leaving them even more vulnerable.
7. Inadequate structures to ensure collective and coordinated action from local communities, civic groups, network of volunteers, inter-agency groups in early warning dissemination and response, pose challenges on-ground.

[3] IAG: Inter Agency Group, a consortium of NGOs, Civil Societies, UN and International Agencies, working in coordination with the OSDMA. See GO-NGO Coordination https://www.osdma.org/capacity-building/go-ngo-coordination.

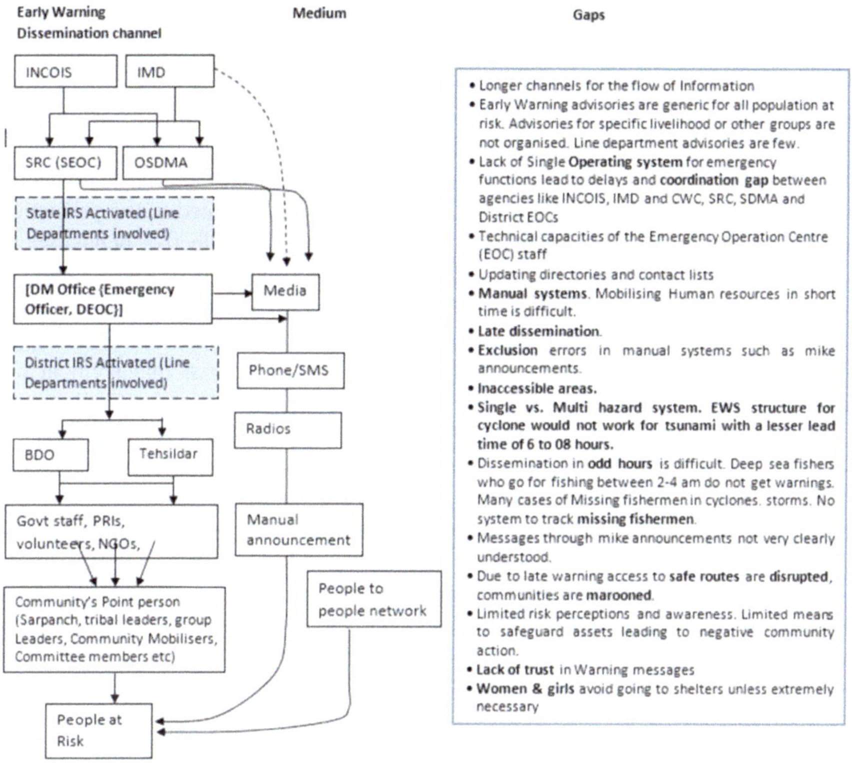

Fig. 7.1 The early warnings dissemination channel in Odisha and the gaps identified. *Source* Adapted by author from the study conducted by Oxfam India

For an early warning system to be effective, it should actively engage different stakeholders, specifically the communities at-risk, to provide the enabling environment that integrates the appropriate technological, regulatory and legal frameworks into local practice as well as meets the adequate operational capacities and defines clearly the responsibilities of all participants (ITU n.d.).

The accessibility and employment of Information and Communication Technologies (ICT) has been the conventional means of community engagement in DRR initiatives that have typically followed a hazard-centred, technology-led approach (Hicks et al. 2019). However, the decades of emphasis on mainstreaming DRR within academia and practice, has led to an increased advocacy for even more people-centred approaches that establish the communities at-risk, at the centre of DRR. It is understood that to strengthen the disaster resilience of a community, it is necessary to empower and engage the most vulnerable individuals in the decision-making processes, thereby building participation, awareness and capacities towards long-term preparedness and mitigation strategies. To take this a step further, the involvement of the at-risk communities in the mapping of risk and collection of grassroots' level socio-economic and environmental data would benefit both those at-risk as well as the scholars and practitioners who utilize this data to better formulate systems and policy frameworks.

The efficiency of an EWS is also be attributed to how well social factors are integrated within the system to create a platform for positive community action. Not all timely warnings result in timely evacuation or early action. Factors such as a lack of risk perception, a lack of trust in the warning messages, concerns related to the protection of livelihood and household assets or issues related to the protection of women and children at evacuation sites determine the collective behaviour of a community during disasters. Field experiences have shown that a well-prepared community with some assistance from trained village-based committee members can avert and absorb the shocks of disasters. Village-based institutions developed through intensive community processes act as the vanguards in disaster risk management and risk reduction. However, Oxfam India's experience suggests that the challenge to tackle attrition of many volunteers will always exist at the last mile; in developing countries and poorer countries. The community volunteers' sustained engagement is also affected by dynamic factors such as occupational and seasonal migration, marriage, and declining health or interests. The systems and structures established by local NGOs, CBOs, SHGs tend to be more sustainable, and their capacities should be enhanced to manage the volunteer base in the respective regions for effective risk management. Technical innovations which can potentially reduce their operational costs in managing volunteer base are also directly beneficial and will also keep them engaged and associated with potential hazards.

7.5 Democratising EWS: Learnings from Oxfam India Projects

In addition to the gap assessment study in Odisha, learnings from Oxfam's community-based early warning initiatives in the transboundary basins of the Brahmaputra and Sharda Rivers also provide valuable insights. They reiterated, many of the socio-economic factors and gaps in EWS highlighted through in the findings of the study. In addition, these learnings helped identify the opportunities that informal community-based networks can offer in strengthening the EWS systems at the local grassroots level. For example, Rampur Lahi, a village in Madhepura, Bihar was severely affected during the 2008—Kosi floods. Many families from the village ignored the Government's early warning advisories as they did not perceive that the floods would affect their village. Not complying with the advisories put their lives at extreme risk, as they were trapped in the floods and could only be evacuated in a bamboo raft. In a tragic turn of events, Nunulal Sardar, a villager and rescue volunteer who built a bamboo raft to aid evacuations, died in the floods. Similar was the case during the 2013 Subarnarekha floods in Odisha (Prakash 2017). In Chadnamkhana a riparian hamlet of Paschimbar Gram Panchayat (Jaleswar block) of Baleswar district, eight families lost their assets, houses and land due to flood erosion. All 30 houses in the hamlet were either completely or partially damaged. During Oxfam's humanitarian assistance in this hamlet, the community shared that they did not receive early warning messages, and recollected that the flooding was sudden, leaving them barely any time to evacuate. This hamlet is just 20 kms downstream of the flood monitoring station at Tambakhuri in Rajghat in Balasore, yet it experienced a last mile connectivity issue pertaining to early warning dissemination. Likewise, during the Saralbhanga floods (2014, 2016–2017, 2018) in West Banglajoda in Kokrajhar, Assam, where a riparian village on the borders of Bhutan recalls the floods experienced by the village in each of the years being very sudden, with warnings coming with lead times of less than an hour. Consequently, the villagers could manage to save very few of their assets. The intensity of the flashfloods and severity of the damages made search and rescue extremely challenging even for the best equipped army, with an eight-year-old girl losing her life in the 2014 floods.

These incidents reiterate that an effective end-to-end early warning system is built on all four pillars including (1) risk assessment, (2) monitoring and warning services (3) dissemination and communication and (4) response capability. Oxfam through its projects has laid a larger focus on communities' participation at all steps of the early warning system, adopting a citizen science approach to strengthen the same vis-à-vis disaster preparedness and support Government in the last mile through participatory actions.

Democratising EWS entails the utilisation of a Citizen Science approach of environmental and ecological sciences which harness public participation to inform research and knowledge production. It has become increasingly relevant with the availability of IT infrastructure for gathering, reporting, and storing data. Moreover, improved literacy rates and education levels have enabled a larger proportion of local populations to participate and contribute to knowledge formation (Frais et al. 2022). Oxfam India has also utilized this approach in other thematic areas, such as community-based surveillance of water contamination in the Sharda Basin in Lakhimpur, Uttar Pradesh. The approach led to an improved community understanding of factors leading to water contamination in the region, positively affecting their involvement in evidence-based advocacy on water quality.

Given the central role of a community's participation in emergency preparedness and minimizing disaster risks, as well as their stakeholdership in the management of their local natural resources and the hazard risks associated, it follows that such an approach would be instrumental in enhancing the effectiveness of early warning systems as well as socio-economic ones. The on-ground realities and insights provided by the diverse voices of different demographics help tailor the network of early warning dissemination to their unique needs. The goal is to equip at-risk communities with the necessary procedures and information in preparedness, for them to be able to take timely and positive actions to save lives and assets. Loose ends at any stage would adversely affect the EWS' efficiency and contribute to the potential human and economic losses that are incurred during and in the aftermath of disasters. Thus, democratising early warning systems is recognizing the critical role of communities in every facet of the EWS, and utilizing their inputs along with their participation to better translate early warnings to early actions.

Oxfam India's projects on strengthening community-based early warning systems were first implemented in Odisha in 2015 and the idea was later expanded within other programs (like TROSA)[4] in Assam and Uttar Pradesh from 2018 to 2021. This section presents three case studies from Oxfam India's interventions with local groups to address challenges faced by at-risk riparian communities. These cases are:

1. The Volunteer Network Management System (VNMS), in Subarnarekha, Brahmani-Baitarani and Mahanadi River Basin in Odisha;
2. Transboundary People-to-People Early Warning Network between (Assam) India and Bhutan in the Brahmaputra River Basin; and
3. *Adhikar-Vaani*, utilizing an early warning dissemination platform for improved access to social protection systems in Assam and Uttar Pradesh.

7.6 Methodology

Across the three case studies discussed in the following sections, Oxfam India's methodology for setting up the community-based early warning systems has remained consistent; evolving and varying in implementation against the contexts of intra-state and inter-state community relations as well as technological innovations to better facilitate these systems. Elements in each of these interventions are complementary and contribute to one holistic design for community-based early warning systems.

Utilizing the existing social networks within these riparian communities was the foundation for these interventions, harnessing the people-to-people connections and their shared experiences with the local natural resources and the associated risks. Oxfam's work strove to further build up relations between these communities as in the case of those settled in transboundary river basins. Each of these, from the VNMS to *Adhikar Vaani* have placed the communities at the heart of the intervention, recognizing their stakes in and promoting their rights to the water resources as well as the corresponding hazard risks involved. Adopting a citizen science approach in these projects engaged communities in the risk assessment and analysis; simultaneously improving community coordination on wider dissemination of early warning information generated by the government agencies and supporting timely and positive community action.

Oxfam India selected these specific river basins for its projects based on the respective riparian communities' exposure to hazards and vulnerabilities. However, the geographical selection was also influenced by Oxfam's strategy to work only among the focus states of Bihar, Jharkhand, Chhattisgarh, Odisha, Assam, and Uttar Pradesh (further based on socio-economic vulnerabilities and development needs). Due to the frequency and severity of floods in these basins, and Oxfam's resource capacity,

[4] TROSA 2017–2021 (Transboundary Rivers of South Asia) is Oxfam's regional water governance programme supporting poverty reduction initiatives in the Ganges–Brahmaputra-Meghna (GBM) and Salween basins.

Odisha, Assam, and Uttar Pradesh were prioritized for these project interventions. For the VNMS intervention in Odisha, communities exposed to recurring floods and coastal hazards, particularly from Subranarekha basin in Balasore, Brahmani-Baitarni Basin in Jajpur and Mahanadi basin in Puri were identified for project interventions. Oxfam India later implemented Transboundary Rivers of South Asia (TROSA) project (2017–2021) through which it engaged with the riparian communities in the Brahmaputra River basin in Assam (and Bhutan) and the Mahakali River Basin in Uttar Pradesh (and Nepal).

With an overall goal of zero casualties and minimizing the loss of assets in disasters, Oxfam India focused on (1) improving technologies for detection, communication and dissemination of early warning messages and advisories at the last mile; and (2) improving coordinated response capacities at the last mile. Key program strategies towards this goal were to:

- Invest for advances in last mile warning communication and dissemination technologies to ensure that timely warning reaches the entire population at-risk, including those in remote locations.
- Strengthen structures for last mile dissemination to complement the forecasting systems.
- Invest in improved participation of local communities, civic groups, and traditional structures in the early warning systems.
- Ensure that all at-risk groups, particularly those from socio-economically marginalized groups, have access to Information in disaster situations and people take informed and timely actions to minimize the loss of lives and property.

At an operational level, steps taken by Oxfam India to establish the foundations of a robust community-based early warning systems in all the target river basins included the following:

1. Engaging with local CSOs, NGOs and community to prepare the Hazard Risk Vulnerability Assessment (HRVA) maps of the targeted region in the river basin. Identification and ranking of the most vulnerable villages in the selected region.
2. Identifying volunteers in the targeted villages, providing training to these volunteers to support
3. local government in dissemination of warning especially reaching the last mile communities to ensure timely action on advisories.
4. Strengthening cultural ties, social relations between upstream and downstream communities to strengthen people-to-people cooperation for early warning.
5. Fostering technological innovations to improve real-time warning dissemination to all at-risk populations.
6. Using early warning dissemination systems during non-disaster periods to improve access to social protection schemes, and thereby keeping the warnings dissemination systems active.

7.7 The Volunteer Network Management System (VNMS), in Subarnarekha, Brahmani-Baitarani and Mahanadi River Basin in Odisha

As elaborated earlier, Odisha is a disaster-prone state. Besides the risks along its coastline, Odisha contains a large part of the 1,41,589 sq km Mahanadi River basin, constituting nearly 4.3% of India's geographical area, shared with Chhattisgarh, and parts of Jharkhand and eastern Maharashtra. Mahanadi is among the peninsular rivers and a major river system in India. It drains into Chilka lake in Odisha, which forms the part of the basin along with six other minor rivers, namely Rushikuliya, Brahmani Baitarni, Budhobalanga, Vamsdhara, and Subarnarekha.

The Volunteer Network Management System in Odisha was designed to organise last mile stakeholders including the CSOs, NGOs, SHGs, and trained village volunteers became points of contact for government authorities to disseminate early warning information and Government advisories to the most vulnerable and remote communities and villages, and supported communities in timely actions and evacuations, during emergencies. The system was aided by technology for a wider outreach in real time. Oxfam India engaged with networks of civil society, with Unnayan as the nodal agency in Balasore, Pallishree in Jajpur and SOLAR in Puri. Participatory HRVA analysis was carried out, in Subarnarekha and Gudikhal Rivers in Balasore, Brahmani-Baitarni and Nuanadi in Jajpur and in Puri—Daya, Luna, and Makara—key distributaries of Mahanadi River. This was a long-drawn exercise requiring awareness generation, local level training and participation.

After the HRVA, 2112 volunteers were identified from 173 villages, with an indirect outreach that covered 34,331 households. Local communities' participation in such a risk assessment provides critical inputs such as information on the local history of disasters and its impact evidenced in different villages. Based on the community consultations and with the technical assistance from the local NGO partners, risk maps were developed. This was followed by preparation of Village disaster management plans and volunteers were trained, to support the community along with identifying the most vulnerable individuals and households during emergencies.

The structure of the system calls for nodal NGOs to coordinate with the local administration and the flood monitoring stations to monitor the flood situation; and once warning or advisories are issued, the communication and dissemination facets of the EWS were aided by the online VNMS platform developed for real-time information dissemination. The VNMS system helped in more systematic engagement with the remotest settlements at every stage of disaster management, from preparedness to response. Community-led HRVA analysis was particularly useful in expanding the risk knowledge and awareness of the community as well as their continued involvement in supporting the wider dissemination of Government warnings and advisories which ensured timely action from the at-risk population (see Annexure 1 for the User Interface of the online VNMS platform).

7.8 Transboundary People-to-People Early Warning Network Between (Assam) India and Bhutan in the Brahmaputra River Basin

Lower Assam is one of the disaster hotspots with around 56 transboundary rivers flowing from the north into the mighty Brahmaputra. The Saralbhanga (also called Swrmanga in India) is one of 56 rivers that flows from Bhutan into the part of the Brahmaputra in Assam, India (Yashwant 2018). This is a lifeline for thousands of people living on the downstream of the river, yet it is also a source of havoc and devastation owing to the flash floods that strike whenever there is incessant rain on the mountains of Bhutan or a cloud burst occurs. It is dotted with c*har* islands in the districts of Dhubri and South Salmara. Chars are the riverine islands formed through silt deposits which house a large population, which often get partially or fully submerged in the monsoon floods, displacing communities year after year. Amidst disaster vulnerabilities, the pace of development has been extremely slow in these regions, further marginalizing and impoverishing these communities.

Oxfam India implemented the Transboundary Rivers of South Asia (TROSA) project from 2017–2021, engaging with riparian communities on issues of water governance. In this project, Oxfam India along with its NGO partner NERSWN[5] worked with the riparian communities in the Saralbhanga River Basin in Kokrajhar, Assam.

The Saralbhanga flows through India to meet the Brahmaputra, coming in from Bhutan, where it is called the *Sarbhang Khola Chu* River.

The river is a lifeline for the *Jamfwi* (alternatively *Dong*) irrigation system, that waters the small fields of paddy, vegetables and areca nuts of the downstream communities of Nepalese, Bodo and Santhal settlers in Kokrajhar district of Assam in India. This simple irrigation system involves diverting water from the Saralbhanga via a small check dam into a network of slender canals crisscrossing the farms around the Ultapani forests.

The aftermath of the flashfloods of 2017 in the Saralbhanga, Aai and Manas Rivers, left behind a trail of devastation in downstream villages like Geluphu and Sarpang. The Bhutanese administration began constructing flood mitigation measures to protect villages within their borders; a cause for concern for the downstream communities in India. Changes in hydrological cycles of the river, whether natural (cloud bursts, drought, landslides etc.) or manmade (dams, flood mitigation measures, boulder mining etc.) in Bhutan severely impact downstream communities in India causing floods on one hand and/or drying up of water in the Dongs on the other.

Civil Society in India with support from Bhutan India Friendship Association (BIFA) engaged with the local administration in Bhutan to share the issues the downstream communities. The *Sarpang Dzongdag's* (Administrative head of district) took

[5] North-East Research and Social Work Networking (NERSWN) is a 'secular, not-for-profit, gender-just, nonpolitical organization' that is led and managed by youth, working towards 'sustainable & holistic change' in the North-East Region of India since 2004–2005. See NERSWN (2005).

a decision to prevent any structural measures in Saralbhanga riverbed that could potentially damage the Dong system. This fostered the cooperation between civil societies in India and Bhutan and brought together the basin level on water management at a local level. The informal ways of cooperation and joint solution—finding led by the civil society actors in India and Bhutan further strengthened the people-to-people ties. Informal communication lines have been set up for transboundary early warning from the BIFA which has led to the improved lead time for downstream communities due to timely information from BIFA on impending floods in Aai and Saralbhanga ever since (Yashwant 2018). In 2019, twelve civil society organizations from Assam and Bhutan signed the Kokrajhar 7 Points Call for Action on Inclusive Water Governance[6] to work together for effective management of river risks (Karmakar 2019).

For early warning systems to work, it is important to prepare and train the communities to respond to the warnings in a disciplined manner. To reduce the vulnerability of riverside communities, disaster risk reduction trainings and preparation of Hazard Risk Vulnerability Assessment (HRVA) documents has been piloted in riverside villages along with identification of individuals for Volunteer Network Management Systems (VNMS) to develop Early Warning Systems. The VNMS model was first implemented in Odisha, and it was replicated in Assam. Under the TROSA project, an online platform for the Water Governance Collective Action Network[7] was used to host early warning dissemination in Kokrajhar, Assam. The volunteer networks were organized across the Saralbhanga River basin which helped the civil societies in India to engage more effectively with village volunteers on issues of water management and early warning communication.

Similar work has been done, by Oxfam, in the Mahakali River basin along the India-Nepal border under TROSA. The Mahakali (also the Kali or Sharda) is a transboundary river originating in the Himalayas and flowing along the India-Nepal border, with a river basin (Uttarakhand and Uttar Pradesh in India) susceptible to a range of hydrometeorological and geological risks that enhance flooding in the region and is home to a demographic with a high incidence of poverty and socio-economic disparity. These factors combined with poor land use plans, unplanned urbanization and inefficient governance have aggravated the risk of flooding (Paudel et al. 2013). As part of TROSA, Oxfam India's work in the Mahakali basin has been to support the Government's efforts to reduce the disaster risk and socio-economic vulnerabilities of these transboundary communities by strengthening the people-to-people network upstream (Nepal) and downstream, as well as widening the access to information to ensure that outreach extends to the most marginalized, at the last mile.

The Transboundary People-to-People Networks in both the Saralbhanga River Basin in Assam and the Mahakali Basin in Uttar Pradesh intervention necessitate bilateral coordination at a national level as well as in the capacity of upstream and

[6] Two-day consultations on Trans-Boundary Cooperation for Effective Management of Water Risks between Bhutan-India (20th–21st June 2019), jointly organized by NERSWN, Oxfam India, TROSA and Aranyak in Kokrajhar, Assam. See Sentinel Digital Desk (2019).

[7] WGCAN, under TROSA (2017–2021) and supported by Oxfam India.

downstream riparian communities. As evidenced in the Kokrajhar case study, NGOs and CSOs on either side of these geopolitical borders play a critical role in safeguarding the interests of the communities, supporting the Governments, and ensuring a positive sum outcome as far as possible and fostering stronger relations between neighbouring countries.

7.9 Adhikar-Vaani: Using Early Warning Dissemination Platform for Improved Access to Social Protection Systems in Assam and Uttar Pradesh

Adhikar-Vaani[8] was launched a month into the lockdown of 2020, to provide a vernacular communication channel for CSOs, NGOs, and government representatives to provide critical information on social protection schemes and vaccine awareness during the early waves of the COVID-19 pandemic. It was built on the WGCAN platform and continues to function as a medium for public awareness on social protection and community rights; it can also be utilized as an early warning system for wider dissemination to the last mile.

The effectiveness of this service lies in the convenience of drafting the messages as well as the ease of wide dissemination as they can be sent in the local vernacular, both in text and audio formats. During the pandemic, even in the case of disaster events, Adhikar-Vaani became an easy and useful medium for the local administrations to send voice-recorded advisories about disasters which struck during Covid years. In addition to early warning messages and advisories, this platform also helped improve public health promotion, and social protection schemes.

As an audio messaging service, Adhikar-Vaani seeks to facilitate social development through citizen engagement. It provides an interface between Government, CSOs and the end-beneficiaries of social protection schemes. The target group for the platform were elected panchayat representatives, citizen science volunteers, CSOs or frontline workers. It was envisioned as a solution to connect various demographics, communities, households or individuals who have been marginalized or left behind in development due to poor access to government schemes, and conversely also helped government services to expand their reach.

The services were also used by Civil Society Organizations to reach out to their communities' network as well. The portal, thus, enabled progress towards achieving equity and inclusion. In the context of Covid-19, Adhikar-Vaani served to reach out to vulnerable groups with information on governmental and non-governmental relief measures, and along with the development messaging for WGCAN, the portal also provided information on Covid-19 related measures, including vaccinations and public health, domestic violence, psychosocial support and other issues. The portal provides audio messages in many languages to reach out to people across the geography of India. The messages are also saved on the website to be accessed and

[8] Adhikar-Vaani, Audio Message Portal on the WGCAN platform.

heard at any time and can also be shared with others to widen the dissemination and reach. This is a cost-effective solution to the information asymmetry observed in the sector of social protection and welfare.

7.10 Comparing the EWS Interventions Undertaken by Oxfam India

The following figures illustrate the forms of early warnings messages disseminated through the VNMS in Odisha in 2017, and the Adhikar-Vaani platform. In the case of the latter, examples of audio advisories and their vernacular transcriptions from the UP State Disaster Management Authority (in Hindi) and from the District Disaster Management Authority in Kokrajhar (in Assamese) are shared.

As mentioned previously, Oxfam's interventions have remained consistent in their foundations yet evolved and contextualised to better facilitate the functioning of an early warning system that can translate to early actions and better integrate DRR mainstreaming at the grassroots (see Annexure Fig. 7.2a, b for an illustration of text SMS [in Odia] through the VNMS in Odisha and audio advisories from the UPSDMA and the DDMA in Kokrajhar, Assam through Adhikar-Vaani respectively).

Table 7.1 illustrates the complementarity of the three different early warning interventions led by Oxfam India. They also reflect the evolution and improvements made in Oxfam's community-based early warning systems. A significant component in this process and structure is the HRVA analysis through which the target villages and volunteers were identified; making it useful to reach the most vulnerable households with ease. Linking the EWS system to social protection schemes further improved its relevance in periods of 'normalcy' and aided the resilience of disaster affected communities in recovering better.

7.11 Conclusion

Despite their unique cultural differentiations, riparian communities across India and the South Asian region face similar risks vis-à-vis their disaster vulnerabilities, as well as their dependence on their local rivers for their lives and livelihoods. Their socio-economic and environmental instability, particularly in transboundary contexts, have local as well as geopolitical ramifications. The socio-economic vulnerabilities of the most marginalized demographics are set to further deteriorate with the increasing risks, intensities and frequency of natural hazard events. Climate-change impacts are increasingly threatening the lives and livelihoods of those settled in these ecologically fragile geographies for decades if not centuries. The development gains made, the infrastructure established, and the disaster management measures that have been in place so far are inadequate to meet the changing risk landscape and pose a

Table 7.1 The key elements of EWS explored through some of Oxfam's interventions in India

Checklist	VNMS	Transboundary EWS	Adhikar-Vaani
Risk assessment	Community led, with technical support from Nodal NGOs	Upper and lower riparian communities' cooperation on river risk management. Community-led HRVA conducted locally in Assam	Drawn from the VNMS platform
Monitoring and warning service	Through relevant institutions and local administration	Upper to lower riparian communities	Through relevant institutions and local administration
Dissemination and communication	Last mile volunteers to support local administration in wider dissemination of advisories	People-to-people network (upper to lower riparian communities)	Local administration using Adhikar-Vaani platform for recording and disseminating advisories
Response capability	Community participation leads to respecting warnings and advisories. Communities engaged in disaster preparedness and response plan. Support from trained volunteers	Community participation leads to respecting warnings and advisories communities engaged in disaster preparedness and response plan. Support from trained volunteers	Alternative media, including audio and text in vernacular languages. Advisories from local departments are more specific and impact based and respected by locals. Improved access to social protection schemes and therefore improved community confidence in the system

Source Oxfam India websites

concern for policymakers and governments across the different levels of the nation-state. Adaptability to the present and the potential of the future are essential in the DRR-integration and community-preparedness measures taken. Strengthening the network of governments, CSOs, NGOs, communities can empower those at the last mile with access to information.

The human and economic costs will only continue to escalate without the institution of adequate, contextually appropriate, and adaptive measures and mechanisms that bolster the resilience of communities through emergency preparedness and accurate early warning systems that translate into early actions. As the examples from the case studies above highlight, a citizen science approach to early warning systems, which places the at-risk and affected communities and their interests at the centre of all decision-making and participation is one that is truly just in assisting them to realize their rights and development aspirations. Inclusivity in participation is a critical enabler of ensuring that no one is left behind, and that the intersectional vulnerabilities and risk exposures of different demographics is considered in the process. Adapting and utilizing technological innovations to streamline universal accessibility to early warnings is then a significant step towards democratising these systems.

Acknowledgements The authors would like to acknowledge the support provided by Ms. Tania Teresa Mathews in developing this paper and also to Professor Janki Andharia for her editorial inputs.

Annexure 1—User Interface of the Online VNMS Platform

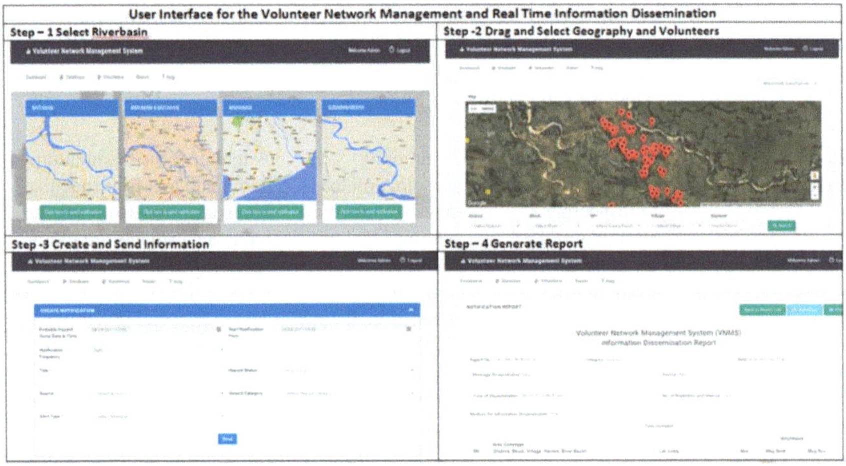

Source https://www.oxfamindia.org/sites/default/files/DRR_AnimeshAprilbooklet_Web.pdf

Annexure 2

An Illustration of text SMS and audio advisories from the UPSDMA and the DDMA in Kokrajhar, Assam through Adhikar-Vaani.[9]

[9] Screenshots of Annexure 2 were taken from *Adhikar Vaani* webpage developed by Oxfam India. The page is inactive now. The page was developed under the project which was led by the author.

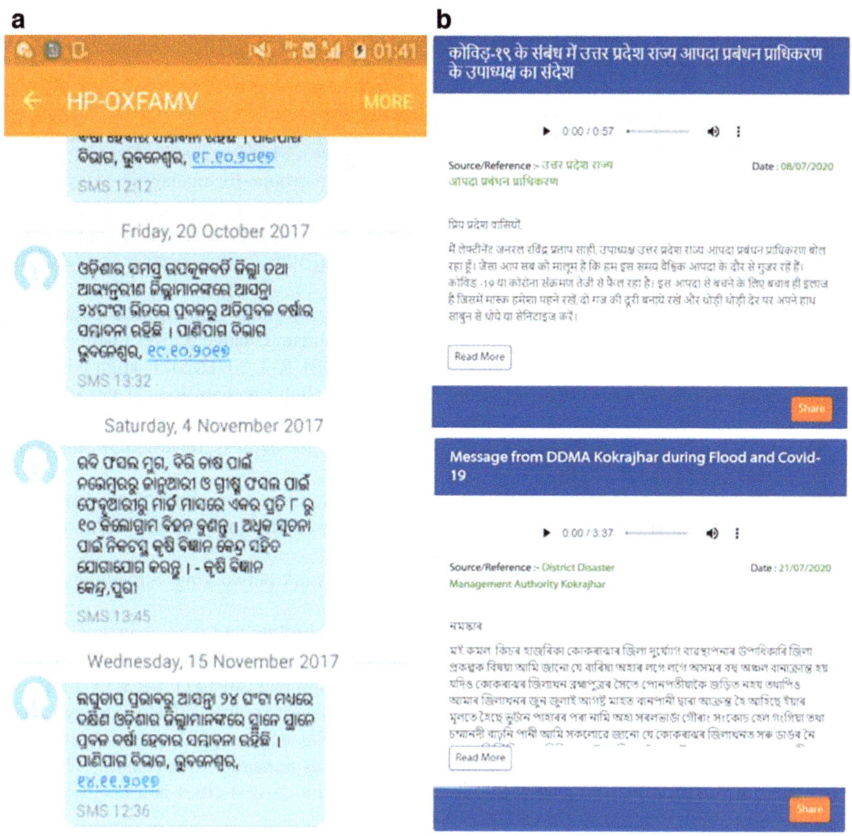

Fig. 7.2 **a** Text SMS (in Odia) through the VNMS in Odisha **b** Audio advisories and vernacular transcriptions from UPSDMA in Hindi (above) and DDMA Kokrajhar in Assamese (below) through Adhikar-Vaani

References

Andharia J (2020) Thinking about disasters: a call for intersectionality and transdisciplinary in disaster studies. In: Andharia J (ed) Disaster studies: exploring intersectionalities in disaster discourse. Springer Nature, Singapore.

Carbon Brief (2018) Mapped: how every part of the world has warmed—and could continue to. Infographics [14 Feb 2022]. https://www.carbonbrief.org/mapped-how-every-part-of-the-world-has-warmed-and-could-continue-to-warm

Eckstein D, Kunzel V, Schafer L (2021) Global climate risk index 2021. Briefing Paper, German Watch. https://www.germanwatch.org/en/19777 (English Version)

Frais D, Hager G et al (2022) Citizen science in environmental and ecological sciences. Nat Rev Methods Primers 2:Article number: 64. https://www.nature.com/articles/s43586-022-00144-4

Global Commission on Adaptation (2019) Adapt now: A global call for leadership on climate resilience. Report, p.14 [13 Sept 2019]. https://www.GlobalCommission_Report_FINAL.pdf (www.gca.org)

Hicks A, Barclay J et al (2019) Global mapping of citizen science projects for disaster risk reduction. Front Earth Sci, Section: Geohazards and Georisks 7. https://doi.org/10.3389/feart.2019.00226
ITU International Telecommunication Union (n.d.) 'Multi-hazard early warning system'. ITU-D Emergency Telecommunications. https://www.itu.int/en/ITU-D/Emergency-Telecommunications/Pages/Multi-hazard-early-warning-systems.aspx
Karmakar S (2019) Indo-Bhutan groups for managing cross-border rivers. Deccan Herald, 22 Jun. https://www.deccanherald.com/india/indo-bhutan-groups-for-managing-cross-border-rivers-742185.html
Krishnan R et al (2020) Assessment of climate change over the Indian region: a report of the Ministry of Earth Sciences (MoES). Government of India. https://www.springer.com/gp/book/9789811543265
MoEF (2021) India: third biennial update report to the United Nations framework convention on climate change. Ministry of Environment, Forest and Climate Change, Government of India. https://unfccc.int/sites/default/files/resource/INDIA_%20BUR-3_20.02.2021_High.pdf
MoES (2020) Statement on climate of India during 2020. India meteorological department—climate research and services. Ministry of Earth Sciences, Government of India. Press Release, 4 Jan 2021. https://mausam.imd.gov.in/backend/assets/press_release_pdf/Statement_of_Climate_of_India-2020.pdf
Mohanty D (2021) How Odisha's model of disaster preparedness came into being. Live-Mint/Hindustan Times, HT Media Limited, 26 May. https://www.preventionweb.net/news/how-odishas-model-disaster-preparedness-came-being
NERSWN (2005) About the Northeast research and social work networking. https://nerswn.org/about. Accessed May–Jul 2023
Paudel D et al (2013) Neonatal health in Nepal: analysis of absolute and relative inequalities and impact of current efforts to reduce neonatal mortality. BMC Public Health 13:Article No. 1239, 28 Dec. https://bmcpublichealth.biomedcentral.com/articles/10.1186/1471-2458-13-1239
Prakash A (2017) Mainstreaming DRR in local development plans to reduce aid dependency. Oxfam India, Blog, 7 Feb for more on Oxfam India' swore in Balasore through its partner Unnayan
Ramesh S (2018) Cyclone Amphan: why India's east coast is a magnet for cyclones. The Print, 13 Nov. https://theprint.in/science/cyclone-amphan-why-indias-east-coast-is-a-magnet-for-cyclones/148539/
Sentinel Digital Desk (2019) Bhutan-India CSO's partnership for inclusive water governance. The Sentinel, Assam, 21 Jun. https://www.sentinelassam.com/north-east-india-news/assam-news/001-bhutan-india-csos-partnership-for-inclusive-water-governance/
The Intergovernmental Panel on Climate Change prepared this Special Report (SROCC) As one of 3 during their Sixth Assessment Cycle, as decided in 2016. The other two Special Reports were the SR1.5 (on Global Warming of 1.5 degrees Celsius) and SRCCL (on Climate Change and Land). https://www.ipcc.ch/srocc/
The NERSWN. https://nerswn.org/
UNDP Press Release (2015) A roof still standing: building disaster resilience in Odisha, Jan. https://reliefweb.int/report/india/roof-still-standing-building-disaster-resilience-odisha
UN News (2021) IPCC report: 'code red' for human driven global heating, warns UN chief. Clim Environ, 9 Aug. https://news.un.org/en/story/2021/08/1097362. Accessed Apr–May 2023
Yashwant S (2018) Villagers in Bhutan and India come together to share river. The Third Pole, Regional Cooperation, 27 Aug. https://www.thethirdpole.net/en/regional-cooperation/villagers-in-bhutan-and-india-come-together-to-share-river

Animesh Prakash is a humanitarian professional with more than fifteen years of service with communities, Government, NGO and UN. He has experience across the spectrum of disaster management, including Climate Change Adaptation, Disaster Risk Reduction and Humanitarian

Response. He is skilled in partnership management, high impact programme delivery, innovation and influencing. His thematic focus of work includes community resilience and risk reduction, risk informed water governance, WASH and early warning. At a senior leadership level, he currently leads the disaster management unit at CARE India. He has previously held CCA, DRR and Humanitarian positions with UN World Food Programme, Oxfam and UNDP. He is an India Disaster Resilience Leadership (IDRL) fellow with TISS. He is recipient of awards for innovation in early warning and DRR from DCHI & Spindle in Netherlands, UNSG & MIT Climate CoLab and UN-OCHA. He has a professional degree in Disaster Management and Human Rights Law.

Pankaj Anand Pankaj is currently the Country Director of Oxfam in Timor-Leste. Prior to this, Pankaj had served Oxfam India for over seven years first as Director of Programme and Advocacy and later as Interim Chief Executive Officer (CEO). In his current role as Country Director, he heads Oxfam's work in Timor-Leste on justice issues relating to Climate, Gender and Economy. Overall, he has 25 years of rich experience cutting across humanitarian and development work. He is one of the editors of this volume and further details are given in the details on editors.

Chapter 8
Seven Times Deeper and Divided into Three: Re-identifying the Lost Focus of the Indus Waters Treaty

D. Saran Prakash

8.1 Introduction

The dearth of water is a matter of rising concern in many countries. The pivotal role that rivers play in energy generation, industries and environment preservation have made rivers crucial for development. For those South Asian countries which are largely driven by agricultural activities, the impact of environmental degradation and poor management of water resources sometimes lead to conflict. In this context, water treaties have exercised an important role and responsibility in peaceful resolution of water-related conflicts. The Indus Waters Treaty (IWT) between India and Pakistan is one example of major international river basin treaty. It is, in fact, quoted as one of the most successful water-sharing mechanism in the recent times.

In 1960, the IWT signed between India and Pakistan, midwifed by the World Bank, gave India three eastern rivers (Sutlej, Beas, and Ravi), whereas Pakistan received the control of the three western rivers (Indus, Chenab and Jhelum). The treaty is considered to have stood up to all its descriptions of being a mechanism providing a moderately reliable framework for the peaceful resolution of water-related dispute. However, the treaty defines absolutely no mechanism to encounter the present-day risks and challenges such as disasters or that of the climate change related impacts. The IWT in particular does not seem to encourage frequent interactions among the riparian countries. In 2010, Pakistan accused India of choking its water supply consistently. And, in 2018, Pakistan again alleged India of violating the IWT with regard to the Kishanganga Dam, and lodged a complaint with the World Bank. India is being called a hydro-hegemon and the paper tries to make meaning of this by discussing the contentions of both countries. It highlights the role of democracy and the implications of some of the political developments on transboundary water cooperation.

D. S. Prakash (✉)
Jamsetji Tata School of Disaster Studies, Tata Institute of Social Sciences, Mumbai, India
e-mail: saranprakashd@gmail.com

J. Andharia and P. Anand (eds.), *Shifting Lands, Flowing Waters*, Disaster Studies and Management, https://doi.org/10.1007/978-981-96-7691-0_8

The importance of democracy for cooperation in transboundary river treaties, is underlined to assess whether the Treaty is catering to the changing social and economic issues. The paper concludes with two critical questions—is IWT to be seen as an example of successful conflict resolution or a tool for conflict management? And in the wake of the abrogation of Article 370,[1] if India does withdraw from IWT, what impacts are to be anticipated for the riparian communities?

8.2 Transboundary Water Resources: Cooperation and Conflict

While conflicts over water may be traced back to thousands of years, importance over this concern has been pressed only in recent years. There are two major factors behind this, one is that media has exercised a significant role in highlighting that the war of the next century will be fought over water for national sovereignty and security. The second factor is the rising concern over water scarcity, particularly in arid and semi-arid regions. Resultantly, the discussions regarding water resources focus more on war threats over transboundary water conflicts, while there is little debate about cooperation over international river basins.

That being the case, what are the enabling factors which make these conflicts fall short of war? Does it have something to do with the power imbalance between co-riparians? In this context, Zeitoun and Warner (2008) argue that if there is no war over shared waters, it does not mean that there would be no conflict [either].

Another significant question to think about is: what makes the study of transboundary water resources more challenging and longitudinal in nature? Transboundary water resources are of two types: groundwater resources and surface water resources. It is comparatively harder to deal with water resources because water is continuously in motion: issues of control, jurisdiction and sovereignty are much more complicated as compared to static land resources and this task is even more difficult when water resources are shared by a number of political entities (Kliot et al. 2001). The Register of International Rivers, in 1978, listed 214 international transboundary river basins in the world (UN. Department of Technical Cooperation for Development. Natural Resources and Energy Division 1978). According to McCracken and Wolf (2019) no formal body has undertaken the delineation of the world's international river basins since 1978. Prior to the most recent update in 2018, 57 inventories of transboundary river basins had been conducted. For instance, Wolf et al. (1999) carried out an inventory in 1999, identifying and listing 261 international rivers, covering 45.3% of the Earth's land surface. Furthermore, Kliot et al. (2001) extended the number of transboundary river basins to 280. However, it is important to note that the latest update in 2018 reported a total of 310 international river basins, covering 47.1% of the Earth's land surface (excluding Antarctica), according to the

[1] Provisions of Article 370, which gave special status to the erstwhile state of Jammu and Kashmir, was abrogated by the Central Government of India on August 5, 2019 (The Economic Times 2019).

Transboundary Freshwater Dispute Database (TFDD) at Oregon State University (TFDD 2018).

Even though conflict over water has not exactly led to full-fledged war until now, conflict has been a tendency of human behaviour throughout history. Particularly since the twentieth century, water has been a significant feature of various conflicts rather than [serving] their sole purpose (Gleick 2006). However, with increasing pressures, especially from rapid population growth, urbanization and climate change, some scholars have argued that this century may in fact witness a birth of full-scale wars for fresh water (Gleick 1993).

Most of the tensions appear especially in parts of southern and central Asia, central Europe and the Middle East where the evidence of water-related conflicts extends back to 5000 years. The probability of increased water related conflicts evolving into an issue of high politics in these areas is soaring, because water is a scarce resource that is important for economic and agricultural development (Gleick 1993). Zeitoun and Warner (2008) state that control over shared water has taken the shape of politics because fresh water is vital for a healthy economic growth and it is not new or much popular term among the water practitioners. Cooper et al. (1999) discuss the roles of upstream and downstream riparians in the shared water resources in Asia, Africa, and South America that hydropolitics in these areas already started from the concern of the downstream riparian about the actions of upstream riparian that may reduce the quantity or quality of the water in the shared water resource. Allan (2001) and Wolf (1995) discuss the two main factors, fast-growing population and economic development, which would make the water issue more political and controversial. Wallace Stegner in Gleick (1993) notes that if you get control over water, you can easily get hold of that land which depends upon it.

8.3 Hydropower Development, Treaties and Disputes

Hydropower development has contributed to one of the largest expansions of dams seen in history through the 1970s. According to some forecasts, as many hydropower dams will be built in the next three decades as were built in the last century, essentially doubling global hydropower capacity, this is a scenario that needs to be watched carefully. Emerging economies, in particular, are under extraordinary pressure to harness the power offered by their natural resources (Opperman et al. 2015).

The quantity of water available in the world, has to meet the rising demands over time, as the competition among water users intensifies. This is nowhere more destabilizing than in river basins that stretch across political boundaries.[2] Nevertheless, experience holds testimony to the fact that in many situations, rather than inviting an open conflict, the need for water sharing can generate unexpected cooperation.

[2] According to an earlier definition by National Geographic, political boundary is an imaginary line separating one political unit, such as a country or a state, from another. This has been updated and can be found at https://education.nationalgeographic.org/resource/boundary/.

In spite of the complex problems, records show that water disputes can be handled diplomatically. In the past half a century, the world has seen only 37 acute disputes involving violence, compared to 150 treaties that have been signed. The history of international water treaties dates as far back as 2500BC, when the two Sumerian city-states of Lagash and Umma crafted an agreement ending a water dispute along the Tigris River which is often said to be the first treaty of any kind.

Furthermore, legal agreements on water sharing have been negotiated and maintained even as conflicts have persisted over other issues. For example, UN (2014) mentions that Cambodia, Laos, Thailand and Vietnam, have been able to cooperate since 1957 within the framework of the Mekong River Commission, and they had technical exchanges throughout the Vietnam War. Similarly, since 1955 Israel and Jordan, have held regular talks on the sharing of the Jordan River, even as they were until recently, in a legal state of war. A framework for the Nile River Basin, home to 160 million people and shared among 10 countries, was agreed in February 1999 in order to fight poverty and spur economic development in the region by promoting equitable use of, and benefits from, common water resources. The nine Niger River Basin countries have agreed on a framework for a similar partnership. Each of these cases reflects two important elements of international water resources cooperation: the requirement of an institution which can effectively develop and regulate the process of engagement over time; and well-funded third-party support trusted by all factions.

Clearly, international treaties have proven to be fairly helpful in preventing the intensification of conflicts over water resources. However, in current times, for South Asian countries which are largely driven by agricultural activities, it is observed that the impact of environmental degradation and poor management of water resources sometimes lead to conflict. India and Pakistan are a case in point and this paper focuses on the Indus River treaty.

8.4 Indus River System and the Treaty

8.4.1 India-Pakistan Relations: History of Indus River System

Agriculture is known to be the lifeblood of Pakistan's economy because it contributes to about 18.5% of GDP, and provides livelihood to 64% rural inhabitants and employs 38.5% of the total national labour force. The sector is also directly and indirectly linked with other sectors of the economy and plays a significant role in socio-economic development of the country (Abedullah 2020). At the same time, Pakistan's agriculture largely employs flood methods of irrigation to grow water intensive crops. Despite having more glaciers than anywhere else in the world and its location in the Indus River Basin, Pakistan is at the risk of acute water scarcity. Its surface and groundwater sources are both increasingly stressed and severe drought conditions persist in parts of the country due to lack of rain (Sleet 2019).

Similarly, the rising demands of water resources by India's increasing population and diminishing quality of existing water resources due to pollution and additional requirements of serving India's spiralling industrial and agricultural growth, have led to a situation where the consumption of water is rapidly increasing while the supply of fresh water remains more or less constant (Dhawan 2017). Since the census of 2011, India entered the league of water deficient nations (Sen 2018). And, agriculture, which is India's primary economic activity is directly linked to the availability of water. Krishnan (2018) pointed out that, according to several studies, 54% of India's agricultural land is battling water scarcity. Most of the land relies directly on the country's water resources. A Government study reported that 15 years is all that it will take for India to reach the level of distress where the country will have only half amount of water needed for the purpose of survival (QRIUS 2018).

Historically when the British left India, the country was torn apart and partitioned and Pakistan was created. Critical water resources that had been shared in British India, now needed a framework of cooperation. Globally, in similar instances partition became the source of transnational water conflicts owing to the suddenness, the arbitrariness and the brutality with which a new country is created. The Indus River System emerged as one of the most contentious one due to the prevailing socio-political and religious tensions in the region, between India and Pakistan.

The title of this paper draws from the popular fable of the Indus River (Kohari 2014). It is believed that when the water first arrived in valley where the mighty Indus River courses through, panic gripped the villagers who had deep attachment to their homelands. As the waters surged through the valley, they faced a Hobson's choice—flee or find refuge in the trees as the water inundated their lands. The devastating impact of the river's power left them with no option but to abandon their belongings and ancestral lands in response to the torrential flow. The tale passed down from generations, indicates that the Indus River was seven times deeper than the length of a rope used in a *charpoy* (a traditional cot with a wooden frame and woven with coir). It symbolizes the profound reverence and awe they hold for the mighty river. As the story unfolds, the river's depth becomes a metaphor for the complexities and power of transboundary water resources that can shape the fate of communities.

Geographically, India, which is the largest country in South Asia,[3] shares its river waters with four South Asian countries-Pakistan, Bangladesh, Nepal, and Bhutan. Located in the north western region of India, Pakistan shares the rivers of the Indus Basin with India. The Indus Basin system is composed of mainly six rivers namely, Indus, Jhelum, Chenab, Ravi, Sutlej, and Beas. These rivers are perennial in nature, which make the Indus plains ideal for the building of large canal irrigation system (Ahmad 2011). And that is why, the Britishers began here the construction of the largest canal networks in the world. During the British Rule this network was manageable under one central command and with the partition in 1947, the Indus basin was divided between two sovereign independent states. The boundary line between the

[3] https://www.britannica.com/place/india.

Fig. 8.1 Map of the Indus River Basin showing major tributaries (Indus, Jhelum, Chenab, Ravi, Beas, and Sutlej) and key cities across Pakistan, India, Afghanistan, and China, illustrating the transition from northern mountains to southern plains. *Source* Pepper (2018), Map of river Indus and its flow through India along with its tributaries (retrieved from Wikimedia Commons, https://commons.wikimedia.org/w/index.php?curid=66085475)

two states was drawn right across the Indus Basin, rendering Pakistan the status of being the lower riparian state. Of the 37 million acres of land irrigated by the canals from the Indus and its tributaries, over 30 million acres fell in Pakistan (Ahmad 2011). See Fig. 8.1.

The Kashmir Valley is the head of the three largest tributaries: Indus, Jhelum, and Chenab Rivers. The first of these two form the Kashmir Valley, and the latter reaches Indian Punjab after going through Kashmir. The other three-Ravi, Sutlej, and Beas, flow from the valleys of Himachal Pradesh, and later join the Punjab water networks. Evidently, all Pakistani waters rise across the Line of Control in Kashmir, where India happens to bask in the status of being the upper riparian (Wirsing 2008).

When Pakistan perceived its livelihood threatened by the prospect of Indian control over the tributaries feeding the Pakistani portion of the Indus Basin, the two governments signed a Standstill Agreement in December 1947 to maintain pre-Partition water flows (Biswas 1992). According to this Agreement, India was not to restrict the natural flow of rivers into Pakistan from the headworks located in India. However, after the Agreement expired in April 1948, India stopped the flow of the Doab and Dipalpur canals into Pakistan (The Round Table 2008). This action highlighted Pakistan's dependency on India for water, particularly in the context of the Kashmir dispute, and triggered a serious water dispute that significantly heightened tensions between the two countries.

8.4.2 The Indus Waters Treaty, 1960

It was in this backdrop of tension and skirmishes over disputed waters and failed attempts at carrying out peaceful dialogues between the two states at loggerheads that the Indus Waters Treaty (IWT) was incubated towards an equitable and peaceful outcome between two neighbouring newly born sovereign countries of India and Pakistan. A prominent name responsible for the idea of an IWT is that of Mr. David Lilienthal,[4] the former chairperson of Tennessee Valley Authority and of the U.S. Atomic Energy Commission. The words of this Mr. Lilienthal in the year 1951 concretized the grounding for the Indus Water Treaty (cited in Gulhati 1973):

> India and Pakistan were on the verge of war over Kashmir. There seemed to be no possibility of negotiating this issue until tensions abated. One way to reduce hostility... would be to concentrate on other issues where cooperation was possible. Progress in these areas would promote a sense of community between the two nations which might, in time, lead to a Kashmir settlement. Accordingly, I proposed that India and Pakistan work out a program jointly to develop and jointly to operate the Indus Basin River system, upon which both nations were dependent for irrigation. With new damns and irrigation canals, the Indus and its tributaries could be made to yield additional water each country needed for increased food production. In the article, I had suggested that the World Bank might use its good offices to bring the parties to agreement, and help in the financing of an Indus Development program. (Gulhati 1973, p. 93)

Lilienthal's recommendation was received well by the World Bank, and the respective governments of India and Pakistan. Following a study of two years by a technical group comprising engineers from India, Pakistan and the World Bank, that aimed at preparing a comprehensive plan for development on a joint basis, in the year 1954, General Wheeler was appointed by the World Bank to put together a proposal for the same, to be considered by both governments. It contained the following elements:

(i). The waters of the three Eastern Rivers (Ravi, Beas, and Sutlej) should be for the use by India;
(ii). The waters of the three Western Rivers (Indus, Jhelum, and Chenab) should be for the use by Pakistan;
(iii). There should be a transition period, during which, Pakistan would construct a system of link canals to transfer waters from Western Rivers to replace the irrigation uses in Pakistan, hitherto met from Eastern Rivers; and
(iv). India should pay the cost of constructing these replacement canals.

With a few reservations, the proposal was accepted by India. However, Pakistan would not accept the proposal unless substantial changes were made regarding the reservoir storage in the replacement plan to meet irrigation uses in Pakistan. Then followed almost four years of negotiations in Washington between the Indian Delegation which was led by Mr. N. D. Gulhati[5] who was the Additional Secretary to the Government of India, Ministry of Irrigation and Power, and the Pakistani Delegation led by Mr. G. Mueenuddin, who was Secretary to the Government of Pakistan,

---

[4] https://www.atomicheritage.org/profile/david-e-lilienthal.

[5] https://www.icid.org/about_ndg.html.

Ministry of Fuel, Power and Natural Resources (The World Bank 1960). The World Bank was represented by Mr. Iliff, the then Vice-President of the International Bank for Reconstruction and Development (IBRD) to sign the Treaty on behalf of the World Bank. He was assisted by a small group of technical experts led by General Raymond A. Wheeler, a senior World Bank engineer.

In May 1959, in the course of the visit by the delegation of the World Bank to New Delhi and Karachi, agreement was reached on the general principles on which, the water treaty was to be rested. This included the system of works to be constructed as part of the settlement arrangements and the financial contribution to be made by India. In the process, it became definite that the cost of financing the system of works in India and in Pakistan was way higher than the capacity of India and Pakistan. Therefore, the World Bank shouldered the responsibility of formulating the plan envisaging financial participation by a number of other friendly governments interested in contributing to bringing about a settlement to this troublesome and contentious water dispute. Finally, in late August 1960 the texts of the Indus Water Treaty and of the Indus Basin Development Fund Agreement were finalized, and on 19 September 1960, in Karachi, the Indus Water Treaty was officially signed with 12 Articles and 8 Appendices.[6]

8.4.3 A Brief Analysis of the Treaty

In a press release issued on 19 September 1960, the World Bank recognized the significance of the Treaty (The World Bank 1960). "*[the treaty] marks the end of a critical and long-standing dispute and opens the way for the development of water resources in the Indus basin, on which depends the livelihood of some 50 million in India and Pakistan*" (p. 1). The Treaty was also appreciated by the leaders of both countries (Ahmad 2011):

> The Treaty was an event of great historic significance for the two countries concerned… and for the whole world.
>
> —Ayub Khan, President of Pakistan

> We are going to have many benefits out of this agreement but important than the material advantages are the psychological and emotional benefit[s].
>
> —Pt. Jawaharlal Nehru, Prime Minister of India

Also, a Permanent Indus Commission was set up composed of two persons—one appointed by the Indian Government and the other one by the Pakistani Government. The Commission was to be responsible for the following tasks:

1. Implementing the provisions of the Treaty and to seek to reconcile any points of disagreement;

[6] https://treaties.un.org/doc/Publication/UNTs/Volume%20419/volume-419-I-6032-English.pdf.

2. Making a general tour of inspection of all the works on rivers, once every five years.
3. On the request of either of the Commissioners, the Commission was to visit any particular work in either country;
4. Reporting at least once each year to each Government; and

Where differences and disputes could not be resolved by agreement between the Commissioners, the Treaty established a machinery for resort to a "Neutral Expert" for a final decision on technical questions and for resort, in certain circumstances to a Court of Arbitration.[7]

The Indus River system, originally comprising six rivers, was divided into three rivers each for India and Pakistan through the Indus Waters Treaty. This division, prompted by the desire to harness and manage the river's waters, aimed to balance the interests of both nations and prevent outright conflict over the shared resource. The bilateral pact of the IWT has been hailed as an exemplary model of conflict resolution in international politics. The Treaty has ridden out a series of hostilities between India and Pakistan, including the full-scale wars of 1965 and 1971, and the more limited war of Kargil in 1999. All this was, in fact taken as an indication for continued cooperation despite escalated tensions and deteriorating relations.

8.5 Mixed Reception—Is India a "Hydro-Hegemon?"

Two months after the India-Pakistan independence, Kashmir remained independent. What is the Kashmir conflict all about? In 1947, the countries of India and Pakistan came into existence, the partition decided the fate of the millions and various state rulers had to choose which country they would belong to. However, the people of Kashmir faced an uncertain future. Maharaja Hari Singh, the Hindu ruler of Muslim-majority Kashmir opted out of choosing either India or Pakistan and remained independent for two months. He signed a "Standstill Agreement" with Pakistan and sent a similar agreement to India, which favoured continuing discussions. But that changed in October 1947, when the Maharaja confiscated armaments from Muslims who were formerly conscripted into British Army. The weapons were distributed to local Hindu village defence forces and provoked an uprising supported by the Pashtun tribesmen from Pakistan. The tribesman was believed to be acting without Pakistan's formal approval. The Maharaja sought military assistance from India; but India's Governor General contended that it would be dangerous to send troops to a neutral state unless Kashmir first offered to accede to it. To supress the uprising, the Hindu ruler's army along with the RSS (a Hindu extremist organization) organized a pogrom against Muslims in Jammu. There is a dispute over the number of Muslims that were massacred; but it could be anywhere between 20,000 to over 100,000 (Saraf 2015). This massacre was a part of widespread violence in the subcontinent in which reportedly 20,000 Hindus and Sikhs were also killed. The Hindu rulers signed a temporary

[7] https://mea.gov.in/bilateral-documents.htm?dtl/6439/Indus.

arrangement ceding Kashmir to India on October 26, 1947. Pakistan disputed the Maharaja's accession claiming that he has no right to sign an agreement with India when a standstill agreement with Pakistan was still in force. On 27 October 1947, Indian troops landed in Kashmir to fight the rebel forces and the Pashtun tribesmen. This led to the first India-Pakistan war. During the war, India's then Prime Minister, Pandit Jawaharlal Nehru promised a referendum, in which the people of the state of Kashmir would be given the authority to decide their fate. A little over two months later, India referred the dispute the United Nations. A Resolution was passed on 13 August 1948 asking both nations to withdraw their forces. Once that happened, a referendum was to be held, allowing the people of Kashmir to decide their political future. But troops were never withdrawn and the referendum never happened. On 1 January 1949, a ceasefire was agreed upon and Kashmir became a disputed territory. A ceasefire line became a de facto border splitting Kashmir in two. Over the following seventy years, India and Pakistan fought three wars making it one of the most militarized zone in the world.[8]

For several reasons, despite the IWT being a bilateral Treaty between the two co-riparians, India has often been accused of being a hydro-hegemon (Hanasz 2014). A basic understanding of the concept of hydro-hegemony suggests that power asymmetry may lead to an inequitable outcome that will result in low intensity conflict to form positive/negative hegemony; further the hydro-hegemon country uses different strategies like military force, coercion pressure, treaties, incentives, securitization and sanctioned discourse to form the interaction over transboundary waters that it prefers (Zeitoun and Warner 2006).

Wolf (1998) discusses the pros and cons of being up-streamer and down-streamer. The author explains that neither of these are in the position to wage a war over water. This is because on the one hand, up-streamer may already be in a better position by location to have control over water resources through different tactics and strategies. And, on the other hand, it would be an imprudence of the down-streamer to attack over the storage or quality related projects of up-streamer because it would result in a wall of water rushing back or result in even worse quality of water than before for the down-streamer.

While the people in general, of both countries, welcomed the Indus Water Treaty, it was also criticized in some quarters within both countries. According to the critics in India, New Delhi had surrendered its rivers to Pakistan. The fact India was to shoulder the cost of replacement works was perceived as an unfair burden. Similarly,

[8] As of 6 August 2019, the Government of India made significant changes to the status of Jammu and Kashmir, a region that has been a subject of dispute among India, Pakistan, and China since 1947. The government revoked the special status, or autonomy, that had been granted to the state under Article 370 of the Indian Constitution. Prior to the revocation, Article 370 provided a unique status to Jammu and Kashmir, granting it considerable autonomy in various matters except for defense, communications, finance, and foreign affairs. It also allowed the state to have its own constitution and flag. However, on 6 August 2019, the government decided to alter this arrangement and reorganize the state into two separate union territories: Jammu and Kashmir, and Ladakh. Jammu and Kashmir became a union territory with its own legislative assembly, while Ladakh became a separate union territory without a legislative assembly. The reorganization aimed to bring these regions under the direct governance of the Indian central government (BBC 2019).

in Pakistan, resentment was expressed on the loss of three Eastern rivers (Ahmad 2011). As various perceptions and analysis on the Treaty have evolved, a detailed scrutiny of the debates read in the context of socio-political realities and environmental considerations, a few pitfalls have been identified, which will shed light on the reason behind perceived role of India being a hydro-hegemon.

Firstly, since even before the Treaty was brought into existence, the three rivers—Sutlej, Beas and Ravi, were known to be the part of a conventional "*sailab*[9]" irrigation. Subsequent developments on these rivers by India disrupted sailab irrigation, as a result of which Pakistan experienced a fluctuating irrigation system and continues to face its effects. This drying up of the *sailab* is in fact seen as India exercising an undue might as the upper riparian.

Secondly, this inconsistency in the flow of Eastern Rivers as a result of the Indian control over them has become an axis point in the hydro-political dispute between Pakistan and India because the tributaries on the Pakistani end of the Eastern Rivers have dried up. India floods them only according to its need, creating a problem for Pakistan.

Thirdly, it is to be known that a storage reservoir has a smaller life than the natural flow of water and storage creation also, happens to be politically motivated (Khalid 2013). The Treaty, in this regard, does not allow for the mutual development of the Indus Basin. It rather thickens the conflicts and treats both parties as adversaries (Kokab and Nawaz 2013).

What is also unsettling is the fact that according to the Treaty, the allocation of water resources is based on quantitative parameters. This is unusual because generally, such allocations should be based on tributary location. While the Treaty states which tributary is dedicated for which nation there is close to no mention or provision regarding the mechanisms that dictate the usage of groundwater. As a result, complications arising because of climate change and the resultant changes in domestic demand, remain or in fact deepen the conflicting situation.

Further, water has become a "strategic tool" of diplomatic policy between the two countries. What it instead should be, is a matter of concern for the governments to ensure the sustainability of the Indus Basin and the riparian communities. Therefore, based on the text of the IWT, the outcomes tilt in the favour of only one nation, thus making India appear as the hydro-hegemon. Another important area of concern is the provision of use of western rivers by India for non-irrigation ends. India, as the upper riparian, can exercise the right to monitor and make amends to the waters of the Indus basin. This concern received its momentum from the construction of the Tulbul Navigation Project by India in the year 1984. Further, the construction of Kishanganga project of 2007 has brought the validity of the IWT under haze because the project was just 160 kms upstream of Muzaffarabad in Pakistan.

When Baglihar Project was conceived in the year 1992, to be begun on Chenab River in Doda district, in the Chenab valley in India, the IWT was brought under

[9] The oldest method of irrigation in Punjab is flood irrigation, locally known as *sailab*. In this method which is restricted to the active flood plains, crops (mainly wheat) are planted after the recession of the summer flood waters (Arnow 1961).

major risk. Under the Article IX, following Pakistan's complaints, a technical team headed by Raymond Lafitte, a Swiss neutral expert, was appointed in 2005, to look into this matter.[10] In order to come to a resolution, the Vienna Convention on the Law of Treaties (1969)[11] was applied. The analysis of the situation was done with the "rules of science" and 'state of the art practices'. This meant that, all of a sudden, the IWT was interpreted in the light of new standards and norms; and that, the Treaty was to adjust as per the emergent realities of the Indus Basin (Akhtar 2010). The support of the team rested in India's favour. This indicated a new interpretation of the Treaty: the construction of dams and hydroelectric projects by India, if anything, according to Pakistan, displayed the strong position of India as the upper riparian, thus strengthening Pakistan's narrative of hydro-hegemony.

Post this judgement followed a critical realization that whenever India decides to launch a project anywhere in the Indus Basin, a neutral expert is assigned for judgement and amendments to be made, if any. Obviously, as the legal procedures take time, India gets done with the formal construction process. This has led to a sense of victimization and helplessness in Pakistan, and [rightful] doubts hovering around the IWT (Bukhari 2011).

The only time when a judgement by the neutral expert tilted clearly in favour of Pakistan was when construction of 330 MW project on Kishanganga River was announced and India donned a strong violation of Article IV and Article VII of the IWT. The said Articles restrain India from expanding its catchment area in any natural/artificial drainage. Kishanganga Hydroelectric Plant Project, aiming to increase the catchment area, was clearly bypassing the content of these Articles. Following Pakistan's complaint, the construction of the Plant and the progress of the Project was halted by the Hague Permanent Court of Arbitration in 2011. Nevertheless, this move of India did come across as a prime example of hydro-hegemony whereby one country's hydroelectric project had the potential to hamper the hydroelectric setup of another state.

On the other side, Brahma Chellaney, a professor of strategic studies at CPR,[12] brings to light the flip side of the hydro-hegemony argument. He points out that, India is anything but a hydro-hegemon. Elaborating on this standpoint, he offers three reasons:

1. India holds a unique status as a riparian. This means that, it is the only country in South Asia which falls in all three categories—an upper riparian, a middle riparian and a lower riparian. This means that India has a direct stake in all the significant river basins in South Asia.
2. Asia has a total of 57 transnational water basins (Chellaney 2014). However, the only treaties in Asia with specifically laid out sharing formulas on cross-border river flows are the ones over which India has signed, with its two lower riparian neighbours—Pakistan and Bangladesh. Each of these two treaties brings new

[10] http://www.jkspdc.nic.in/beta/baghlihar_power_project.html.

[11] https://treaties.un.org/doc/Publication/UNTS/Volume%201155/volume-1155-I-18232-English.pdf.

[12] https://www.cprindia.org/people/brahma-chellaney.

principles of international water law. While the 1966 the Ganges Treaty with Bangladesh set a new standard by guaranteeing the delivery of specific water quantities in critical rice season, the IWT for its part stands out as the world's most generous water-sharing arrangement.

3. It is to be highlighted that India indeed stands in a position of being potentially affected by water-related actions of upstream countries especially China and Nepal. In fact, no other country in South Asia is more vulnerable to China's reengineering transboundary flows than India, because it alone receives half of all waters that flow from the territories which are controlled by China.[13]

Therefore, these points, highlight the other side of the discourse of India's hydro-hegemony. While this debate has been going on, what is of more significance currently is to understand that India and Pakistan, having survived over 70 years of hot and cold relationship, what is the actual impact on the IWT?

8.6 What Is the Survival of Indus Waters Treaty Dependent Upon, Today?

The literature centred around conflicts over water resources between the riparians in general underscores that water holds the capacity of triggering violent and social conflict. However, the level and the intensity at which such conflicts may arise is still unclear. With increasing concerns that climate change is exercising on water resources, many hydrologists, climatologists and water experts have evaluated the risks of prospective water wars with reference to historical evidence. The debate now requires a focus on impacts that hydro conflicts and consequent moves of the parties involved, will have on the riparian communities.

With increasing strain on the availability of water in Pakistan and in India, the IWT too has come under stress, with some questioning its very existence. While this is mainly driven by the growing demand and decreasing availability of fresh water resources, and the degree of dependence of riparian communities on the trans boundary water resources, the Treaty today, must be understood also in the context of the politics of current times, which seem to dictate the fate of the Treaty and that of the riparian communities as well.

An analysis of the developments over the years, shows that the Treaty has served as a tool of political division as opposed to being a mode of equitable distribution of

[13] India's concerns regarding water-related actions of upstream countries, particularly China and Nepal, are significant due to the country's vulnerability to potential impacts on transboundary flows. Among the countries in South Asia, India is uniquely affected by China's reengineering of water flows because it receives approximately half of all waters that originate in territories controlled by China (Chellaney 2019). China has been undertaking various water infrastructure projects in its territories, including dam construction, river diversions, and hydropower developments (Wenjing et al. 2022). These activities have the potential to influence the downstream flow of rivers that originate in China and flow into India, particularly in regions like the Brahmaputra and the Indus River basins (Giordano and Wahal 2022).

water resources, as conceived originally, and perhaps also with the potential of aiming for conflict resolution. This is to say that, both India and Pakistan have been using it to further their propaganda and respective nationalist agendas, thus creating a huge gap to be filled with integrated approaches for cooperation and conflict management, and later, resolution. Not only has this strained the diplomatic relationships between the two countries even further, but it has also exercised an unfavourable impact by not being able to deal with rising concerns brought about by climate change (Mustafa 2010).

It appears clear that while the IWT is appreciated by many for having provided a way of reconciliation for the two countries in conflict, there is a long room for improvement to bring an ultimate conflict resolution to pass. Why is this so? India's benefits as the upper riparian have led to Pakistan claiming India's violation of the rules of IWT. Not only has this brought a lot questions on the Treaty, but it has also brought the very existence of the Treaty under jeopardy.

From one perspective, the survival of the Treaty can be said to be based on the modifications in agriculture and irrigation communities of both Pakistani Punjab and Indian Punjab. Both sides of Punjab grow food products mostly for the purpose of exporting to overseas markets. Water intensive agricultural products like rice are exported. As they do this, they are, in a way exporting water in virtual form and putting their water resources under strain. Though fundamentally, Pakistan needs to restructure its economy, what should not be missed out here is the fact that Pakistan is growing food even for water rich regions. In this context, the European Union is the top buyer of Pakistani rice. Agriculture accounts to 94% of all the water consumed in Pakistan. This is already creating growing water stress in Pakistan.

In addition, the survival of the IWT is linked to political factors also. In the year 2018, following the Uri attacks of 2016 (Negi 2016), the Indian Prime Minister Narendra Modi made the statement, "*blood and water cannot flow together*" (Express Web Desk 2016). This was to mean that, in the light of violence that erupted between the two countries, on the border, expecting that India will continue to remain adversely unaffected, is an unrealistic assumption, especially with regard to sharing and allocations of the Indus waters. Here, the key issue between the two countries is, in fact, that of state-sponsored terrorism by Pakistan against India.

Two things are crucial for the survival of any water treaty in general: rights and obligations. In any treaty, the comparative benefits and burdens should be such that the advantages for each party must outweigh the burdens. Or else, the country that sees itself as a victim and as weaker, may fail to comply with its obligations, or completely withdraw from the treaty.

With regard to the signing of the IWT, other than the contention that it was signed by the Prime Minister of India, instead of the President of India who is the Head of State (Goradia 2019), what was also problematic was the absence of democratic involvement of the Kashmiris, who were to be affected by this Treaty.

8.7 Democracy and the Missing Kashmiris

The three western rivers, mentioned earlier—Indus, Chenab and Jhelum are reserved for the use of Pakistan flow through Kashmir. A great deal of politics that has become a part of the IWT is in fact about the territory of the erstwhile state of Jammu and Kashmir and India's and Pakistan's claims on the territory.

Chellaney (2019) notes in a podcast interview that, basin-wide collaboration is not possible with any kind and level of autocracy (Svensson & Chellaney, 2019). There has been considerable amount of displeasure in Kashmir, since a long time, about India negotiating on these rivers without consulting the local state government. The IWT completely avoided discussing the disputed region of Kashmir in the negotiations and resultantly, the final draft took the shape that it did. However, the bulk of the Treaty directly affects the physical land mass of Kashmir and thereby, the Kashmiri people. For that matter, the lack of representation provided a push for the Jammu and Kashmir Assembly to initiate a resolution in the year 2003, rejecting the Indus Water Treaty, and also hire a consultant to quantify their losses (Saadat 2018).

Further, the concern relating to the democratic participation involving the people of Kashmir in general, and the riparian communities of both sides of the border in particular, needs to be discussed in the light of newly emerged haze over the sustainability of the Treaty, with regard to the more recent abrogation of the Article 370, and the declaration of Jammu and Kashmir, and Ladakh as new Union Territories, and full-fledged parts of India. This brought into popular doubt whether India will possibly withdraw from IWT. The question to ask here is, when the Treaty itself is a bilateral one, can India unilaterally withdraw from the Treaty. Is that even a possibility?

8.8 Can India Withdraw from IWT?

In this matter, the Treaty per se has not laid down any such provision for either country to unilaterally walk out of the pact. Article XII of the Treaty states:

> The provisions of this Treaty, or, the provisions of this Treaty as modified under the provisions of Paragraph (3), shall continue in force until terminated by a duly ratified treaty concluded for that purpose between the two governments.

In this case, if India still moves ahead to withdraw itself from the Treaty, it is required to abide by the 1969 Vienna Convention on the Law of Treaties. Also, and nevertheless, what India can do is reduce the flow of water to Pakistan by utilizing the provisions of the Treaty. But, with drastic measures, also come along repercussions. Factually, 80% of Pakistan's 21.5 million hectares of farmland is irrigated by river Indus and its tributaries (Stewart et al. 2018). Reducing the water flow of rivers to Pakistan will indeed trigger havoc in Punjab and Sindh provinces of Pakistan. Besides, India, at present, does not possess sufficient infrastructure to use the additional water available. This means that, any attempt at stopping waters to Pakistan would result in flooding areas in Kashmir, leading to disastrous consequences. Politically, this move of India may further aggravate the elements of terror in the country.[14]

8.9 Abrogation of Article 370

As mentioned earlier, the sharing of water of Jammu and Kashmir with Pakistan by India under the IWT had fostered popular grievance in Jammu, Kashmir and Ladakh. The abrogation of Article 370 by the Indian government, and the consequent sudden backlash from underdevelopment in Jammu and Kashmir has in fact prompted India to empower on several modestly sized run-up rivers, hydro-power projects in Jammu and Kashmir to address chronic shortages there. Run-up river hydro-power projects are permitted in the Indus Treaty. In this sense, can India, without actually withdrawing from the Treaty and without violating any of the terms, ensure that water availability and sharing with its neighbour does not hamper the economic development of Jammu and Kashmir? This remains a contentious question.

8.10 Conclusion and Scope Ahead

> Fierce national competition over water resources has prompted fears that water issues contain the seeds of violent conflict. …If all the world's peoples work together, a secure and sustainable water future can be ours.[15]
>
> —Kofi Annan, Seventh Secretary-General of the United Nations

So far, this paper, has articulated the enabling factors that brought the IWT into existence. Having presented the nature of status of each country party to the Treaty,

---

[14] This is in reference to the longstanding political and territorial disputes between India and Pakistan, particularly over the region of Kashmir. Kashmir has been a contentious area, claimed by both countries in its entirety. This dispute has led to armed conflicts, insurgencies, and acts of terrorism perpetrated by various groups. If India were to take measures to reduce water flow to Pakistan, it could be interpreted as an aggressive move or an act of hostility, further escalating the existing tensions between the two nations. In response to such actions, there is a possibility that extremist and terrorist groups within Pakistan may exploit the situation to advance their agendas. These groups, which may have links or sympathies with elements of the Pakistani establishment, could use the water dispute as a pretext to incite violence and destabilize the region. They may attempt to launch attacks within India or in the disputed territory of Kashmir to retaliate against what they perceive as an attack on their interests or their country. Moreover, the move could also trigger anti-India sentiments among some sections of the Pakistani population, potentially leading to protests and unrest. This could create a fertile ground for extremist recruitment and radicalization, further exacerbating the terrorism threat in the region. Therefore, this means that any provocative action regarding water flow by India could be exploited by terrorist groups and may lead to a surge in violence and terrorism-related activities in the region. The situation demands careful handling and diplomacy to avoid further exacerbating the already complex political and security landscape.

[15] https://www.un.org/press/en/2002/sgsm8139.doc.htm.

and the contentions around India being called a hydro-hegemon, the reasons which should be sustaining IWT even today, has brought the very crucial factor of the role of democracy in this Treaty. Having left the discussion of the deeper impact of the abrogation of Article 370 by the Indian Government, for further scope of study, this paper now moves towards the concluding section.

How much significance and what relevance, after all, does the IWT hold in current times especially against the backdrop of political developments? One way of addressing this concern is by attempting to comprehend that the inadequacy of IWT, if it were to fall into place, would have come from the dearth of focus on challenges arising out of climate change, uncertain environmental flows, and the exigencies of growing demand. The IWT, drafted nearly 8 decades ago, lays out guidance in the event of adverse climate effects, like flooding. In such a situation, each country must inform the other country beforehand if it holds any information or confirmation of excessive flows of water in the river basins owing to abnormal rainfall or similar climate change concern. Paragraph 8 of Article IV of the IWT states the same in the following words:

> Each Party agrees to communicate to the other Party, as far in advance as practicable, any information it may have in regard to such extraordinary discharges of water from reservoirs and flood flows as may affect the other Party.

In order to mitigate the potential risks hovering on the riparian communities, it is needed that such communication is made a norm between India and Pakistan. It is for this purpose that, agile, trustworthy and effective cooperation is established between these co-riparians.

However, this is not sufficient. With the increasing challenges posed by climate change, the complications too have multiplied. It should be highlighted that, the IWT was signed in 1951, and the Environmental Impact Assessment (EIA) came into force in 1972 through Principle 21 of the Stockholm Declaration of 1972.[16] As a result, several challenges and risks related to the availability of water in the river basin and the impact of climate change were not addressed in the originally signed Agreement. Therefore, it becomes pressingly relevant that the terms of IWT are renegotiated keeping in view the challenges of the current scenario. For that to happen, it is required that a stable institutional structure is brought into existence, comprising a combination of experts of (for example) geo-politics, disaster studies, hydrology etc., who can address the rising concerns adequately and monitor the situation continuously in a scientific manner, keeping in focus disaster risks, disaster prevention and disaster mitigation.

This requires a careful consideration of the changing roles of South East Asian countries, owing to the rise of China in the overall geo-political environment in the region (Pal 2021). China is known to be tapping into the resources of all international rivers (Jia 2017). On top of it, China has graduated from building large dams to mega dams. While it is not claimed in this text that the country may use water as a weapon against its neighbouring countries, the fact that the China holds the very capacity

[16] https://legal.un.org/avl/ha/dunche/dunche.html.

to do that, may render the downstream countries weak, cannot be ignored. This is a concern because in such cases, the emerging risks may have to be dealt with, with a different kind of strategy altogether, when the downstream countries may find themselves rather helpless. For instance, China has built 11 mega dams on the Mekong River (Economic Times 2021), which is known as the arterial waterway of Southeast Asia. These dams have led to recurrent drought downriver and have turned the Mekong Basin into a security and environmental hotspot (Eyler 2020). Against this backdrop, China's plan of building dam on Brahmaputra (Varma 2021) stands as a hovering concern for India with reference to how it may influence the national interests of the country.

Michel and Pandya (2009) noted that for countries to participate in a cooperative framework, benefit sharing must offer rewards greater than those of unilateral action. In its Preamble, the IWT does put forth a number of possible grounds for cooperation and benefit seeking. The treaty's focus on cooperative management and peaceful resolution of water disputes must occupy centre stage. Re-identifying the lost focus of the IWT and revisiting the original intentions and implications of the treaty, taking on board the evolving challenges posed by water scarcity, population growth, and climate change in both countries, is perhaps the way forward. Resting the base on Article VII of the IWT, both countries still have the scope to undertake joint research, in order to consider the need of re-orienting the focus of the Treaty to account for the emerging needs.

8.11 Postscript

This chapter is being published in the backdrop of 2025 terrorist attacks on tourists in Pahalgam (The Economic Times 2025), an event that significantly shifted the dynamics of Indo-Pakistani relations. Following the attacks, India announced a suspension of the operational mechanisms of the IWT (Gilani 2025)—a major political move, especially given the Treaty's survival through multiple wars and extended periods of diplomatic tension. For years, strategic analysts have pointed out the uneven structure of the IWT. While India was able to fully use the eastern rivers through large-scale projects like the Bhakra and Pong dams (Mukul 2025), it was heavily restricted when it came to the western rivers. With the suspension of the Treaty, India is no longer required to inform Pakistan about new projects, share hydrological data, or allow on-site inspections. Though India currently has limited capacity to store or redirect water from the western rivers, the suspension opens a door for long-term strategic development. This shift has serious implications not just for India's water and energy security but also for Pakistan's increasing vulnerability as a downstream state.

Within the broader context of this paper, which reflects on the Treaty's endurance and the challenges it faces in today's shifting climate and political realities, the 2025 suspension raises pressing questions. Can a mid-twentieth century Agreement still function in a world marked by intensifying political hostility, growing water scarcity, and rapidly changing regional priorities? The moment invites a rethinking of how transboundary water-sharing agreements can be made more equitable, flexible, and responsive to crises—especially in regions where shared natural resources have historically been shaped more by rivalry than by collaboration.

References

Abedullah (2020) Issues and strategies to revitalize the agriculture sector of Pakistan. Islamabad, Pakistan, Jan 10

Ahmad A (2011) Indus Waters Treaty: a dispassionate analysis. Policy Perspect 8:73–83

Akhtar S (2010) Emerging challenges to Indus Waters Treaty. Abasyn J Soc Sci, 3–66

Allan JA (2001) The middle east water question—hydropolitics and global economy. I.B. Tauris, London

Arnow T (1961) Geological survey water-supply paper 1608-A. Government Printing Office, U.S

BBC (2019) Article 370: what happened with Kashmir and why it matters. Retrieved from https://www.bbc.com/news/world-asia-india-49234708

Biswas AK (1992) Indus Water Treaty: the negotiating process. Water Int 17:201–209

Bukhari A (2011) Setback in the water dispute. DAWN, Karachi

Chellaney B (2014) China, Asia's water hegemon. Mint, Sept 16, available at https://www.livemint.com/Opinion/rM0WMFDmz8ADEeXGJjkEMP/China-Asias-water-hegemon.html

Chellaney B (2019) India is behind in the water wars with China & Pakistan, it must get its act together. The Print

Cooper R, Elhance A, William A (1999) Hydropolotics in the third world: conflict and cooperation in international river basins. Foreign Affairs

Dhawan V (2017) Water and agriculture in India: background paper for South Asia expert panel during the Global Forum for Food and Agriculture (GFFA). OAV—German Asia-Pacific Business Association, Hamburg

Eyler B (2020) Science shows Chinese dams are devastating the Mekong. Foreign Policy

Express Web Desk (2016) Blood and water cannot flow together: PM Modi at Indus Water Treaty meeting. The Indian Express, Sept 27, available at https://indianexpress.com/article/india/india-news-india/indus-water-treaty-blood-and-water-cant-flow-together-pm-modi-pakistan-uri-attack/

Gilani I (2025) Indus Waters Treaty suspension: India-Pakistan water crisis 2025. Frontline, Apr 25. https://frontline.thehindu.com/politics/indus-waters-treaty-suspension-india-pakistan-water-crisis-2025/article69500145.ece

Giordano M, Wahal A (2022) The Water wars myth: India, China and the Brahmaputra. United States of Institute of Peace

Gleick PH (1993) Water and conflict: fresh water resources and international security. MIT Press, pp 79–112

Gleick PH (2006) Water conflict chronology. Pacific Institute for Studies in Development, Environment and Security, California

Goradia P (2019) Nehru blundered by signing Indus Water Treaty, Mar 16

Gulhati ND (1973) Indus Waters Treaty: An exercise in international mediation. Allied Publishers, New Delhi

Hanasz P (2014) Power flows: hydro-hegemony and water conflicts in South Asia. Secur Chall 10:95–112

Jia SX (2017) China's international transboundary river politics: security and diplomacy of shared water resources. Transboundary Water Governance and Conflict Resolution in Asia

Khalid I (2013) Pakistan foreign policy: evolution, development, and strategies. Peace Publication

Kliot N, Shamir U, Shlomi D (2001) Institutions for management of transboundary water resources: their nature, characteristics and shortcomings. Water Policy 3:229–255

Kokab R, Nawaz A (2013) Indus Water Treaty: need for review. Asian J Soc Sci Humanit 2:210–218

Krishnan M (2018, Feb 21) Restoring India's depleting river—an uphill task

McCracken M, Wolf AT (2019) Updating the register of international river basins of the world. Int J Water Res Dev 35(5):732–782. https://doi.org/10.1080/07900627.2019.1572497

Michel D, Pandya A (2009) Troubled waters: climate change, hydropolitics and transboundary resources. The Henry L. Stimson Center, Washington, DC

Mukul S (2025) Pahalgam terror attack: Indus Waters Treaty on hold, but when can India turn off Pakistan's tap completely? India Today, Apr 26. https://www.indiatoday.in/india/story/pahalgam-terror-attack-stop-indus-waters-treaty-on-hold-when-can-india-turn-off-pakistan-tap-completely-2715324-2025-04-26

Mustafa D (2010) Hydropolitics in Pakistan's Indus Basin. United States Institute of Peace, Washington, DC

Negi M (2016) Uri attack: an inside story of how it happened. India Today, Sept 19, available at https://www.indiatoday.in/india/story/uri-attack-inside-story-pashtun-map-pakistani-ammunition-jash-e-mohammed-341761-2016-09-18

Opperman J, Grill G, Hartmann J (2015) The power of rivers: finding balance between energy and conservation in hydropower development. The Nature Conservatory, Washington, DC

Pal D (2021) China's influence in South Asia: vulnerabilities and resilience in four countries. Retrieved from Carnegie Endowment for International Peace, Oct 21. https://carnegieendowment.org/2021/10/13/china-s-influence-in-south-asia-vulnerabilities-and-resilience-in-four-countries-pub-85552

Pepper K (2018) Map of river Indus and its flow through India along with its tributaries. Available at https://commons.wikimedia.org/; also at https://commons.wikimedia.org/w/index.php?curid=66085475

QRIUS (2018). Dying rivers foretell a grave future for India and its agrarian economy. Gurgaon, Haryana, India, Mar 6

Saadat BK (2018) Indo-Pak and hydro-politics. Srinagar, Kashmir, India, Jul 26

Saraf M (2015) Kashmiris fight for freedom. National Institute of Kashmir Studies, Mirpur

Sen S (2018) If 80% water consumption in India is for agriculture, why is it unregulated and inefficient? New Delhi, India, May 3

Sleet P (2019) Water resources in Pakistan: scarce, polluted and poorly governed. Nedlands, WA, Australia, Jan 31

Stewart J, Podger GM, Ahmad MU, Shah SM (2018) Indus River System Model (IRSM)—a planning tool to explore water management options in Pakistan: model conceptualization, configuration and calibration. DFAT-CSIRO Indus SDIP Project

Svensson, J. (Host) & Chellaney, B. (Guest) (2019, August 20) India's hydro-diplomacy with Pakistan, Bangladesh and China [Audio podcast episode]. In Global Water Forum. https://www.globalwaterforum.org/2019/08/30/brahma-chellaney-indias-hydro-diplomacy-with-pakistan-bangladesh-and-china/

TFDD Transboundary Freshwater Dispute Database (TFDD) 2018 Transboundary freshwater spatial database. Oregon State University. Retrieved from https://transboundarywaters.science.oregonstate.edu/content/transboundary-freshwater-spatial-database

The Economic Times (2019) Abrogation of Article 370 unconstitutional, people of J&K bypassed: petitioners to SC. The Economic Times, Dec 10, available at https://economictimes.indiatimes.com/news/politics-and-nation/sc-commences-hearing-on-pleas-challenging-abrogation-of-article-370/articleshow/72454345.cms?from=mdr

The Economic Times (2021) After Mekong, China focuses to take control over Brahmaputra's water. The Economic Times, Feb 11, available at https://economictimes.indiatimes.com/news/international/world-news/after-mekong-china-focuses-to-take-control-over-brahmaputras-water/articleshow/80851084.cms?from=mdr

The Economic Times (2025) Pahalgam attack: 28 dead, several injured in targeted firing; survivors recall horrendous act as govt mulls action plan, Apr 23. https://economictimes.indiatimes.com/news/india/pahalgam-attack-28-dead-several-injured-in-targeted-firing-survivors-recall-horrendous-act-as-govt-mulls-action-plan/articleshow/120540638.cms
The Round Table (2008) Pakistan: end of the water dispute. The Commonwealth J Int Aff, 72–75
The World Bank (1960) Indus Water Treaty. World Aff, 99–101
UN. Department of Technical Cooperation for Development. Natural Resources and Energy Division (1978) Register of international rivers. Pergamon Press for the United Nations, New York
United Nations (2014). International decade for action 'water for life' 2005–2015. Retrieved from United Nations Department of Economic and Social Affairs (UNDESA), Oct 23. https://www.un.org/waterforlifedecade/transboundary_waters.shtml
Varma K (2021) China approves building dam on Brahmaputra in Tibet, near Arunachal Pradesh border. The Wire
Wenjing Z, Hong J, Rogers S (2022) The next phase of China's water infrastructure: a national water grid. China Dialogue
Wirsing RG (2008) The Kashmir territorial dispute: the Indus runs through it. The Brown J World Aff, 225–240
Wolf A (1995) The Asian water forum: summary report. Glob Environ Change: Human Policy Dimensions 5:149–154
Wolf AT (1998) Conflict and cooperation along international waterways. Water Policy 1:251–265
Wolf AT, Natharius JA, Danielson JJ, Ward BS, Pender JK (1999) International river basins of the world. Int J Water Res Dev 15:387–427
Zeitoun M, Warner JF (2006) Hydro-hegemony—a framework for analysis of trans-boundary water conflicts. Water Policy 8:435–460
Zeitoun M, Warner JF (2008) International relations theory and water do mix: a response to Furlong's troubled waters, hydro-hegemony and international water relations. Polit Geogr 27:802–810

D. Saran Prakash is a doctoral researcher at the Jamsetji Tata School of Disaster Studies, Tata Institute of Social Sciences (TISS), Mumbai, where she studies women's roles in shaping food security during disasters. She holds an MA in Disaster Management from TISS and brings together academic grounding with rich field-based experience in disaster risk reduction and climate change adaptation. Saran has contributed to culturally rooted resilience initiatives, including a 2023 project with the Risk and Resilience Institute (RRI), supported by ICCROM and funded by the Swedish Postcode Foundation, which examined the intersections of culture, climate adaptation, and disaster risk management in Jodhpur. Earlier, in 2020, she worked on an RRI initiative in Assam that focused on traditional flood risk management practices. This effort led to the publication of Milijuli, which was presented at the Understanding Risk Forum 2020 (UR2020).

She is an active member of the Confederation of Risk Reduction Professionals (CRRP), India, and has organized and spoken at webinars in partnership with the National Institute of Disaster Management (NIDM). In 2022, she represented CRRP India at the Global Platform for Disaster Risk Reduction (GPDRR) in Bali. In 2021, she served on the organizing committee of an International Roundtable hosted by Oxfam India, JTSDS-TISS, and NIDM.

Her research interests include disaster risk reduction, climate change adaptation, gender and disasters, women's agency, disasters and food (in) security, indigenous knowledge for Disaster Risk Reduction and Management, community-based disaster management, and the varied dimensions of agency, vulnerability and coping.

Chapter 9
Citizen Science and Community Engagement in Transboundary Areas: Learnings from the Mahakali Basin in the Indo-Nepal Border Region

Rajan Subedi and Mukunda Upadhyay

9.1 Introduction

Unplanned urbanization and industrialization have emerged as major factors responsible for transboundary environmental problems such as wastewater mismanagement, improper solid waste discharge, increase in plastic and air pollution and climate change (Fernandez 2002; Hoel 2002; Zhao et al. 2012; Marks et al. 2020; Abas et al. 2019). Burning of coal in Poland (Löfstedt 1998), heavy metal pollution in ASEAN countries (Ding 2019), and transport related air pollution in Europe (Berge et al. 1999) are examples of environmental problems across shared borders. Since they transcend geographic boundaries political and local level co-operation is needed to augment these problems (Fernandez 2002; Hoel 2002). The transboundary water assessment programme (UNEP-DHI and UNEP in 2016) indicates that 218 out of 286 transboundary river basins are at risk of high pollution levels which are projected to increase. The Ganga–Brahmaputra-Meghna (GBM) basin in South Asia lies in the top three hot spot transboundary basin of the world with increasing exposure to floods and droughts (UNEP-DHI and UNEP 2016). Many riverine communities across the planet are deprived of access and control over their riverine resources and their problems are further exacerbated with the change in the river structure (Matanzima 2020; Babatunde 2020). Similar challenges are faced by several South Asian communities because of inadequate water governance frameworks and regimes (Shan et al. 2020). This paper is based on the systematic engagement of the authors through the years 2020–2021 through the initiative taken by Oxfam in the Mahakali river basin that

R. Subedi (✉)
Oxfam, Nepal
e-mail: RaSubedi@oxfam.org.uk

M. Upadhyay
Assam, India

J. Andharia and P. Anand (eds.), *Shifting Lands, Flowing Waters*, Disaster Studies and Management, https://doi.org/10.1007/978-981-96-7691-0_9

traverses Nepal and India. The issues, the process, the learnings offer insights that may be of value to other regions facing similar challenges.

9.2 The Transboundary Mahakali River and the Water Problems

Nepal's eastern, western, and southern borders are attached with India and all the rivers of Nepal drain into India. Most of the western border of Nepal with India is administratively separated by Mahakali river.

The Mahakali river flows from Kalapani at an elevation of 3600 m (see Fig. 9.1). The river flows through Sudurpaschim Pradesh of Nepal covering Darchula, Baitadi, Dadeldhura, and Kanchapur districts of Nepal. In India, the river flows through the Kumaon region of Uttarkhand and the plains of Uttar Pradesh before finally assimilating into the Ghagra. Five major tributaries and their major sub-tributaries join Mahakali river in Upper Mahakali basin above Sarada barrage. The average discharge of Mahakali river is about 658 m^3/s at Sarada Barrage while the flow fluctuates heavily during monsoon and winter season. There are altogether 9 glacial lakes and 164 glaciers within Mahakali river basin in Nepal (CBS 2019). Glacial lake outburst floods (GLOF), flash-flood, landslides and floods are major hazards within the basin.

Besides, this river has been facing several environmental problems such as water pollution, increase in flooding and landslides, haphazard sand mining, soil erosion

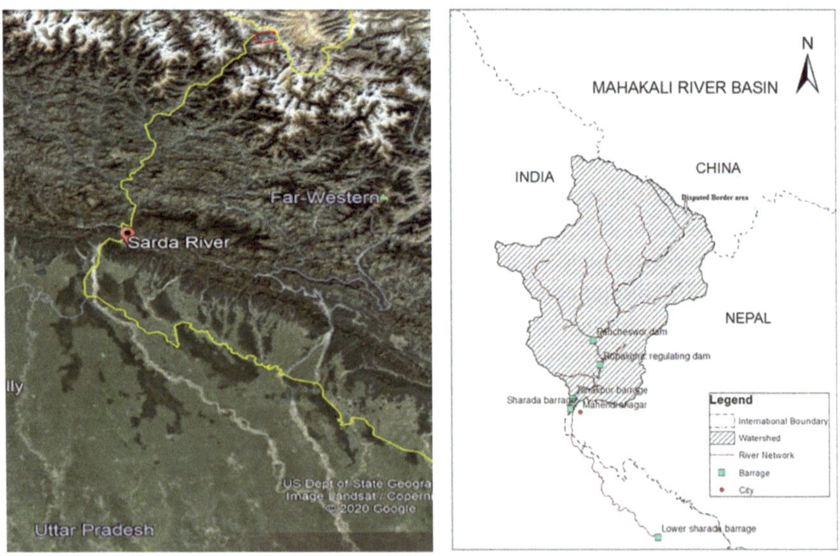

Fig. 9.1 Mahakali river basin upstream of Sarada Barrage. *Source* Bagale and Adhikari (2020)

(Chalise et al. 2019) and drying up of water sources in some areas (Subedi and Gupta 2019). Dumping of solid waste in the banks of rivers of Mahakali by businesses, local municipalities, hotels, and communities living around the river basins (Dhangadi Declaration into action, 2020) is also observed and contributes to deteriorating water quality. More than 1070 households are dependent on the Mahakali river for drinking water as they do not have any other alternative to access the same. This has a direct impact on the health of the local population who suffer from all sorts of ailments attributes to drinking contaminated water. Clearly the cumulative adverse impacts of all of these factors causing riverine pollution are very high. The increasing demand of sand for road construction, and other infrastructural development lead to higher extraction of sand from the river and other sources in South Asia that is leading to serious environmental problems in the rivers of South Asia (Larson 2018). The haphazard mining in rivers has also led to environmental impacts like increased turbidity of water, increase soil erosion, shrinking of riverbeds, adverse effects on the hydrological functions etc. (Gavriletea 2017).

There is not much systematic research focussing on the challenges faced by the riparian community in the Mahakali basin especially on those issues concerning their water needs in relation to their livelihoods and socio-economic development. Little effort has been made to understand the interconnected complexities associated with transboundary uncertainties influencing the lived experiences of the community in the basin. These are further complicated in the name of political sensitivity in the transboundary areas. This has resulted in limited options for the communities to interact with the government officials. According to a study conducted by Upadhyay and Mehzabbeen (2020), the lower Mahakali Basin has been severely impacted by water pollution due to the dumping of industrial waste in the external environment. This has resulted not only in the contamination of water but has also altered livelihood patterns in the Mahakali. Upadhyay and Mehzabbeen (2020) have also highlighted how transboundary communities in both India and Nepal were forced to abandon fishing and adopt other means of earning because of state conflicts, irresponsible industrial contamination and unscientific fishing by private actors. Even though the Mahakali river forms a geographical boundary between India and Nepal, centuries old cultural ties between the different communities living across the border, have created healthy social and economic ties which provide the foundation for strengthening transboundary water governance.

In the recent decades, the Indo-Nepal and Indo-Bangladesh border communities of South Asia have been educating the world with new perspectives on the idea of co-operation. In a process of cultural repair, communities are uniting together to undo the historical injustice meted out to them by synthetic borders created by the imperialist British forces. A new age of diplomacy has seemingly begun in South Asia in which science is being used as a tool in hydro diplomacy—a science of the people, by the people and with the people which teaches states, the art of water governance and diplomacy. The application of science, for solving local water problems carries the potential to sensitize communities to take stewardship of their shared resources. This paper highlights such efforts wherein citizen science has been used as a tool to promote peace and regional harmony.

9.3 Understanding the Citizen Science Approach

Citizen Science is the community's involvement in scientific inquiry for ensuring effective water governance. It could be referred to as citizen involvement in deciphering facts and scientific evidences for better advocacy and stakeholder engagement. In the current context citizen science has been used to engage the community in a constant process of monitoring water quality through a co-ordinated approach largely aimed at increasing the efficiency of basin level water governance. In the entire exercise the major focus has primarily been on augmenting the sense of community ownership over shared resources. While technical knowledge on water quality monitoring and environmental ecology has also been a focus, a shared sense of resilience and environmental sustainability at a basin level should be the expected outcome from such citizen science exercise.

Citizen science as a tool and method of strengthening transboundary co-operation began in 2018 with the aim to harness the power of culture to strengthen water governance in the Mahakali region. A very comprehensive framework was designed by the experts working in the field of water governance in both India and Nepal. This began first by institutionalising the communities in both the countries through networks and educating them on critical water issues in the region. Two community-based organisations-the Women Empowerment Centre (WEC) in Nepal and Sharda Nadi Nagrik Manch (SNNM) in India played a phenomenal role in identifying the core issues associated with water in the whole basin. In the initial stage, only three water quality parameters (pH, BOD, DO, TDS)[1] were tested with the help of citizen scientists in Nepal and India. In the later stages, important parameters like Iron, TDS, Dissolved Oxygen, Biological Oxygen Demand, Fluoride, Faecal coliform, and Phosphate etc. were also added. Further, information and data on waste management sites and sewage disposal practices were collected based on which deliberations were organised for further dialogues on water governance. At the initial stage, both WEC and SNNM worked independently in their respective upper and lower stretches of the Mahakali. Later as community dialogues intensified, joint training and consultations were organised to devise basin level citizen science plans. Within the ambit of citizen science, communities also developed their own early warning dissemination maps, identified areas of major contamination as well as executed joint water quality testing.

The total length of the river was divided into various zones both in the upper and lower basin. The approximate length of Mahakali from Darchula bridge to Bhimdutta is around 160 km. About 16 points at the longitudinal section of 10 kms were selected in the upper basin. The major areas in the upper Mahakali basin where the citizen science exercise and trainings were conducted are—Sribagad, Kakada, and Dattu in Darchula Meeting point of Chamelia and Sera, Dimbarghat, Kusmautghat, Teerkhadeni, Parshuram Dham, meeting Point of Mahakali and Rangoon River, Khalla, Badibasti, Bhujela Khola, and Kutiya kabar. In the lower basin, the major areas for intervention was the Sharda river and SooteaNala(a tributary of the Sharda

[1] pH is a measure of the concentration of hydrogen ions, the acidity or alkalinity of water. BOD: Biological Oxygen Demand, DO: Dissolved Oxygen.

River). Citizen Science exercises which entailed meetings to discuss community-based scientific inquiries, were conducted in Govindnagar, Palia Kalan, Paraspur, Srinagar, and Chauri Kalan. Since 2018, members from Women Empowerment Centres (Subedi and Shakya 2018) in Nepal and community volunteers in India were trained on 'citizen science techniques' including assessment of water quality in different sampling sites.

9.4 Actors Leading the Citizen Science Initiatives in the Mahakali Basin

In Nepal citizen science exercises were conducted by leaders from the Women Empowerment Centres (WEC) and the data and reports were analysed with the support of social mobilisers and public-spirited individuals. The women leaders design the framework and the mode and tools of assessment with the support of the local CSO's. After collecting the data and information, the women leaders began to educate and advocate with the local administration on specific parameters and critical concerns based on evidence gathered. In the past, Nepali women at the community level show cased extraordinary leadership skills in the field of citizen science leading the local government, and the private sectors such as hotel associations and even sand mining contractors to acknowledge the deteriorating water quality and water ecology of the Mahakali. On the Indian side of the Mahakali, efforts were made by the Sharda Nadi Nagrik Manch (SNNM) to collect information on water quality issues in the lower Basin in India. Some of the most important areas for collection of information are Chauri kalan, Govind Nagar, Sri nagar, areas surrounding the contaminated Sutiya Nala and areas around the local sugar factories.

Upadhyay and Mehzabbeen (2020) evaluated the lived experiences of the community in and around the sugar industries located in the Mahakali Basin. They found that some of the rivulets which were flowing freely in the past are now contaminated to a level where people have now abandoned fishing and they speculate that if such conditions continue then some of the surface water sources may have been completely destroyed by the industries. Among the major industry-affected areas in the lower basin are Govind Nagar, Paraspur, and Bada Gaonthe pH values were found to be abnormally high. The central idea of these practices was to bring the community together on a common platform so that water quality parameters and their dynamics are understood by the community in a simple clear manner.

In order to inculcate a culture of ownership, the SNNM members used citizen science approaches by periodically and consistently collecting information from various sources for a continuous period of time. One group of volunteers would train the other and the process would continue and they would map the information for better understanding of the local water issues. The intent is not only to produce information, but also to develop a sense of ownership so that through the process of self-testing, one realizes the importance of water governance. The approach adopted aims to converge citizen science with local culture. Data collection requires precision, time, and patience. Also, since volunteers are from various background, the testing of water is carried out when people agree on a common free time. Women leaders generally take the lead in collecting and recording information to be further discussed with the local authorities. Some of the contamination in the lower basin is obvious as the communities have reported dying fish population and a foul smell coming from water. The purpose of citizen science is to organise people on a joint platform for a more organized advocacy.

9.5 Key Issues Identified by the Community

Various water quality issues were identified by the community and shared at different transboundary platforms led by Women Empowerment Centre (WEC) Nepal and Sharda Nadi Nagrik Manch (SNNM). Such platforms provide opportunities to discuss critical water related issues, share observations and provide solutions. When such forums act, people from different areas get to know the challenges in both upstream and downstream. Although the Mahakali citizen science framework initially focussed only on water quality, later joint groups focussed on collaborating on dissemination of early warnings. In fact, when the upstream community leaders met the downstream ones for the first time, they were filled with surprise when they heard the challenges of people living in the lower basin. In the Lohaghat transboundary consultation, the downstream community leaders interacted on critical basin-related issues. Many such dialogues helped upstream and downstream cultures to know each other in a spirit of collaboration. Citizen science approach helps strengthen the '*roti-beti*' relation between India and Nepal, and offers an excellent scope effective water governance in the future.

Interested community members were trained on different water parameters and offered systematic explanation about what different water parameters indicate. One of the major areas for intervention as identified by the citizen science groups was the presence of Iron in different locations. Even the report of the Central Ground Water Board (CGWB), India reveals the presence of iron, nitrate and arsenic in water sources in Lakhimpur Kheri district through which the Sharda river flows to meet the Ghagra (Pandey 2013). However, the current water related citizen science exercise does not take into account arsenic contamination as the tools required for its testing are not available within the current framework. Nevertheless, the communities in both the upper and lower basin were trained on testing Nitrate in their water sources. Nitrate testing in the community is a tedious process and unlike other parameter testing, it requires some time if the test is carried out manually. In fact, high concentration of Nitrate can result in thyroid related problems and may also be carcinogenic, if found beyond the permissible limit (Parvizishad et al. 2017).

One of the major problems in some parts of the Mahakali basin is the presence of faecal coliform (E. Coli) in the water. It is generally found in the digestive tracts of warm-blooded animals and do not live long outside the host body. For example, one of the most contaminated sites with regard to E. Coli is Rangoon in upper Mahakali basin along with other areas. The reason for presence of faecal coliform is related to direct sewage connection to Rangoon river in Nepal. Presence of faecal coliform was seen in other areas as well especially in the lower parts of the basin. Since the testing is being carried out in a joint manner by the volunteers, continuous awareness on hydro hygiene is also being carried out through consultative meetings and discussions. When community members themselves examine the presence of harmful pathogens in the water, their sense of ownership of the water quality improves. This is an important step because it helps to ensure that the water is safe to drink and use. During the various citizen science exercises, it was observed that in the mountains, the most contaminated points are small springs or streams which bring water to the river.

9.6 Dialogue, Sensitisation, and Impact

The efforts of the WEC and SNNM were very successful in identifying the existing challenges in waste management in the Mahakali basin. The concerns of community in the Sharda basin with regard to industrial pollution was highlighted by Upadhyay (2020a) as well who analysed the experiences of the community living near such contaminated sites. His analysis on the irregular waste management of the local sugarcane factories showed how such contamination is detrimental to the general health, livelihood, community drainage and the overall ecology of the Sharda basin. In the downstream districts of Nepal and India, sewage and other waste which are channelled into the tributaries have converted them into undeclared sewers. Although pollution of water sources in the upper basin is less than the lower basin, it was found through different citizen science exercise that the environmental threats posed by contaminated water is increasing even in the upper stretches. The basin level citizen science exercises revealed that dumping sites located near major water sources have been affecting both the surface and ground water in the area. It has also impacted the water quality in the overall basin. In fact, if one visits the twin towns of Dharchula and Darchula, one can witness a poor waste management system, detrimental to the local ecological health.

Few civil society organizations in the upper stretches have been trying to bring people together through the process of citizen science to initiate dialogues on water quality. A similar model is being adopted in areas in and around Tanakpur as well where the local hotel associations along with the Women Empowerment Centre and Sharda Nadi Nagrik Manch are putting efforts to initiate consultations on the Tankapur-Mahendranagar waste management issue. The Bhimdatta Mayor (Nepal) also committed to take special initiatives to bring stakeholders in the Mahakali basin together for improving waste management in the area using citizen science approach. Similar initiatives are required to be taken at the Darchula and Dharchula urban spaces as well wherein waste governance should be carried out through citizen science. The impact of the dialogue was such that in 2021, the hotel association of Dadeldhura committed to develop a Sanitation plan in which they would focus on sustainable water and sewage management. Previously two eco-friendly hotels were opened in 2020 because of the constant advocacy. The businessmen drew inspiration from the leaders from the Women Empowerment Centre in Nepal.

In 2021 there were multiple rounds of transboundary dialogues in the Mahakali after which citizens of Palia, Lakhimpur kheri, Tanakpur, and Kanchanpur Nepal joined hands to establish a citizen science observatory. The local rafting agency also joined hands to support these public-spirited individuals to take special initiatives in Mahasheer (a type of fish) conservation, promotion of cross border trade and regular practice of citizen science. Their co-operation has been appreciated by the local administration and politicians. This paper advocates the replication of such unique stakeholder co-operation in other parts of the world where UN agenda 21 has been implemented in its true spirit.

In Nepal, the Women Empowerment Centres have been consistently engaged in multiple dialogues with the local municipalities and the Hotel associations of Kanchanpur and Dadeldhura on different environmental concerns in the Mahakali urban ecology. The WEC also played a phenomenal role in the development of a Sand-mining plan in Bhimdatta municipality (Subedi and Gupta 2019) providing their support in understanding the risks faced by the Mahakali ecology due to illegal mining and extraction by non-state actors. On the Indian side, the members from the Sharda Nadi Nagrik Manch through their Water Quality campaigns mobilised the masses on the existing condition of poor water quality in the basin.

The Mahakali *samwad* (*dialogue*) also became an important platform for both the WEC and SNNM to present the pertinent water issues in the basin. The Mahakali *samwad* led by the Indo-Nepal Joint Action Forum (INJAF) has been able to bring local governments, Civil Society Organisations (CSOs), academia and communities of both India and Nepal to commit to the conservation of the shared resources in the Mahakali river. The citizen science approach helped to share the evidence-based information and appealed for a co-ordinated response to the changing climatic terrain of the basin (Subedi and Gupta 2019). The success of the *samwad* and the citizen science efforts of WEC and SNNM could be discerned from the fact that local community leaders and governments in border municipalities appreciated the joint ownership of water resources in Nepal and India.

9.7 Citizen Science and Cultural Sustainability

Citizen science should be considered as an important component in the framework of Agenda 21[2] which advocates for the implementation of local citizen science plans. Upadhyay (2020a) has shown how citizen science can be an effective tool

[2] United Nations Conference on Environment & Development Rio de Janerio, Brazil, 3 to 14 June 1992 AGENDA 21, available at https://sustainabledevelopment.un.org/content/documents/Agenda21.pdf.

to bring sustainability of water resources in river basin through cultural sustainability: actions which intend to preserve basin cultures for ensuring sustainability of natural resources. We would like to emphasise here that sustainable development without people's involvement in decision making may not prove beneficial since culture has an instrumental role to play in water governance. In the Mahakali summit held in March 2020, the Bhimdatta local administration very clearly explained how the local administration is actually implementing the vision of the United Nation's Agenda 21. The Mayor himself was very clear in the instrumental role of culture in determining the hurdles in maintaining a standard water quality. The local administration of Bhimdatta also emphasised the need for a co-ordinated approach to be adopted by the borders' administrative towns for effective integrated water resources governance—assessment of water resources, sanitation and drinking water, impacts of climate change on water, conservation of aquatic eco-systems and most importantly measuring water standards through a futuristic lens. It also acknowledges the importance of various indigenous practices and cultural knowledge in the sustainable development framework and gradual incorporation of environment friendly technologies in the national and sub-national culture. In the historic Pithoragarh consultation, the local administration appealed to their counterparts in for continuous dialogue focussing on resolving issues around water governance in the border areas.

The authors have witnessed that all these efforts and co-operation was possible due to the co-ordinated citizen science exercises in the basin. Evidences generated through citizen scientists became the basis of deliberations between the local administrations, panchayats, communities, civil society groups and private actors. Further, this has changed the narrative and discourse from 'water sharing' to 'water conservation', creating an enabling environment for future dialogues and commitment for a joint action. One of the most important achievements due to this was the establishment of the joint citizen science observatory in the areas of Tanakpur, Pithoragarh, and Dharchula which at this stage is informal and maintained by local livelihood and citizen groups. Such efforts made by the civil society groups now need to be endorsed by the state.

In this connection, it is proposed that the local administration at the districts in India and Nepal may have joint consultation with the Bhimdatta administration which have been endeavouring to implement Agenda 21 imbibing the value of citizen science. The Lohaghat civil society dialogue and Pithoragarh consultation of 2021 are some of the major historical footsteps taken towards a more comprehensive transboundary dialogue in the Indo-Nepal border areas. The local municipalities surrounding the Mahakali in Nepal have been pioneers in demonstrating good models of water governance. Some of their major actions are shared in Box 1.

Box 1 Citizen Science Efforts for Improving Local Water Governance Strategies in Bhimdatta Municipality, Nepal

Because of the efforts of the Women Empowerment Centre and different citizen science efforts, stakeholders in the local municipalities in the Mahakali Basin in Nepal successfully advocated with the local administration to implement the following:

(a) Endorsement of a Sand-mining plan by Bhimdatta municipality and Budget Allocation for ensuring ecological sustainability in the Mahakali Basin.
(b) Budget allocation by hotel associations for Waste Management plan and endorsement of the plan by Bhimdatta municipality.
(c) Support to transboundary communities to uplift the water from Mahakali river for agricultural families in Dasarathchand municipality. This initiative will also benefit Indians in an indirect way through better food grain production.
(d) Announcement of 'one house one tap' program for the riverine communities by Pashuramdham and Dasarathchand Municipality of Dadeldhura and Baitadi districts respectively under which communities' rights to water and life are being secured.
(e) Joint Intervention by the private sector of both India and Nepal through subsidised diversified livelihood opportunities for youths in transboundary areas. Some of the petty business holders have also promoted gravel and sand art to empower the local women.
(f) Establishment of Joint Citizen Science observatory by the private actors in Tanakpur, India for consultative discussions, water quality monitoring and hazard monitoring.

Source Oxfam projects

9.8 Conclusion and Recommendations

In the past there have been few dialogues between the experts, communities and the local governments in both India and Nepal wherein people have committed to ensure water quality in the basin. Citizen Science as an integral component has been a central binding force connecting various stakeholders. As of now, the local governments in Nepal have joined hands with the civil society setting an example for their neighbouring nations. The Nepal Model of Sustainability requires replication in countries like India where community participation seems to be limited in decision making. In the past, the few discussions on water quality and citizen science in Dhangadi, Lohaghat, Pithoragarh, Dharchula, Tanakpur, MahendraNagar, and Lakhimpur Kheri have united people of the Mahakali in one thread and people have been appealing for co-operation from their states for a better tomorrow. Citizen Science therefore is not merely collection of data but a medium to foster collaboration, generate awareness among locals and government level decision makers as also help bring peace in South Asia.

The deep realization by transboundary communities that the conservation approach to sustainability is the key to solving major problems was manifested during the Lohaghat consultation and series of Mahakali *samwad* when both the communities came forward in series of dialogues starting from the Dhangadhi consultation. The impact was seen in policy decision making in Nepal which stood as a model in South Asia. Citizen science was found to be an effective tool which can change policies. At another level, the civil society also seems very interested in taking leadership to imbibe a sense of ownership on common property resources. The joint citizen science test being conducted by the people of India and Nepal in the Mahakali is a ray of hope of a sustainable future.

Through this paper, we would like to emphasise on mainstreaming citizen science in water governance at a South Asian level, not just as a tool for monitoring water quality but also developing a sense of ownership among the community. In fact, citizen science can be so effective that it has the potential to change the very face of transboundary governance in the region. We can create models of good governance by learning from each other fostering peace and co-operation. In this we should not forget to imbibe the culture of respecting traditional practices of water governance which are kept hidden in the pages of history. Policy makers should now rethink the Eurocentric model of governance and revisit South Asian tradition of citizen science and include them in their local sustainability models. We finally assert that Nepal and India's local governments' recognition of citizen science framework will create promising opportunities for strengthening transboundary water governance in the Mahakali basin.

References

Abas N, Saleem MS, Kalair E, Khan N (2019) Cooperative control of regional transboundary air pollutants. Environ Syst Res 8(1):1–14. https://doi.org/10.1186/s40068-019-0138-0

Babatunde AO (2020) Oil pollution and water conflicts in the riverine communities in Nigeria's Niger Delta region: challenges for and elements of problem-solving strategies. J Contemp Afr Stud 38(2):274–293. https://doi.org/10.1080/02589001.2020.1730310

Bagale DR, Adhikari KD (2020) Mahakali treaty: delay in implementation and resulting impacts from Nepal's perspective. Water Policy 22(4):658–669. https://doi.org/10.2166/wp.2020.141

Berge E, Bartnicki J, Olendrzynski K, Tsyro SG (1999) Long-term trends in emissions and transboundary transport of acidifying air pollution in Europe. J Environ Manage 57(1):31–50. https://doi.org/10.1006/jema.1999.0275

Central Bureau of Statistics (CBS) (2019) Environment statistics of Nepal. Kathmandu, Nepal: National Planning Commission, Government of Nepal. Available at: https://www.climatenepal.org.np/sites/default/files/doc_resources/Environment_Statistics_of_Nepal_2019.pdf

Chalise D, Kumar L, Spalevic V, Skataric G (2019) Estimation of sediment yield and maximum outflow using the IntErO model in the Sarada river basin of Nepal. Water 11(5):952. https://doi.org/10.3390/w11050952

Ding Y (2019) Heavy metal pollution and transboundary issues in ASEAN countries. Water Policy 21(5):1096–1106. https://doi.org/10.2166/wp.2019.003

Fernandez L (2002) Solving water pollution problems along the US-Mexico border. Environ Dev Econ 7(4):715–732. Retrieved Oct 13, from http://www.jstor.org/stable/44379446
Gavriletea M (2017) Environmental impacts of sand exploitation. Analysis of sand market. Sustain 9(7):1118. https://doi.org/10.3390/su9071118
Hoel M (2002) 32 Transboundary environmental problems. Handbook of environmental and resource economics, 472.
Larson C (2018) Asia's hunger for sand takes toll on ecology. Science 359(6379):964–965. https://doi.org/10.1126/science.359.6379.964
Löfstedt RE (1998) Transboundary environmental problems: the case of the burning of coal in Poland for heating and electricity purposes. Glob Environ Chang 8(4):329–340. https://doi.org/10.1016/s0959-3780(98)80001-0
Marks D, Miller MA, Vassanadumrongdee S (2020) The geopolitical economy of Thailand's marine plastic pollution crisis. Asia Pac Viewp 61(2):266–282. https://doi.org/10.1111/apv.12255
Matanzima J (2020) Large dams: long term impacts on riverine communities and free flowing rivers. Water Int, 1–3. https://doi.org/10.1080/02508060.2020.1808321
Pandey DS (2013) Ground water scenario of Lakhimpur Kheri, 2012–13. Government of Uttar Pradesh. Report available at https://cgwb.gov.in/District_Profile/UP/Lakhimpur-kheri.pdf
Parvizishad M, Dalvand A, Mahvi A, Goodarzi F (2017) A review of adverse effects and benefits of nitrate and nitrite in drinking water and food on human health. Health Scope. https://doi.org/10.5812/jhealthscope.14164
Shan V, Singh SK, Haritash AK (2020) Water crisis in the Asian countries: status and future trends. In: Resilience, Response, and Risk in Water Systems, pp 173–194. https://doi.org/10.1007/978-981-15-4668-610
Subedi R, Gupta S (2019). MahakaliSambhad-a civil society organization lead dialogue in Mahakali basin, Nov. Oxfam. https://oi-files-cng-prod.s3.amazonaws.com/nepal.oxfam.org/s3fs-public/file_attachments/Mahakali%20Sambad%20English.pdf
Subedi R, Shakya S (2018) Emerging women leadership in transboundary water governance. Oxfam. https://cng-cdn.oxfam.org/nepal.oxfam.org/s3fs-public/file_attachments/WEC-Brochure.pdf
United Nations Conference on Environment & Development Rio de Janerio, Brazil, 3 to 14 June 1992 AGENDA 21, available at https://sustainabledevelopment.un.org/content/documents/Agenda21.pdf
UNEP-DHI, UNEP (2016) Transboundary river basins: status and trends, summary for policy makers. United Nations Environment Programme (UNEP). http://twap-rivers.org/assets/GEF_TWAPRB_SPM_compressed.pdf
Upadhyay M (2020a) Pragmatic approaches to transboundary river governance: exploring pathways for co-operation on shared water resources. World Water Policy 6(1):101–114.
Upadhyay M (2020b) Water footprint and stewardship: narratives from the Sharda basin, transboundary rivers of South Asia. Oxfam India Publication, Supported by the Government of Sweden, accessible at https://d1ns4ht6ytuzzo.cloudfront.net/oxfamdata/oxfamdatapublic/2020-12/PolicyBrief-TROSA.pdf?g6TLX1t1.wEzaDGas26z23z_njBX4Co8
Upadhyay M, Mehzabbeen E (2020) Exploring water Stewardship in the Mahakali. Oxfam India Publication, Supported by the Government of Sweden.
Zhao L, Qian Y, Huang R, Li C, Xue J, Hu Y (2012) Model of transfer tax on transboundary water pollution in China's river basin. Oper Res Lett 40(3):218–222. https://doi.org/10.1016/j.orl.2012.01.011

Rajan Subedi is the Program Manager—River Basins at Oxfam in Nepal. He holds a master's in environmental science and brings over 15 years of experience in climate change, transboundary water governance, and disaster risk reduction. Rajan has led multi-country programs, facilitated stakeholder dialogues, and represented Oxfam at regional and global platforms. He is passionate about ecological justice and continuous learning in the development sector.

Mukunda Upadhyay Mukunda Upadhyay is a seasoned Disaster Management Professional based in India, with years of extensive experience in designing and implementing disaster risk reduction (DRR) policies and programs. He specializes in strengthening systems for effective emergency responses, developing technological solutions for project management and response governance, and designing recovery frameworks for long-term resilience. He has been credited with creating systems that enhance disaster preparedness, response, recovery, and resilience outcomes envisioned by various state disaster management authorities across different states in India, including Uttar Pradesh, Assam, Nagaland, Meghalaya, Tripura, Sikkim, Arunachal Pradesh, and others. Additionally, he has been involved in various community-based risk management initiatives, working closely with riparian communities, local governance systems, and civil societies in the Ganga–Brahmaputra-Meghna region. His structural and non-structural models, designed and created to support vulnerable communities in disaster-prone regions, have received significant appreciation. He has also been instrumental in various transboundary water governance initiatives in the Mahakali basin, emphasizing citizen science, early warning systems, cultural sustainability, and promoting a sense of stewardship among various critical actors. In the past, he took the lead in organising global and regional dialogues on water stewardship, riparian community rights, water footprints, and transboundary co-operation. His role in designing technology-embedded platforms for project management, disaster governance, and community-based transboundary early warning systems has effectively contributed to achieving desired system outcomes.

Apart from programmatic endeavours, he also mentors scholars on critical disaster risk reduction research and is an invited lecturer at several Administrative Staff colleges, universities, and national and international forums. He is an alumnus of the prestigious Jawaharlal Nehru University (JNU), the International Institute for Population Sciences (IIPS), and Assam University (AU).

Chapter 10
International Court of Justice on Transboundary Harm: Reflections on South Asian Transboundary Water Governance

Saumya Kumar

10.1 Introduction

The International Court of Justice (ICJ) is an important global judicial body created to ensure a sovereign State's respect for international rule of law. It identifies one body of international rules and principles and interpret them universally (Lauterpacht 1982). It is referred to as a central judicial body of the international community that works not only on settling disputes but also towards the development of the principle of international law governing the international society in general.

A very important principle of international law—is sovereignty. Manifesting as States' superior and inherent right to manage the environment and natural resources within their territory, this principle of absolute territorial sovereignty has transitioned to where limited territorial sovereignty is more visible, specifically in treaties between South Asian countries on transboundary water. The acceptance of limited territorial sovereignty has emerged with the recognition of the principle of transboundary harm. The paper will identify essential features of these treaties in the context of political tensions around them.

The ICJ has considered many other international agreements on transboundary water to prevent harm. It has also focused on the practice between the states or the legal instruments between the parties. It has developed vital principles and practices to prevent transboundary harm in the context of transboundary rivers. This paper will identify these principles and practices to provide a framework to build up the no-harm principle for its application in the South Asian region. The use of the no-harm principle has the potential to enrich the discourse on transboundary water governance.

S. Kumar (✉)
Jamsetji Tata School of Disaster Studies, Tata Institute of Social Sciences, Mumbai, India
e-mail: saumya.kumar@tiss.ac.in

J. Andharia and P. Anand (eds.), *Shifting Lands, Flowing Waters*, Disaster Studies and Management, https://doi.org/10.1007/978-981-96-7691-0_10

10.2 Absolute Sovereignty and Introduction of Transboundary Harm

States have complete freedom to use and manage natural resources within their territory. Control over natural resources emanates from the fundamental idea of sovereignty. It is reflected in early international principles like Permanent Sovereignty over Natural Resources enshrined in the United Nations General Assembly Resolution in 1962 (Tyagi 2015). States back then considered that they had a superior and inherent right to manage the environment and natural resources within their territory. During the debates on the 1962 Resolution, some countries stated that the idea of sovereignty presupposes the right to use natural resources. At the same time, a more recent interpretation of the provision questioned the balance between the rights and focused also on the duties of the States emerging from these provisions (Chimni 1998). On similar lines, a water dispute between the USA and Mexico led to labelling absolute territorial sovereignty as the Harmon Doctrine. The doctrine allows absolute supremacy to the state to use the natural resources, specifically water, within its territory without considering the needs of other states (Salman 2007). This principle has been used mainly by many upper riparian countries to claim that they have no obligation to maintain an adequate supply or quality of water for the lower riparian countries.

However, this idea of sovereignty fails to consider a crucial point: the possibility of transboundary harm. The principle of transboundary harm covers any human activity which results in harm or any action caused with reasonably proximate causal relation to human conduct whereby the impact is suffered across the border (Rao 2002). The formal introduction of transboundary harm as a distinct concept in environmental law was undertaken with the adoption of the Draft Articles on the Prevention of Transboundary Harm in 2001. In the context of this Draft Agreement, it was understood to refer to environmental damage that originates in one state and affects the territory of another where the damage is beyond a particular threshold level of severity (Xue 2003).

Major accidents, especially in Europe, created a need to challenge the idea of absolute territorial sovereignty. The water used to douse the fire in the Sandoz Factory in Switzerland in 1986 drained into the Rhine River with dissolved chemicals, which led to massive destruction of the river's ecology, which cut across many countries, before entering the sea (Lewis 1986). Although the Swiss domestic framework was competent to ensure preventive mechanisms and the private party's liability for Swiss citizens, no framework was available for information sharing and accountability for the damage caused across international boundaries without the initiation of the procedure by the State (Brüggemann and Halfon 1990). The failure of the Swiss Government to contain the pollutants led to severe harm downstream. The downstream countries could not hold Switzerland responsible, as there was no international or even bilateral agreement which created any responsibility for Switzerland to notify the downstream states. Many have argued that the other countries affected by the pollutants, in this case, were unwilling to hold Switzerland responsible, as they all had a history of similar events in the past (Schwabach 1989). A similar legal vacuum came up after the Chernobyl Nuclear meltdown in the USSR in 1986, which led to radiation exposure in the USSR and many parts of Europe. Here again, even though the USSR was aware of the leak, the information about the accident was not released to other countries soon enough, which could have allowed them to undertake preventive measures (Malone 1987). The delay in sharing information about the incident with other countries is often categorised as one of the many consequences of the political uncertainty of the Cold War. In any case, it highlights the importance of information sharing as a crucial aspect of preventing harm in neighbouring countries.

Way back, the Trail Smelter case of 1932 is considered a landmark case articulating the need to recognise the obligation of states to prevent transboundary harm, much before the Draft Articles on Transboundary Harm, 2001, were introduced. The case was about air pollution originating in Canada from the fumes released by smelters since 1905, which affected the farms in Washington State, USA. After failed negotiations and mediation by the International Joint Commission, the matter was submitted for arbitration in 1932 to an arbitration Tribunal specifically constituted to settle the dispute under the Boundary Waters Treaty of 1909 (Read 1963).

The arbitration had to determine whether there was a transboundary harm for which Canada was responsible. The Tribunal decided that states cannot allow their territory to be used in any way which could cause injury to another country. This well-known decision covered public and private activities (Rubin 2006). It also included any harm caused by air pollution or other injury. It laid down the foundation of the no-harm rule, which became the conceptual anchor for International Environmental Law after this case (Miller 2006). Although a crucial decision for environmental law, the Trail Smelter has been criticised for not prioritising the ecological damage caused by Canada, focusing more on the protection and balancing of sovereign interests, and emphasising the establishment of causal links between the harm caused and the action of Canada.

The gaps in determining accountability in the Sandoz Factory case and the Chernobyl Accident, both in 1986, created the need to revisit the idea of absolute territorial sovereignty. The importance of recognising the no-harm rule provided the basis on which different elements of transboundary harm have evolved as an exception to complete sovereignty. As stated earlier, the paper will elaborate upon various aspects of recognising transboundary harm, especially in the context of transboundary water governance, focusing on South Asia.

10.3 The Emergence of Limited Territorial Sovereignty and Transboundary Harm

Transboundary water sharing has been a matter of multiple debates in modern times. The international community had seen several issues and challenges in applying the Harmon Doctrine, where the upper riparian states were always in a position of power

(Just 1998). This was followed by the concept of limited territorial sovereignty, which is applied when the riparian State is not allowed to utilise the water of an international river in a way which causes significant harm to the reasonable utilisation by other riparian States (E/ESCWA/SDPD/2012/Glossary). It is interpreted as the right of territorial sovereignty and the corollary duty not to cause significant harm to the sovereign rights of other States (Thorson 2008). In the Corfu Channel Case (Corfu Channel, United Kingdom v Albania, Judgment, Compensation 1949) the International Court of Justice (ICJ), though unrelated to transboundary rivers, laid the groundwork for the no-harm principle. Over a period of time, the concept of no-harm was also adopted in the context of environmental damage in transboundary regions where the impact of human activity creates damage beyond a particular threshold level of severity (Xue 2003).

In the context of water sharing, introducing and developing the principle of limited territorial sovereignty with the introduction of the transboundary harm principle brought in much-needed accountability and dialogue between the upper and lower riparian countries (Kliot 2001). Limited territorial sovereignty restrained states' use of water resources by respecting that their actions should not harm other countries. These principles were operationalised by qualifying water use with equitable and reasonable use. This was first mentioned in the Salzburg Session of 1961 at the Institute of International Law, followed by substantive inputs from the International Law Association in 1966 (Qureshi 2016). These were followed by the report of the 6th Committee of the General Assembly and the Helsinki Rules, 1966, leading to the codification of the equitable and reasonable use and the no-harm rule in the UN International Watercourses Convention (UNWC), 1997 (Salman 2007). The UNWC codified that transboundary water should be shared equitably and reasonably (Article 5(2)), respecting the no-harm principle, strengthening the dominance of limited territorial sovereignty. The purpose of the UNWC is clearly stated in Article 8, which provides the general obligation of the State to cooperate based on sovereign equality, territorial integrity, mutual benefit and good faith to attain optimal utilisation and adequate protection of an international watercourse. Scholars have argued that the UNWC and the more recent case laws are moving from limited territorial sovereignty to the principle of community of interest where all the riparian states are at equal levels engaged in using the entire river (Vučić 2019). Part III of the UNWC is considered an essential component where the principle of equitable and reasonable use is projected as procedural obligations, providing avenues for sharing information (Article 14).

The concept of limited territorial sovereignty has helped to understand the relationship between riparian states. Many bilateral treaties on water in South Asia utilise the no-harm principle in different forms (see Table 10.1). The challenge with the UNWC and the South Asian treaties on transboundary water is that limited territorial sovereignty and the no-harm principle are not effectively implemented. The standards of equitable and reasonable use in UNWC needs to be supplemented with minimum standards. These minimum standards can be provided by bilateral transboundary water treaties tailored to the specific requirements of the river and its regional context, which can contribute towards operationalizing the no-harm principle more effectively.

Table 10.1 Dimensions of water sharing bilateral treaties in four countries of South Asia

	Indus Water Treaty India—Pakistan	Mahakali River Treaty India—Nepal	Ganga River Treaty Bangladesh—India
Use	Drainage, conservation of soil erosion, dredging as long as, as far as practicable no material damage occurs to the other party	Specific amount of water to be shared + power plants	Formula for water sharing at the Farakka Barrage
Principle for use	Restricted use—Domestic use, non-consumptive use, agricultural use and generation of hydro-electricity (of the other river)	Article IX(1) Principles of equality, fairness and no-harm—guide the relationship	Optimum utilisation of the water by providing a fair and just solution
No-Harm	No material damage to the other party for both self and other side use Pollution—as far as practically possible	Specific provision	Principle of equity, fair play and no-harm in cases of emergency
Institution	Permanent Indus Commission	Mahakali Commission	

10.4 Contestations on Transboundary Water in South Asia

This section provides an overview of the transboundary water treaties in South Asia and identifies their critical features. The overview will help examine how limited territorial sovereignty and the principle of transboundary harm are conceptualised in the context of transboundary water management.

The South Asian region has many massive river systems that transverse significant national boundaries. River systems like the Indus, Ganga Brahmaputra Meghna, and the Mekong are home to an enormous population. The region evinces great diversity in ecology, biodiversity, environment, culture and ethnicity and is connected to the livelihood of people in the South Asian countries of India, Pakistan, Bangladesh, Bhutan, Nepal and Myanmar. Most of these countries are also dominated by unplanned urbanisation, with cities like Delhi, Kathmandu, Kolkata, Dhaka, and Lahore mushrooming with large populations and limited spatial planning. The increasing population, intensive agricultural activities and rising impacts of climate change are expected to exacerbate the vulnerabilities of people living in the region (Salman 2003). Further, the entire Himalayan-Hindu Kush belt and the plains are subject to severe hydrological hazards like floods and droughts. These hazards are magnified by the absence of coordination mechanisms between countries managing dams, leading to multiple cases of flash floods in the Terai region or even droughts in many lower riparian areas. The coordination and management of water resources are proving to be a severe challenge in this region. Cooperation becomes even tougher with the influence of pre-existing disputes, especially relating to boundaries (Myint 2012). The rivers in the higher areas of China, Nepal, India and Bhutan are fast-moving and are seen as a source of attractive hydroelectric power, which the countries are keen to utilise. This invariably affects the quality and quantity of water downstream. With the increased urbanised settlement and lax government regulations (Ahmed 1999) despite multiple treaties, the management of transboundary water resources in the region is a challenge (Uprety 2011).

The Indus River system consists of multiple rivers traversing India and Pakistan. It is governed by the Indus Water Treaty (IWT), 1960, negotiated for the common interest of optimum river development across the two countries. The Indus Water Treaty divides the Indus tributaries into eastern and western rivers, where India's unrestricted use of the eastern rivers is assigned. With respect to western rivers, India has the right to set up projects for domestic use, non-consumptive use, agricultural use and generation of hydroelectricity on the river of the west without impeding the flow of the river into Pakistan. Pakistan has unrestricted use of the western river, as well as the domestic and non-consumptive use of the eastern river. The treaty has been considered a unique format as it allows for division based on geographical location and not the volume of water. Some have argued that the too-technical treaty ignores the socio-economic realities of the communities living along the river affected by climate change and the fluctuating diplomatic relations (Zawahri 2009).

In many cases, the definitions and the scope of activities allowed on the country's river have evolved over some time (Akhter 2010), like the kind of dams allowed as non-consumptive use as seen in the case of the Kishenganga Arbitral Award (Award in the Arbitration regarding the Indus Waters Kishenganga between Pakistan and India 2013). In this award, the dispute was over whether India's proposed dam could be considered non-consumptive use on the western rivers, as such service was not explicitly covered under the said term in the IWT. Non-consumptive use is one of the basic allowances given in the treaty, which includes a wide range of functions like—navigation, floating of timber, flood protection, fishing wildlife and other beneficial purposes. The no-harm principle, although not very structured or defined in the IWT, is reflected in places like Article IV (2) of the IWT, which allows parties to non-consumptive use of the river without materially changing the flow to prejudice the use of the river by the other party. Further, Article IV(3) of IWT allows a wide range of activities on the rivers assigned to each state for drainage, soil erosion conservation, and dredging as long as, and as far as practicable, no material damage occurs to the other party. The provision relating to the prevention of pollution is restrictive as given in Article IV(10) of IWT, creates the obligation to prevent undue pollution which would have adversely affected the usage of the river (as far as practically possible) with a special emphasis that the reasonableness of the prevention will be based on customary practices.

The limited scope of the no-harm principle restricts consideration of environmental disputes under the treaty, as was seen in the Kishanganga arbitral award, where the ecological impact of the project was seen as one of the factors to be considered for the implementation of the project. Still, the matter was left to the parties to address (Paras 94 and 110). In its final award, the Court was not keen to pursue the customary law of precautionary approach as a specific treaty prevents the Court from becoming a policymaker in determining the balance between acceptable environmental change and other priorities (Para 112).

As regards the water sharing between India and Nepal, the implementation of the Gandak and Kosi agreements created political turmoil in Nepal, which led to a specific provision in the 1990 Constitution in Nepal on treaties on water sharing requiring a two-third majority approval of the representatives (Subedi 1999). This was followed by the Mahakali River Treaty (MRT) in 1996 as a framework treaty on equality, mutual benefit and no-harm. The provisions gave control over the supply of the Sarada Barrage to Nepal and the power of the Tanakpur Barrage in Nepal to India without prejudice to Nepal's sovereign right in exchange for a predetermined supply of water from the Barrage and a specific amount of electricity. The treaty allowed the construction of the integrated Pancheshwar multipurpose project, where both parties had equal rights over the water and the power stations on the plant. This treaty was expected to be the foundation for future state agreements.

Unfortunately, the implementation of the MRT has been limited, with many scholars referring to it as a defunct treaty as its performance is subject to the implementation of the Pancheshwar Project (Bagale 2020). In addition, the treaty brought a significant issue of diplomatic tussle to the forefront—regarding the definition of the term 'existing consumptive use' in Article 3, on which the distribution amount was to be calculated. The Indian interpretation (being an upper riparian state) of the term was more expansive, including the water from the second auxiliary Sarada Canal, while Nepal restricted the term to water only from the main canal (Desai 1998). In the Mahakali Treaty, although the bare minimum river flow in the lean months was provided, there was a gaping hole in the issue of determining the flow where the bare minimum flow was not maintained (Desai 1998). This would allow India to determine the flow in the lean months during severe scarcity. The principle of equity, fairness and no-harm (Article IX (1) of the MRT) is expected to guide the relationship between the two countries, even though there is no mechanism to do so as the Mahakali Commission has limited power to address disputes (Subedi 1999). The treaty provides for an equal partnership concerning sharing some resources with the objective of maximum net benefit. This provision may direct the focus towards the community of interest approach. Still, there appears to be no substantive provision allowing its practical implementation, as the entire river system is divided based on multiple projects where different agreements are anticipated. The no-harm principle is alluded to in Article 7, which provides that the parties agree not to obstruct or divert the water to affect its natural flow adversely.

The Ganga River Treaty, 1996 (GRT) between Bangladesh and India was preceded by a wide range of setbacks due to multiple reasons. The construction of the Farakka barrage and the uncertainty about the amount of water released affected the flow of the Bhagirathi in Bangladesh. It was a matter of grave concern during the negotiation process (Pandey 2012). The 1996 Treaty established a formula providing a specific amount to be released from the Farakka barrage in the dry and wet seasons. Any deviation from this amount was allowed only for reasonable use. It has been challenging to enforce the main mandate, as India was expected to release at least 90% of Bangladesh's share (Thomas 2017). The treaty provides a fair and just solution for optimum water utilisation. It applies the principle of equity, fair play and no-harm in emergencies such as the times when the water level goes very low (50,000 cusecs) or future agreements based on the treaty (Article IX). For dispute resolution, the Indo-Bangladesh Joint River Commission is constituted for referral with the two governments giving the freedom to the governments to decide if they want to go for arbitration.

An overview on the three key bilateral treaties indicates that while the use of water in the treaties was based on the negotiations carried out, they all respect the idea of limited territorial sovereignty by employing terms like restricted use, optimum use, and, most importantly, using terms like no-harm, no material damage. This indicates that the idea of limited territorial sovereignty is respected by upholding the no-harm principle. Unfortunately, besides the undeniable political issue, as seen in the Kishenganga Arbitral Award, there is no standard mechanism for operationalising the no-harm principle. The persistent multiple disputes on the nature of the hydroelectric plant in IWT, the construction of the barrage in MRT and the amount of water shared in dry months in GRT over the years indicate that, besides the political issues, the operationalisation of these principles is vague and uncertain and needs more clarity. In this context, different methods developed by the ICJ to uphold the limited territorial sovereignty by respecting the no-harm principle are explored in the next section.

10.5 ICJ on Transboundary Harm

In light of the diverse ways in which the idea of transboundary harm is suggested in the different treaties in South Asia, it is useful to understand the ways in which the ICJ has contributed towards developing this principle further. Decisions like the Corfu Channel Case (Corfu Channel, United Kingdom v Albania, Judgment, Compensation 1949) and The Threat or Use of Nuclear Weapons Case (Legality of the Threat or Use of Nuclear-Weapons, Advisory Opinion 1996) have considered transboundary harm in the context of sovereignty and use of force. These decisions created the exception to absolute territorial sovereignty by deciding that a state cannot allow its territory to be used to harm another country. The cases of the ICJ have helped expand the different kinds of harm and have given new life to environmental disputes. The ICJ issues on transboundary environment cases have evolved in three phases (Tignino 2020). The first series of cases examined existing sovereignty principles employed to explain transboundary harm, like the Corfu Channel case or the Gabčíkovo-Nagymaros Project case. While the second phase encompasses cases like the Pulp Mills case focused on developing principles of environmental protection, the third phase focuses more on building the procedural requirements for environmental protection. These cases highlight how crucial aspects of transboundary harm and due diligence are structured and the third phase is valuable to map out alternatives for operationalizing the no-harm principle in South Asia.

10.5.1 Gabcikovo-Nagymaros Case (Gabcikovo-Nagymaros Project 1997)

The case before the ICJ was based on the obligations mentioned in the Budapest Treaty, 1977, between Hungary and Slovakia for constructing dams and gates on the Danube River to improve flood management, navigation, energy production and water availability in urban cities. Slovakia completed the construction of a portion of the project in Gabčíkovo, after which Hungary refused to construct the dam at Nagymaros in 1989 due to environmental concerns. Since the project had come to a standstill, Slovakia, in its attempt to negate any harmful impact of the stalemate, unilaterally diverted the water of the river to an alternate barrage system, which affected the quantity of water available to Hungary, especially the water supply of Budapest. The dispute focused on whether Hungary had the right to delay and terminate the Treaty based on its environmental concerns and whether Slovakia's unilateral diversion action was legal under the Treaty and general principles of international law.

Regarding the legal status of the termination of the treaty, Hungary maintained that suspending the project was a logical necessity due to environmental concerns. Slovakia argued that the scientific evidence did not establish the ecological risk, allowing necessity as an exception. The Court identified that the state of condition arises out of a grave and imminent peril that should not affect the other state's essential interest. Hungary argued that the project's construction was a severe and imminent danger to the environment; therefore, suspending and abandoning the Project was necessary. The Court concluded that the environmental concerns raised by Hungary were significant but were not enough to establish a grave and imminent peril for Hungary's environment as the facts indicated a mere apprehension of a possible danger. The ICJ concluded that Hungary's suspension and abandonment of the Project was unnecessary, and therefore, Hungary failed to fulfil the treaty's obligation.

The ICJ considered the legal validity of Slovakia's countermeasure of diverting the water by assessing whether the action was taken in response to a prior international wrongful act with due notification about discontinuation followed by a reversible countermeasure proportional to the injury suffered. The ICJ found that Slovakia's unilateral withdrawal and diversion of the Danube River responded to the wrongful act of abandoning the project after regular communication between the States. However, it led to the deprivation of natural resources in Hungary, which was not proportional to the injury suffered by Slovakia. The countermeasure was seen as an act infringing upon Hungary's territorial integrity and sovereignty, as the alleged claim over Slovakia's equitable share of the river was more than the injury suffered by Slovakia. This decision of the ICJ is perceived as one of the first cases bringing environmental concerns to the forefront of transboundary harm, although it was not a significant issue. It has been argued that this case expanded the restricted aspect of the Corfu Channel case on no-harm to a broader stage and paved the way for applying transboundary harm in environmental cases (Nakamichi 1997). While some scholars find this decision restrictive and conservative, many others have argued that it came long before some of the critical international policies on environmental law to explain its restrictive scope (Stec 1998). The ICJ's failure to consider the Precautionary Principle concerning Hungary's reason to abandon the project has been touted as a missed opportunity to allow the intrusion of sovereignty. The decision has been questioned on identifying the threshold level for harm (Stec 1998). The act of diverting the Danube River was assessed as a countermeasure, citing approximate application of proportionality as the standard to conclude that the diversion is wrong. However, the well-known standard of equitable and reasonable sharing of transboundary resources (well known in the Helsinki Rules and often cited as a part of customary law) was not considered to assess the threshold level of the transboundary harm caused by the diversion. This case is one of the first examples when the ICJ is expected to operationalize the no-harm principle, paving the way for building up the jurisprudence on transboundary harm and environmental concern within the ambit of harm which has not been considered in the South Asian treaties.

10.5.2 Pulp Mills Case

The dispute between Argentina and Uruguay (Pulp Mills on the River Uruguay, Argentina v Uruguay, Order, Provisional Measures 2006) was a culmination of a long-drawn tension between the two countries on the construction of two pulp mills by Uruguay near its border with Argentina along the River Uruguay. Argentina challenged the construction as it contested that it violated Uruguay's obligation to ensure optimum and rational utilisation of the river, protect the aquatic environment, and prevent pollution under the 1975 Statute of the River Uruguay (Statute). In addition, Argentina claimed that the Statute required Uruguay to comply with the procedural requirements of sharing information regarding projects affecting the river. Uruguay denied that there was any harm or risk of harm to the river or its ecosystem.

As per the 1975 Statute of the River Uruguay, the Administrative Commission of the Uruguay River (CARU) was created to facilitate optimal and rational utilisation of the river as per the treaty's obligations. A significant aspect of the decision was the treaty's scope, as Uruguay interpreted that the treaty covered only the harm caused to the river's water and not any pollution. The ICJ relied on Article 60 of the 1975 Statute to conclude that the jurisdiction of the ICJ was restricted to the activities mentioned under the provision, which did not include noise pollution.

Concerning the procedural issues, Argentina claimed that Uruguay had delayed sharing the reports of the environmental impact assessment with the CARU, violating the procedural obligations. The ICJ acknowledged applying the equitable and reasonable use principle with the no-harm principle as a part of international law governing transboundary water sharing. The case brought to the forefront the linkage between procedural and substantive obligations for implementing the equitable and reasonable usage with no-harm principle in the context of a complex treaty. The Court stated that there is a functional link between procedural and substantive obligations as they complement each other for a common purpose but with relative independence. The substantive obligation of equitable and reasonable utilisation and the related duty of preventing transboundary harm are both crucial and very general in their standard and operationalised by procedural requirements. The violation of procedural requirements could either lead to harm or may not lead to any harm. In cases where there is harm, the procedural violation will be helpful to establish the failure to meet the due diligence standard and the duty to prevent transboundary harm. The Court identified that the purpose of sharing information with the CARU, as mandated in Article 7, was to allow parties to cooperate and negotiate possible changes in the project. The delay in sharing the information by Argentina rendered this provision ineffective, violating the procedural provision of the treaty.

ICJ considered Articles 1 and 27 of the 1975 Statute concerning the substantive obligations to identify that the treaty was created for the optimum and rational utilisation of the river. A balance must be created between the parties' rights and the need to use the river for economic and commercial activities. Article 27 brought in an additional level of scrutiny, requiring the parties to strike a balance between the use of the water and the protection of the river, consistent with the objective of sustainable development. The decision has been hailed as the ICJ's first attempt to consider the nature of sustainable development in the background of international water courses, which can broadly be understood to include certain procedural and substantive obligations.

Argentina contended that the environmental impact assessment undertaken by private entities and submitted to CARU did not meet the international standards and, therefore, cannot be considered a valid document for allowing the construction of the mills. Considering Article 41 with due diligence, the ICJ recognised that the State has the responsibility to adopt appropriate rules and measures within the framework of the domestic legal forum to protect and preserve the aquatic environment and to prevent pollution and also a certain level of vigilance in the enforcement and exercise of administrative control including monitoring safeguarding the rights of the parties. The court was clear that the due diligence requirements are imposed upon the state under the customary duty of preventing transboundary harm. The court specifically identified the undertaking of an environmental impact assessment as a non-negotiable due diligence requirement. Further, the statute created a procedural requirement to cooperate in good faith, including the duty to notify in cases of transboundary harm. Reliance was placed on Article 7 of the 1975 Statue to establish the requirement to inform a neighbouring state of a project or activity with potential transboundary effects as soon as possible.

Examining the key international agreements and standards, the Court observed that since both parties are not signatories to any international environmental Impact Assessment (EIA) agreements, they can determine the requirement process within their domestic legislation. This point has been crucial in the debate on the application of soft law and hard law in International Law. Recently, many sectors have seen an increasing growth of soft law regimes. The ICJ clarifies that these soft law provisions do not create a binding obligation. However, it guides States to structure their domestic laws or even guide bilateral agreements, followed by a treaty or agreement. In this case, the absence of any hard law on the environmental impact aspect made it impossible for the Court to comment on the validity of the data, allowing the domestic legislation to prevail, which, as some have argued, led to restricting any benchmark for transboundary environmental impact assessment (Zhu 2012). The Court considers Article 40 to elaborate that an EIA is necessary to protect and preserve the aquatic environment from activities that may cause transboundary harm. The court does indicate that the EIA must be undertaken prior to the implementation of thc project and should be considered before any decision to authorise the proposed project. An essential aspect of the decision is that it indicates that EIA must be followed when necessary by continuous monitoring of the project and its impact on the environment throughout the project's life as it is a necessary extension of the State's obligation of due diligence and prevention of transboundary harm. Considering the actual harm done to the river, the ICJ considers scientific data from both parties to conclude that the amount of dissolved oxygen phosphorus, and phenolic substance did not affect the quality of the river.

This is an important case which provides an example of how the ICJ flagged optimum and rational utilisation to cover environmental damage and provided a framework of due diligence process to be undertaken as a part of procedural obligation, thus offering a justification for information sharing and conducting an EIA. Similar terms have been used in the three South Asian treaties covered in the previous section of the paper, and they open up the possibility of creating a procedural framework.

10.5.3 Dredging by Nicaragua and Road Construction in Costa Rica Along the San Juan River (Costa Rica v. Nicaragua)

The two critical issues in this dispute relevant to this paper are the dredging of the San Juan River by Nicaragua and the construction of a road by Costa Rica, polluting the river in Nicaragua. The main contention in this case was on the territorial limits between the two parties. Costa Rica presented that, as per Article 9 of the 1858 Treaty of Limits, Nicaragua cannot use the San Juan River to carry out hostile acts or carry any act which causes damage to Costa Rican territory. Further, the presence of armed forces in the disputed territory and the dredging of the river by Nicaragua violated the sovereignty and territorial integrity of Costa Rica. The Ramsar Convention was included to indicate that Nicaragua had an obligation to consult with Costa Rica to coordinate future policies and regulations concerning the conservation of wetlands and their flora and fauna. The ICJ concluded that the right bank of the channel in which Nicaragua dredged is not part of the boundary. Therefore, the disputed territory belongs to Costa Rica, and the activities undertaken by Nicaragua in that territory (like establishing a military presence) were in breach of Costa Rica's territorial sovereignty. Concerning the compensation calculation for the damage caused by the occupation of Nicaragua, the ICJ considered the environmental damage and the cost of environmental recovery in the disputed territory to arrive at the compensation amount.

For the dredging carried out by Nicaragua, the ICJ relied on the Pulp Mills case and observed that each state is free to determine the process and content of the environmental impact assessment based on its domestic legislation. Relying on the report developed by Nicaragua in 2006, the ICJ accepted that the dredging activity would not cause significant transboundary harm. Therefore, there was no need to carry out an environmental impact assessment.

In December 2010, Costa Rica started working to construct a road within the Costa Rican territory along part of its border, which Nicaragua challenged due to water pollution caused by road construction. Costa Rica contended that a state of emergency was declared in the region because of a border dispute. Therefore, Costa Rica was exempted from the obligation to conduct an environmental impact assessment based on their domestic law. Considering the arguments of Costa Rica, the Court identified that although Costa Rica alleges that there was no evidence of any harm in the preliminary assessment, a substantial part of the project would affect the river, which was near a sensitive wetland. Taking note of this, the Court concluded that Costa Rica's construction of the road carried a risk of significant transboundary harm. Therefore, there was an obligation to conduct an EIA. The Court did not find any legal obligation in the CARU for the parties to coordinate, share information or consult with each other. And yet, the Court ordered Costa Rica to consult in good faith with Nicaragua to undertake measures to prevent significant transboundary harm. Concerning the substantive obligations to quantify the actual harm caused by the sediments, the court was not willing to consider the scientific data submitted by Nicaragua as there was no clarity on the quantity of sediment deposited in the river after the construction of the road.

A major significance of the case is to support the proposition made in the Pulp Mills case on the significance of conducting EIAs where there is a risk of transboundary harm. The factual circumstances used in this case from scientific reports to past assessments done indicate the strategy which can be used to substantiate the existence of transboundary harm to invite the necessity of an EIA. The importance given to the procedural aspect of conducting an assessment could set a precedent for states to work towards bringing procedural clarity and incorporating standards required to reduce risks. The scope of the duty of information-sharing consultation is to be derived from the agreements between the states, which can help establish the violation of procedural and, in some cases, substantive obligations.

10.6 Operationalising Transboundary Harm in South Asia

The treaties and agreements considered in this paper on transboundary governance in South Asia focus on the division of water, in many cases, quantifying the exact amount to be released by states. Many political reasons at the time of the negotiations can be attributed to the specificity of these agreements. However, modification or transition may be necessary since many have proven ineffective or never utilised (Thomas 2017).

Most of these agreements are based on principles like optimal use, equity, fair play or equality. The principle of equitable and reasonable use is considered a focal point of customary international law on transboundary water governance, codified in the UNWC 1997 and in multiple decisions of the ICJ. The aspect of preventing transboundary harm or the no-harm principle is reflected in a restrictive or indirect manner in the South Asian treaties like the obligation to maintain natural flow or prevent a material change of the river for the lower riparian states or to carry out preventive measures of environment protection up to a reasonable standard (IWT) or employing the principle of equity, fair play, and no-harm in case of emergency (GWT).

Incorporating due diligence in the context of transboundary harm, as envisaged in the Pulp Mills Case, will help expand and develop some of these procedural requirements. The ICJ decision on the Gabcikovo-Nagymaros Case laid down the foundation of the no-harm principle with substantial inputs from the Trail Smelter case. Connecting the no-harm principle with due diligence, the ICJ established that states must assess the nature of their action and impact on other countries, moving away from the idea of absolute territorial sovereignty and has ensured that the states' obligation to prevent significant transboundary harm is now recognised as a part of customary law mandating due diligence to prevent such harm. Seen mostly as a critical aspect of environmental cases, the principle is finding more relevance in many contemporary issues like cybersecurity (Takano 2018).

The challenging part is that the ambit of the threshold of harm and standards to reduce harm is not very well defined by the ICJ, and it has relied primarily on the practice between the states or the legal instruments in use between the parties (Pulp Mills Case). There needs to be a unified idea of what standards are required to satisfy the minimum threshold to constitute harm. Many international environmental policies consider the precautionary principle to create an obligation on states to prevent foreseeable significant damage or minimise the risk. The Pulp Mills case relied extensively on the scope of the 1975 Statute to establish its jurisdiction only on matters relating to river pollution, not noise pollution.

A close look at the South Asian agreements in the paper indicates that these agreements are very restrictive, and many disputes exist regarding the definition and scope of water use. Like in the Indus Water Treaty, the usage of non-consumptive use and production of hydroelectric power have been the focus of much debate in the Kishanganga Arbitral Award. Where a treaty ratified in 1960 specifically allows the construction of hydroelectric dams with some specifications, the courts and the ICJ may not be willing to disallow these activities even in cases where there is transboundary harm. The relatively newer guidelines on environment protection, or disaster risk reduction, like the Paris Agreement, 2015 or the Sendai Framework, 2015, may not necessarily be considered while identifying minimum standards for assessing a state's mandate of due diligence while interpreting a treaty like the IWT on transboundary water.

A significant contribution of the ICJ's decision, especially the Pulp Mills case and Costa Rica v. Nicaragua case, is the obligation to conduct EIA in cases with a risk of transboundary harm. Most South Asian agreements in the 60s and early 90s have limited provisions concerning environmental protection. This is perhaps because environmental issues were more benign and not as severe as in the present day. In light of this dated and restrictive approach, the decisions of the ICJ mandating EIAs as an essential part of due diligence to avoid transboundary harm are to be welcomed in transboundary agreements. However, the EIAs will be applicable only in cases of environmental harm. While the ICJ has expanded the purview of due diligence, the implementation of respective EIAs is subject to state practices, which vary greatly (Knox 2002). As a consequence, some authors contend that EIA is reduced to being a part of the 'myth system' of international environmental law (Bodansky 1995). This means that while EIAs are legally required, the enforcement in letter and spirit is difficult. The decisions of the ICJ create a necessary provision towards preventing transboundary harm, but the reliance on domestic laws for implementation creates uncertainty. The problem is further compounded by the dated international agreements in South Asia, which vary significantly across the nation-states.

Procedural obligations have also been considered an essential aspect of due diligence. The duty to cooperate is an important procedural duty, including the duty to notify and consult with the concerned state (Handl 1980). As observed in the Costa Rica and Nicaragua case, the scope of duty and procedure of information sharing is to be derived from the agreements between the states, which can help establish the violation of procedural and, in some cases, substantive obligations. Most of the South Asian agreements provide some scope for notification, specifically through the commissions constituted under the agreement. The importance and utility of this procedural obligation depend strongly on the powers given to the commission. In light of this, South Asian countries need to recognise the critical role played by these commissions, reorient them towards conflict prevention, and work on water conservation and disaster risk reduction collaboratively. From primarily articulating the demands made by the states (of the Commission) for the parochial interest of the respective nation-states, these commissions can be transformed into a more robust body working towards addressing the more universal concern of climate change.

In light of the pressing demands of climate change and the impact of increasing frequency of disasters, especially hydrological events like floods, transboundary water governance in South Asia requires a cooperative approach to deal with extreme events. The ICJ decisions provide an insight into the different avenues under which the dated and ineffective agreements in South Asia can be re-oriented towards addressing the compelling needs of the present.

10.7 Concluding Remarks on the International Court of Justice

The treaties on transboundary water governance in South Asia focus on specifying the precise amount of water, governments must release in the year. The specificity of these agreements can be attributed to numerous political considerations at the time of the negotiations. These treaties are based on limited territorial sovereignty and respect the no-harm principle in different forms. The operationalisation of this principle is not very clearly articulated in the treaty and varies from time to time, heavily dependent on the political relationship between the countries. Climate change will unavoidably force states to abandon these limited outcomes of protracted discussions soon and require that the states build a more robust framework for transboundary governance of water and other natural resources.

Given these expected future transitions, the ICJ rulings' contribution to fair and reasonable use is essential to international law on transboundary water governance. It is important to recognize the issues around the structure of the ICJ, and that judges are required only to apply the law, which others had created—implementing the legislator's will, often reduced to being a mouthpiece (Series 2018). On similar lines, as per the ICJ Statute, Article 59, the decisions are binding only on the parties, and there is no recognition of the theory of precedents (Venzke 2018). The ICJ decision is a subsidiary source of international law at par with scholarly decisions (Article 38 of the ICJ Statute). Considering this structure, it is crucial for States to embrace these principles proactively, specifically in light of the global issue of climate change.

While it is true that the judges are the mouthpiece of the legislative body, it is also essential that the law needs to be explained and interpreted before it is pronounced (Waldock 1983). It is well recognised that the ICJ clarifies and shapes the law by its judgements and separate opinions (Jennings 1995). The ICJ identifies the states' responsibility to prevent significant transboundary harm, including the practice of due diligence to prevent such harm, through its judgements. The no-harm principle, no transboundary harm, has existed for some time. With its input, the ICJ was able to clarify the operationalization of these principles and thereby develop practices like due diligence and EIAs as a part of the principle. This way, the decisions mentioned in this paper trace the contribution of the ICJ towards the normative propositions of the state of general international law on transboundary harm.

Considering case laws given by the ICJ on human rights protection, State responsibility and the Law of Seas, it is suggested that the ICJ decisions do contribute towards exploring linkages between specialised areas of international law and related fields by mainstreaming a specialised field into the general principles of international law (Sloan and Tams 2013). A similar trend can be observed in the context of the ICJ's development of international environmental law, where issues on water pollution, environmental harm, and transboundary water governance are connected to general principles of international law, like the no-harm principle. This practice contributes to the development of international environmental law, thereby providing a ground for countries to consider these issues while dealing with the challenges of transboundary harm. For South Asian countries, where political tensions have affected the implementation of the treaties and therefore, adopting state responsibilities in the context of transboundary water could be challenging. Focusing on mainstreaming, for specialised areas like transboundary water governance which so closely linked to increased frequency of hydrological disasters due to climate change, using general principles of international law like the no-harm principle as given by the ICJ to frame the negotiations could be a useful start.

For climate change, it is understood that the global system operates with a rule of negotiation where agreements are punctuation marks of an ongoing process of debate among states (Bodansky 2017). This indicates that even though the state governments have significant power in formulating agreements, judicial opinion can influence the negotiation process by setting the terms of debate, evaluating standards and establishing a framework of principles for negotiations (Bodansky 2017). With its clarification on transboundary harm, the ICJ can be a valuable source to establish a framework for negotiating transboundary water-sharing treaties and institutions in South Asia. The decision of the ICJ is also understood to provide a groundwork for future litigation in terms of building arguments based on the past decisions even though they are not binding (Alvarez-Jiménez 2011).

Transboundary water governance treaties in South Asia are highly specific, reflecting political considerations of their time and have faced similar operationalisation challenges. Climate change pushes states to move beyond these limited agreements and develop a more robust framework. The paper argues that ICJ's role in shaping international law on transboundary water governance and the fair and reasonable use of shared resources is significant. While judges primarily apply the law, they play a crucial role in interpreting and clarifying it. The ICJ's contributions include defining states' responsibilities in preventing transboundary harm and linking specialised areas of international law with general principles. This approach contributes to developing international environmental law and offers guidance for dealing with transboundary challenges. In the context of climate change, the ICJ can provide a framework for negotiations and influence the terms of debate among states. Though not binding, its decisions ought to serve as a foundation for future treaties.

References

Ahmed ID (1999) Water, power and people: a South Asian manifesto on the politics and knowledge of water. Water Nepal 7:113–121

Akhter M (2010) More on the sharing of the Indus Waters. Econ Polit Wkly 45:99–100

Alvarez-Jiménez A (2011) Methods for the identification of customary international law in the International Court of Justice's jurisprudence: 2000–2009. Int Comp Law Q 60(3):681–712

Award in the Arbitration regarding the Indus Waters Kishenganga between Pakistan and India, 2011-01 (Permanent Court of Arbitration December 2013).

Bagale DR (2020) Mahakali Treaty: delay in implementation and resulting impacts from Nepal's perspective. Water Policy 22:658–669

Bodansky D (1995) Customary (and not so customary) international environmental law. Ind J Global Legal Stud 3:105–119
Bodansky D (2017) The role of the international court of justice in addressing climate change: some preliminary reflections. Ariz. St. LJ 49:689
Brüggemann R, Halfon E (1990) Ranking for environmental hazard of the chemicals spilled in the Sandoz accident in November 1986. Sci Total Environ 97:827–837
Chimni BS (1998) The principle of permanent sovereignty over natural resources: toward a radical interpretation. IJIL 208:213–214
Corfu Channel, United Kingdom v Albania, Judgment, Compensation, ICJ Rep 244, ICGJ 201 (International Court of Justice December 15, 1949)
Desai B (1998) Sharing of international water resources: the Ganga and Mahakali River treaties. Asia Pac J Environ Law 3:172
Gabcikovo-Nagumaros Project, 1997 I.C.J. 3 (International Court of Justice February 1997)
Handl G (1980) The environment: international rights and responsibilities. ASIL Proc 74:223
Jennings RY (1995) The international court of justice after fifty years. Am J Int Law 89(3):493–505
Just RN (1998) International water resource conflicts: experience and potential. In: Just RN (ed) Conflict and cooperation on trans-boundary water resources. natural resource management and policy. Springer, pp 1–26
Kliot NS (2001) Institutions for management of transboundary water resources: their nature, characteristics and shortcomings. Water Policy 3:229–255
Knox JH (2002) The myth and reality of transboundary environmental impact assessment. Am J Int Law 96:291–319
Lauterpacht H (1982) The development of international law by the International Court. Cambridge University Press
Legality of the Threat or Use of Nuclear Weapons, Advisory Opinion, ICJ GL No 95, [1996] ICJ Rep 226 (International Court of Justice July 1996)
Lewis P (1986) Huge chemical spill in the Rhine creates havoc in four countries. New York: New York Times
Malone LA (1987) The Chernobyl accident: a case study in international law regulating state responsibility for transboundary nuclear pollution. C J Environ Law 12:203
Miller RA (2006) Surprising parallels between Trail Smelter and the global climate change regime. In: Bratspies RM (ed) Transboundary harm in international law: lessons from the trail smelter arbitration. Cambridge University Press, pp 167–180
Myint T (2012) Governing international rivers: polycentric politics in the Mekong and the Rhine. Edward Elgar
Nakamichi M (1997) The International Court of Justice decision regarding the Gabcikovo-Nagymaros Project. Fordham Environ Law Rev 9:337
Pandey P (2012) Revisiting the politics of the Ganga water dispute between India and Bangladesh. India Q 68:267–281
Pulp Mills on the River Uruguay, Argentina v Uruguay, Order, Provisional Measures, ICJ GL No 135, [2006] ICJ Rep 113, (2006) 45 ILM 1025, ICGJ 2 (ICJ 2006) (International Court of Justice July 2006)
Qureshi WA (2016) Equitable apportionment of shared transboundary river waters: a case study of modifications of the Indus waters treaty. San Diego Int Law J 18:199
Rao P (2002) Prevention of transboundary harm from hazardous activities—a sub-topic of international liability. Environ Policy Law 32:27–32
Read JE (1963) The Trail smelter dispute. Can Yearb Int Law 1:213–229
Rubin AP (2006) Pollution by analogy: the Trail smelter arbitration [abridged]. In: Bratspies RM (ed) Transboundary harm in international law: lessons from the Trail smelter arbitration. Cambridge University Press, pp 46–55
Salman S (2007) The Helsinki Rules, the UN Watercourses Convention and the Berlin Rules: perspectives on international water law. Water Resour Dev 23:625–640

Salman, SMA (2003) Conflict and cooperation on South Asia's international rivers, legal perspective. World Bank, Washington, DC
Schwabach A (1989) The Sandoz spill: the failure of international law to protect the Rhine from pollution. Ecol Law Q 16:443
Series GML (2018) 1. Decisions of the ICJ as sources of international law? Decisions ICJ Sources Int Law? 7
Sloan J Tams CJ (2013) The development of international law by the International Court of Justice. Hague Yearb of Int law 26:216–238. Brill Nijhoff
Stec S (1998) Of Solemn Oaths and obligations: the environmental impact of the ICJ's decision in the case of the Gabcikovo-Nagymaros project. Yearb Int Environ Law 8:41
Subedi SP (1999) Hydro-diplomacy in South Asia: the conclusion of the Mahakali and Ganges River treaties. Am J Int Law 93:953–962
Takano A (2018) Due diligence obligations and transboundary environmental harm: cybersecurity applications. Laws 7:35
Thomas K (2017) The Ganges water treaty: 20 years of cooperation, on India's terms. Water Policy 19:724–740
Thorson EJ (2008). Sharing Himalayan glacial meltwater: the role of territorial sovereignty. Duke J. Comp. & Int'l L. 19:48.
Tignino MB (2020) The role of international case law in implementing the obligation not to cause significant harm. Int Environ Agreem 20:631–648
Tyagi Y (2015) Permanent sovereignty over natural resources. Camb Int Law J 4(3):588–615
Uprety K (2011) Legal aspects of sharing and management of transboundary waters in South Asia: preventing conflicts and promoting cooperation. Hydrol Sci J 56:641–661
Venzke I (2018) The International Court of Justice during the battle for international law (1955–1975)—colonial imprints and possibilities for change. The Battle for International Law in the Decolonization Era (2018), Amsterdam Law School Research Paper, (2018-08)
Vučić M (2019) Access to water in the context of the international watercourse: a theory of the community of interest. Anali Pravnog fakulteta u Beogradu 67:73–94. https://doi.org/10.5937/AnaliPFB1903077V
Waldock H (1983) The international court of justice as seen from bar and bench. Br Yearb Int Law 54(1):1–5
Zawahri NA (2009) India, Pakistan and cooperation along the Indus River system. Water Policy 11:1–20
Zhu X (2012) International Court of Justice's impact on international environmental law: focusing on the Pulp Mills case. Yearb Int Environ Law 23:106–130
Xue H (2003) Transboundary damage in international law (Vol. 27). Cambridge University Press

Saumya Kumar works as Assistant Professor at the Jamsetji Tata School of Disaster Studies, Tata Institute of Social Sciences (TISS), Mumbai. She teaches courses on Humanitarian Action, Transboundary Governance, Peace and Conflict Studies courses for the MA/MSc Programme in Disaster Management. Saumya has been awarded a Fellowship—for the Commonwealth Futures Climate Research, by the Association of Commonwealth Universities in 2024–25. She graduated from NLU-Jodhpur, and possesses a Master's degree from NUJS, Kolkata, specialising in International and Comparative Law. She is also pursuing her doctoral study at TISS on Climate Change Litigation in India.

Chapter 11
Water Issues in Small and Seasonal Transboundary Rivers and Aquifer Systems Across Bihar, India and Nepal

Aparna Unni, Eklavya Prasad, Ajaya Dixit, and Rakesh Mishra

11.1 Introduction

As neighbouring countries, northern part of India is closely interconnected with Nepal. Many of India's major, medium, and small rivers originate in the mountainous catchments in Nepal. Among the major river systems are—the Mahakali, the Karnali, the Gandak and the Kosi. While these are well studied, other medium and small rivers originating in Nepal and flowing downstream into India, have received less attention. This paper therefore focuses on the dynamic, water-related issues concerning small trans-boundary rivers, using illustrations from the border areas in Bihar with the Terai of Nepal.

One predominant water-based risk in these small trans-boundary rivers is of flash floods which have extremely localized impacts. Border areas of Bihar's districts like Pashchim (West—W) Champaran, Purbi (East—E) Champaran, Araria and Kishanganj face repeated damage from flash floods from these *pahadi* (hilly) rivers. Furthermore, unlike riverine floods, flash floods can occur multiple times anytime in the year between January and October. One assessment done in Gaunaha block of Pashchim Champaran noted between 30 and 50 instances of flash floods in a year.

Another shared water-based risk in the region comes from arsenic contamination of groundwater. A spatial and temporal heterogeneity in arsenic concentrations is seen

A. Unni
Kerala, India

E. Prasad (✉)
Megh Pyne Abhiyan, Delhi, India
e-mail: meghpyneabhiyan@gmail.com

A. Dixit
Kathmandu School of Law, Kathmandu, Nepal

R. Mishra
Kathmandu, Nepal

J. Andharia and P. Anand (eds.), *Shifting Lands, Flowing Waters*, Disaster Studies and Management, https://doi.org/10.1007/978-981-96-7691-0_11

in the Terai aquifers of the Gangetic Plain. This raises concern as rural populations on both sides of the border are mainly dependent on groundwater for drinking.

A landscape impacted by regular flash flood from small transboundary rivers and encountering arsenic contamination in groundwater requires a governance framework which will address these issues of water-based risk in a carefully calibrated manner. Such a framework can then be extended to similar landscapes in other parts of the India-Nepal border.

11.2 Typologies of Transboundary Rivers Between Nepal and India

The mountain ranges and forests of Nepal are extremely important watersheds, not just for itself but also for its southern neighbour India. They are the birth place of countless streams and tributaries which feed and form major well-known rivers which traverse downstream into India. It would not be an exaggeration to state that these transboundary rivers and water-sharing agreements surrounding them lie at the heart of bilateral relations between the two countries.

These river systems have been classified into three types based on their origin. First, there are the four big transboundary river systems which arise in the Greater Himalayas that stand at an average elevation of 6000 metres (m), snow-fed and perennial. These are well-known and documented—the Mahakali (Sharda), Karnali (Ghaghara), Narayani (Gandak) and Koshi (Kosi). Then there are the slightly smaller transboundary rivers, perhaps not as well-known and include those perennial and rain-fed rivers which originate in the Mahabharat range (at an altitude of 1500–4500 m) in Nepal. These are the West Rapti (tributary of Ghaghara), Kamla and Bagmati (Kosi tributaries in Bihar), Mechi and Kankai (both tributaries of Mahananda, a major transboundary river between Bangladesh and India) (NCVST 2009).

The third type of transboundary rivers are the much smaller and seasonal, that originate in the Chure, i.e. Siwalik, range of Nepal which is at an elevation of 500–1500 m. The Siwalik is the youngest low elevation range, between the Mahabharat range in the north and the Terai plains in the south that runs continuously from Indus River in Pakistan in the west, throughout the length of Nepal until it reaches Brahmaputra River in India in the east (Ghimire 2016). These innumerable small monsoon-fed seasonal rivers from the Chure range in Nepal, cross the India and Nepal border (in Bihar), acquiring transboundary character. These rivers have extremely variable flow and short length. Individual river course flows through a few neighbouring districts on both sides of the border.

Despite the multitudinous shared river systems, large and small, between the two countries, it is only the four major Himalayan rivers that have hitherto garnered attention. Most analyses of water-sharing agreements between Nepal and India have been confined to the treaties of Kosi (1954, 1966), Gandak (1959, 1964) and Mahakali (1996), all of which focus on hydropower, irrigation and/or flood-control.

The Mahakali treaty also incorporates the 1920 treaty on the Sharda barrage, irrigating parts of India's Uttar Pradesh and parts of Nepal's Kanchanpur district. The Kosi and Gandak project mostly serve Bihar in India and some areas in Nepal.

The state of Bihar in India shares approximately 601 kilometre (km) of international border with Nepal (Tewary 2020). This estimate however, varies. Another estimate puts the length of the international border between Bihar and Nepal to 726 km, the highest compared to the international border along the four Indian states—Uttarakhand (275 km), Uttar Pradesh (551 km), West Bengal (100 km) and Sikkim (99 km) (Goyal 2020). The districts in Bihar along the India-Nepal border are Pashchim (W) Champaran, Purbi (E) Chamapran, Sitamarhi, Madhubani, Supaul, Araria and Kishanganj, while the adjoining districts in Nepal are Parsa, Chitwan, Bara, Mahottari, Sariahi, Rautahat, Dhanusa, Siraha, Saptari, Sunsari, Morang and Jhapa. These border districts encounter numerous challenges such as recurring flash floods, and arsenic contamination of the groundwater threatening people and habitations along the international border.

This paper discusses two predominant water challenges of the region, (i) flash floods in the smaller transboundary rivers and (ii) arsenic contamination in transboundary aquifers. Rather than focus on the entire border between the two countries, the paper will focus on the region contiguous to international border between Bihar and Nepal. The paper also proposes a governance framework to manage the challenges. Though the Chure rivers are central to the well-being of the people living in their proximity, they are largely invisible in the bilateral discussions and policies. This oversight is surprising despite considerable impact of floods and seasonal droughts on the lives and livelihoods of those living along them on both sides of the international border.

The Chure rivers drain an area of about 18,860 square kilometre (km^2) in Nepal which represents around 12% of the county's area (Shrestha and Silwal 2017). As mentioned above, there are no estimates of the discharge by the innumerable small, seasonal transboundary Chure rivers across the international border. Indirect estimates suggest that the Chure rivers have a combined annual run-off of 456 cubic metres/second (m^3/s) (Shrestha and Silwal 2017). In contrast, a large river's discharge is many times this combined run-off. The mean annual flow of the Kosi at Chatara, for example, is 1500 m^3/s.

Even an accurate number of these rivers are hard to find let alone the details on seasonal/annual discharge, sediment load and water quality. MPA's[1] research has identified 148 small and seasonal transboundary rivers across the international border in the seven bordering districts of Bihar (MPA 2020). The information was compiled by holding discussions with local inhabitants, and referring to secondary documentations such as media reports, and Government of Bihar's (GoB) documents.

These are preliminary estimates and it is still challenging to arrive at an authentic number of these smaller transboundary rivers in the seven districts of Bihar. The

[1] MPA: Megh Pyne Abhiyan, a public charitable trust working on water and sanitation issues in the water-stressed rural and urban areas in East India focussing on the flood-prone region of north Bihar, and water deficient and source contaminated areas of Dhanbad district in Jharkhand.

districts are Pashchim (W) Champaran, Purbi (E) Champaran, Sitamarhi, Madhubani, Supaul, Araria and Kishanganj.

Another layer of complexity that exists in pockets across the Bihar-Nepal border is due to arsenic contamination in groundwater. According to the Central Ground Water Board's (CGWB) study on Ground Water Quality in Shallow Aquifers in India, 18 districts in Bihar arc at risk due to arsenic contamination in groundwater (CGWB 2018). A study by A. K. Ghosh[2] also suggests that 18 as arsenic contaminated. While the CGWB study suggests Pashchim (W) and Purbi (E) Champaran on the Bihar-Nepal border affected by arsenic contamination, the study by Ghosh suggests that four border districts on the Bihar-Nepal border—Pashchim (W) Champaran, Purbi (E) Champaran, Supaul and Kishanganj are affected (Jamwal 2015).

In Nepal, the districts with reported arsenic contamination are Sunsari, Saptari, Siraha, Dhanusha, Sariahi, Rautahat, Bara, Nawalparasi, Rupandehi, Kapilbastu, Banke Kailali and Kanchanpur. The contamination level in these districts varies from 10 to 50 microgram/litre (μg/l) and higher than 50 μg/l while Jhapa, Morang, Mahottari, and Parsa have reported of low levels of arsenic concentrations (Thakur et al. 2011).

11.3 Flash Floods Along the India-Nepal Border

Flash floods are the common water hazard associated with the small and seasonal transboundary rivers of this region. The main meteorological phenomena causing intense rainfall are cloudbursts, the monsoon trough, and monsoon depressions (Shrestha 2008). It is crucial to understand this phenomenon for addressing the related problems so that appropriate policy measures and management systems can be put in place supported by on-ground preparations. The main factors triggering flash floods in the border region are intense rainfall and landslide dam outburst floods (Shrestha 2008). This paper will deal largely with intense rainfall.

In 2019, floods triggered by intense rainfall affected a total 14.9 million people in 27 districts of Bihar. Approximately 42% of the total affected population were from the seven districts bordering Nepal. See Table 11.1.

Diverse types of floods affect the population in the border districts of Bihar and flash flood is one of them. This became evident between July 7 and 13, 2019, when heavy rainfall in the border districts and along the India-Nepal border led to flash floods in seven border districts of Bihar. But there is no data to prove the differential impact caused by riverine or flash floods. According to official state data, as of July 18, 2019, heavy rains in Bihar and Nepal affected 4.68 million people in 831 panchayats under 92 blocks in 12 districts of Bihar. Nearly half of the districts affected by this rainfall might have faced flash floods.

[2] Former Chairperson of Bihar State Pollution Control Board and former Professor-in-Charge, Department of Environment and Water Management, A N College, Patna.

Table 11.1 Flood-affected population in the border districts of Bihar in 2019

District	Affected population (in million)
Araria	1.11
Kishanganj	0.23
Madhubani	1.88
Pashchim Champaran	0.009
Purbi Champaran	0.67
Sitamarhi	2.08
Supaul	0.25
Total	6.23

Source Disaster Management Department, Government of Bihar (2019)

In the same year, on the other side of the international border in Nepal, landslides and floods caused by torrential monsoon rains affected 28 eastern districts. Siraha, Sarlahi, Saptari, Mahottari and Rautahat districts in Madhesh Province; and Sunsari, Morang and Udayapur districts in Province 1 were the most severely affected districts. Six of these severely affected districts, namely Siraha, Saptari, Mahottari, Rautahat, Sunsari and Morang, lie on the border with Bihar India. The incessant rainfall in different parts of the country since July 11, 2019, triggered flooding and landslides in the south-eastern districts of Nepal particularly in Province 1 and Province 2. Out of 31 affected districts, the worst affected districts were Mahottari, Dhanusa, Rautahat, Sarlahi and Siraha. About 76,990 families (384,950 people) were affected since last week. According to the situation report from the Nepal Ministry of Home Affairs (MoHA), 78 people lost their lives, 22 people injured and another 22 were missing. In ministry's daily situation report, Nepal Red Cross Society (NRCS) was informed that 8798 families across 31 districts have been displaced by the current floods and landslides (IFRC 2019).

A growing body of evidence indicates that the frequency and intensity of flash floods are increasing in the countries of the Hindu Kush Himalayan region.[3] Flash floods are severe events that occur with little or no warning (Shrestha and Bajracharya 2013). In Bihar, the state Flood Management Improvement Support Centre (FMISC) describes flash floods as "floods from Nepal rainfall, where the lead time is short (8 hours) and recession fast." This definition is based on flood type, source area, warning time, flood duration and recession, and impact on agriculture.

The recurrence of floods in the small and seasonal transboundary rivers in districts of Bihar bordering Nepal has largely gone unnoticed prior to 2017. That year border districts of Bihar faced high impact as these smaller rivers experienced massive flooding. In 2017 Nepal experienced incessant rainfall from August 11 to 14, resulting

[3] In Xichang and Niujuangou in Sichuan, China; Dhemaji District in Assam, India; the Bhote Koshi/Sun Koshi basin (two case studies), Lal Bakaiya, and Madi in Nepal; and Chitral District in Pakistan.

Table 11.2 People affected in 2017 floods from districts along India-Nepal border

District	Affected population (in million)
Araria	1.85
Kishanganj	0.97
Madhubani	0.76
Pashchim Champaran	1.08
Purbi Champaran	2.22
Sitamarhi	2.27
Supaul	0.41
Total	9.56

Source: Disaster Management Department, Government of Bihar (2019)

in widespread floods across 35 of the country's 77 districts. Several districts experienced the heaviest rainfall in over 60 years and about 80% of the land in substantial parts of the Terai was inundated (Govt. of Nepal 2017). Most rivers flowing south brought flooding in districts of Bihar. The Bihar government acknowledged the phenomenon. Excessive rainfall of more than 400 millimetres (mm) within 24 h had led to flooding of the rivers causing damages to districts on both sides of the border. Araria, for instance, which had been facing a drought-like situation suddenly found itself bearing the brunt of the floods. It received exceptionally high rainfall, and "rivulets and streams, which had been dry for decades, sprang to life. A meandering local river called the Parman became a torrential stream" (Sethi 2017).

The floods in 2017 affected a total of 9.56 million people in the districts of Bihar that border Nepal (Govt. of Bihar 2017). See Table 11.2.

11.4 Need for Systematic Data

That flash floods in the small transboundary rivers are a serious water risk which needs to be addressed is established beyond question. The challenge, however, is getting systematic data of the flooded rivers, the timeline of floods, areas, and people impacted. The available data are assorted lone case studies associated with individual rivers and local media reports which mention in passing the small rivers and areas affected by them when they flood. In 2016, Megh Pyne Abhiyan (MPA) conducted a post disaster recovery need assessment in moderate flood conditions from 2007 and 2016 across three small rivers—Chegraha, Amahawa and Pandai in Pashchim Champaran, Bihar. It showed that the flash floods were not limited just to the monsoon period but could happen any time during the year between January and October. The study suggested that the probability of occurrence was the highest during the monsoon months. Between 2007 and 2016, the Chegraha, had faced 30–60 instances of flash floods in a year. These floods damaged standing crop—the kharif paddy and sugarcane, the major cash crop. The farmers in villages such as Harkatwa and Rupwaliya faced increased indebtedness in the aftermath of the floods. In Harkatwa alone, on an annual basis approximately 50 acres of standing paddy gets destroyed because of floods. The total produce projected in 50 acres is around 600 quintals amounting to a loss of Rs 720,000 (@ Rs 1200/quintal). Flash floods cause similar damages to the region's principal cash crop—sugarcane. During the catastrophic floods of 2010, 2000 quintal of standing sugarcane crop in Harkatwa was destroyed, and the village lost close to Rs 500,000. In addition, siltation, destruction of farm bunds and land erosion bring associated damages (MPA 2017).

Table 11.3 attempts to map the damage caused by the smaller transboundary river floods between 2014 and 2019 in Bihar (*Source* Collated by MPA from various local newspaper reports).

One of the most pressing issues is that such data which identifies and quantifies damages from flash floods in the border districts, is hard to come by as there is no systematic effort to maintain and/or gather information on part of any government body. It is, in fact, essential to map other areas that have been and are likely to be impacted by flash floods.

For an effective governance framework to address water-based risks in the border areas of Bihar with the Terai of Nepal, it is essential to include the transboundary aquifers between India and Nepal which are discussed in the next section.

11.5 Characteristics and Challenges of Transboundary Aquifer Between India and Nepal

In the Indian sub-continent groundwater aquifers are internationally shared.. Groundwater resources (as against rivers) however are not easily perceived as having transboundary dynamics. India shares its groundwater aquifers across its international border. At least eight aquifers have been identified as transboundary aquifers (Dhiman and Jain 2010). They are as follows.

1. Aeolian, Alluvial and Tertiary Sandstone Aquifers (India, Pakistan)
2. Upper Tertiary, Quaternary Alluvial aquifers and Bhabar Terai Aquifer (India, Nepal)
3. Alluvial/Deltaic aquifers (India, Bangladesh)
4. Tertiary Sandstone/Siltstone and Proterozoic (Granite, phyllite, quartzite) aquifers (India, China, Pakistan)
5. Tertiary (Tipam Sandstone) aquifers (India, Bangladesh)
6. Older Alluvium aquifer (India, Bhutan)
7. Sandstone and Siltstone aquifers (India, Myanmar)
8. Proterozoic (Granite-gneiss) aquifers (India, Bangladesh)

The aquifer between the India-Nepal border under the Gangetic plain alluvium runs across the border of the states of Uttarakhand, Uttar Pradesh, and Bihar. It is largely recharged from the Bhabar zone of the Siwalik foothills, partly in Nepal. Bhabar zone

Table 11.3 Snapshot of transboundary river floods in districts along the Bihar-Nepal border

Date	River	District	Remarks
August 19, 2014	Bagmati, Bangari, Tilawe, Dudhaura, Motiya, Sariswa	Purbi Champaran	Increase in water levels due to heavy rainfall of three days
August 24, 2014	Jhim	Sitamarhi	Heavy rainfall in night of Aug 24 plus release of water from Nepal causes embankment breach of Jhim river inundates several villages in Sonbarsa block, cuts off communication; Jhim flows over Kanhauli-Nepal road
August 29, 2014	Rato, Lakhandei	Sitamarhi	Leak from repaired breach of embankment of Rato river in Sursand; Lakhandei inundates several villages in Kanhauli
August 29, 2014	Dhaus, Khiroi, Thumhani, Kokri	Madhubani	Water enters several houses in Navgachi village
July 5, 2015	Marha, Jhim	Sitamarhi	Continuous three days of rainfall in Nepal catchment in early July causes rise in water levels of the two rivers; 18 villages in Parihar and 6 villages in Sursand affected
July 15, 2015	Sikrahana, Bagamti, Tilawe, Bangari, Masah, Tiar, Gad	Purbi Champaran	[news of] Rising waters mounting pressure on embankments [no flood]
July 19, 2015	Lalbakeya, Adhwara group, Khiroi	Sitamarhi	Rivers in spate [apart from the major rivers] owing to high rainfall in Nepal
July 20, 2015	Bakra, Noona	Araria	River levels fallen, but erosion increased. Threat of erosion to bridge at Patariya Ghat on Bakra river; Villages like Tira Khardah, Nemua, Pipra are under threat of erosion from Bakra river
August 22, 2015	Oriva, Sikta, Dhatua	Pashchim Champaran	Sikta block is underwater owing to these rivers
August 22, 2015	Rato, Jhim, Lakhandei, Adhwara	Sitamarhi	Rise in levels of water
August 22, 2015	Sikrahana	Purbi Champaran	Rise in level of water has stopped transportation at Phulwariya Ghat

(continued)

Table 11.3 (continued)

Date	River	District	Remarks
August 24, 2015	Tilawe, Sikrahana	Purbi Champaran	Waters entered several villages in northern Suguali
August 25, 2015	Masan	Pashchim Champaran	River erosion at Mahuawa village in Bagaha block; 10 acres of crops underwater
August 25, 2015	Harbora	Pashchim Champaran	Damage to over 50 feet of road in Gaunaha block between Pachrukhiya and Lakshnauti
August 25, 2015	Sikrahana, Dudhaura, Bangari, Tilawe	Purbi Champaran	Rise in water levels threaten several villages in Sugauli and Banjariya blocks
August 25, 2015	Bachhraja	Madhubani	Erosion causing damage to road between Benipatti and Kataiya
August 30, 2015	Bangari	Purbi Champaran	Rise in level of Bangari river has caused breach of embankment between Baluwa and Vritta villages; water has entered these two villages along with Bairiya village in Ramgarhwa block; people evacuated their homes, several acres of crops underwater
September 1, 2015	Mechi, Mahananda, Chenga	Kishanganj	Heavy rain on August 30 and 31 causes sudden rise in levels in rivers; water enters several villages; kutcha houses and crops destroyed
August 30, 2016	Bhaluhi, Masan	Pashchim Champaran	Bhaluhi river has claimed two out of the three rooms of the primary school in Dhayar Don in Ramnagar panchayat; Bridge [10.5 crore rupees worth] between Bahuwari and Shorwa villages in Ramnagar under threat from erosion by Masan River. Schools in several villages of Don area such as Gardi, Gobarhiya, Khairahni, and Barwa Dih were at risk from hilly rivers erosion
July 11, 2019	Oriya	Pashchim Chamaparan	Water of the hilly river gushed into Dharmarpur village, triggering the threat of flood in Dharmarpur and its adjoining village
July 11, 2019	Tengraha river	Pashchim Champaran	Embankment of Jhikari Tengraha river breached in Barwa Khuda village under Laukariya police station circle, Water enters into villages Barwa Khuda, Binwaliya, Bhelayi, Nautanki, Majhuwa, and Narayangarh; parts of neighbouring East Champaran affected
July 12, 2019	River not mentioned	Araria	Madanpur village [15 km east of Araria hq] submerged after incessant rainfall
July 14, 2019	Mahananda, Kankai, and Mechi rivers	Kishanganj	All blocks affected

(continued)

Table 11.3 (continued)

Date	River	District	Remarks
July 14, 2019	Bakra, Parman, Nuna	Araria	Floods in Palasi, Sikti, Kursakanta, Jokihat, blocks caused by incessant rainfall in catchment;
July 14, 2019	Lalbekiya river (Patehi), Pandai (Dhaka), Gandak (Dhaka and Motihari town)	Purbi Champaran	Heavy rain (Purbi Champaran received 54.4 mm, 214.92 mm and 91.59 mm rain on July 10, July 11 and July 12 respectively); Lalbekaya overflowed and entered Jiwali, Padumkher, Yamuna, Saratha, Karamawa, Chandanwara, Phoolwariya, Badrwa, Lakansen, Pachpakri, Murali, and Bhandar villages in Patahi and Dhaka blocks; 30,000 population impacted in Patehi, Dhaka and Phenhara blocks
July 16, 2019	Sariswa, Tilawe, Gadh which flow into Sikharana and Budhi Gandak	Purbi Champaran	As many as 13 panchayats in Ruxaul block affected; hilly rivers in Nepal overflowed—waters gushed into 14 panchayats affecting 23 villages in Dhaka block
July 25, 2019	Singahi river	Pashchim Champaran [Ramnagar block]	Breach of protection bund of Singahi river between Singha and Baluwa Bagalwa villages at 6 places; Incessant rain and heavy discharge of water (1.14 cusecs) at Valmikinagar barrage, water levels have risen in Pandarai, Singha, Sukhaura and Masan rivers; Water gushed into Baluwa Bangali Tola, Sinahi, Sipahi Madhubani, Gudgudi Hariharpur, Katharwa, Fagunahawa, Raibari Mahuwa, Jhar Mahiyi
July 26, 2019	Hargorda, Pandarai, Maniyari rivers	Pashchim Champaran	Mohammadpur, Dhumnagar, Rajpur Tumkayiha panchayat (all villages), Kesariya panchayat (all villages), Chaturbhujwa, Ajuwa, Shikarpur and Bairiya, Tumkariya in Narkatiaganj were affected; approach route to these villages submerged and crops damaged

is generally permeable and unconfined in nature. Coarse sand, gravel, pebble, and cobble constitute the Bhabhar zone and consists mostly of reworked sediments from the Siwalik rocks. Spring line from Bhabhar zone separates Terai zone and consists of alternating layers of sand, silt, clay, and gravel deposits of varying thickness. The Upper Siwalik conglomerate is a potential aquifer unit intermittently found in Siwalik rocks across the border (Dhiman and Jain 2010). Some sections along India-Nepal Terai are known for their artesian conditions especially for auto flowing wells, which forms prolific aquifers. In general, groundwater flow is towards south from Nepal border. In the last few decades, groundwater development in Terai zone has lowered the piezometric head. Thus, aquifer management across the border is important for ensuring that the aquifers are not over-pumped, depleted and that existing auto flowing condition of artisan aquifers maintained.

The regional geological setting of Bihar-Nepal border across the area is peculiar owing to the largely sedimentary rocks in Nepal and alluvial flood plains in Bihar. They give rise to groundwater systems that have large geographical extent, regional dynamics of stocks, flows and quality. The regional system involves multiple aquifers, contiguous in nature, with complex recharge and discharge characteristics. They are collectively shared by a large number of villages on either side of the border, and thus mean the presence of a large number of stakeholders involved in the use and management. This dependence has implications on the aquifers, and so do other human interventions including urbanisation. They have impacts on quantity, water quality, scarcity, and access. Given the trans-boundary nature of the aquifers, management has to be holistic for sustenance of the resource. Therefore, considering aquifers as a common unit for understanding groundwater resource is imperative (MPA et al. 2014).

Arsenic contamination, one of the biggest threats to safe drinking water in rural areas, is another major water-related risk in the border areas of Bihar and Nepal. Over 90% of the population living in the border districts depend on groundwater for drinking and domestic requirements. In Nepal, the World Health Organization (WHO) detected the presence of arsenic in groundwater in the Terai districts was detected for the first time in 1999 (Thakur et al. 2011). Since then, many individuals and organisations have made multiple attempts to understand arsenic contamination in the region. UNICEF and Nepal's Drinking Water Supply and Sanitation Department conducted one of the earliest testing of the groundwater of Terai in 2001. The study found that 3% of the 4000 tested tube wells across 20 districts had arsenic concentrations above 50 μg/l while 10% exceeded the WHO limit of 10 μg/l. A subsequent study led by Nepal's National Arsenic Steering Committee in 2003 reported that over 23.7% of tube wells in 20 districts of Terai had concentration over the WHO limit (NASC and UNICEF 2007). To summarise, arsenic contamination of Nepal's Terai districts has been well-established, with close to 30% of the tube wells having concentrations above the WHO limit. This is a serious issue since around 47% of Nepal's total population lives in Terai region and, as mentioned earlier, 90% of them rely on groundwater as their primary source of water for drinking. Thus approximately 0.5 million people are at risk as they consume water with contamination that is greater than 50 μg/l (Shrestha 2012).

Parasi District, Nepal's Lumbini Province on Nepal's border with Uttar Pradesh close to Bihar is one of the worst arsenic-contamination-affected districts. In one community-based study, in Patkhouli village of the district the highest prevalence (18.6%) of arsenicosis was found: 95.8% of tube wells were contaminated with arsenic. The districts of Rautahat and Siraha, both along the border with Bihar also have high levels of concentration. In these districts 26% of shallow wells failed to meet the WHO's standard of 10 μg/l. In districts of Parasi, Bara, Parsa, Rautahat,

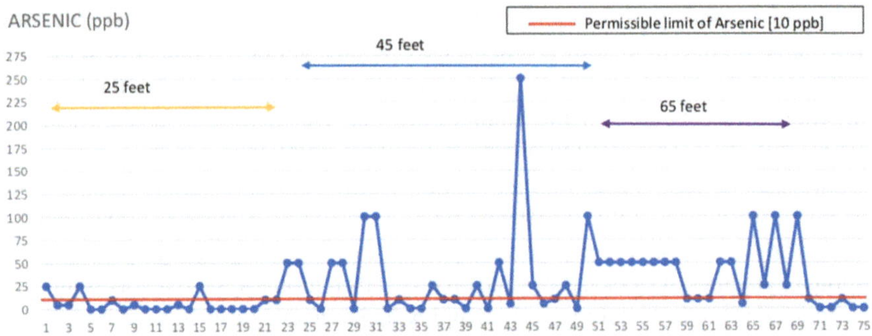

Fig. 11.1 Drinking water profile of a village in W Champaran—mapping presence of arsenic in all water sources[4] (Megh Pyne Abhiyan et al. 2018a, 2018b). *Source* Megh Pyne Abhiyan field study

Rupandehi, and Kapilvastu, the prevalence of arsenicosis has been found to be on average, 2.2% (Shrestha 2012).

On the Bihar side, the Central Ground Water Board (CGWB), recognizes Pashchim (W) Champaran and Purbi (E) Champaran as having problems of arsenic contamination. But Bihar State's own Public Health Engineering Department (PHED), which oversees the supply of safe drinking water in rural areas, does not list any of the bordering districts with Nepal as being arsenic-affected.

Individual case studies, though limited in scope, have established the presence of high levels of arsenic in Bihar's districts bordering Nepal. A study in Khap Tola of West Champaran district close to the international border showed high levels of arsenic (maximum value being 397 μg/l), in excess of the WHO limits of 10 μg/l. The analysis showed that 57% of the samples from private hand pumps in the shallow aquifer zone of 15–35 m have arsenic concentration greater than 50 μg/l (Bhatia et al. 2014). Supaul, another bordering district, has not been recognized by state or central government with arsenic contamination. However, field tests by a local organization between 2013 and 2015 have indicated presence of arsenic in over 100 samples tested in a single ward (Fulkaha Ward number 11 of Kishanpur Dakshin panchayat Kishanpur block) (Megh Pyne Abhiyan et al. 2018a). Incidentally, the official records such as those of the National Rural Drinking Water Program (NRDWP) have listed the same area as having no groundwater contamination. The knowledge of existing groundwater quality in the far-flung rural areas along the border shows differing evidence. This calls first for developing a database from ground up through large-scale testing and building of drinking water profile of villages (see Fig. 11.1 for reference).

[4] The graph was developed as per the water quality analysis of all drinking water sources (in this case, hand pumps accessing water from different depths) in a particular village in Pashchim Champaran district. It was developed by MPA between 2017 and 2018. The line graph shows arsenic contamination in all the hand pumps tested in the village. Points 1–21 along the x-axis in the graph are hand pumps of 25 feet depth. Following this, points up to 52 are hand pumps of 45 feet depth. Points 53–70 indicate hand pumps of 65 feet depth. Beyond this point are hand pumps of 105, 145,

11.6 Contextualizing Water Governance for Small and Seasonal Transboundary Rivers

Small and seasonal transboundary rivers influence and impact communities along the Bihar-Nepal border in multiple manner and a number of factors make them serious. First, there is lack of official acknowledgement of the prevalence of flash floods brought by the small, seasonal transboundary rivers impacting millions in both nations on a regular basis. The impact is not only during the monsoon period. Second, the details of small, seasonal transboundary rivers remain unknown, invisible and unacknowledged by the state and policy makers. Thirdly, there is absence of collecting localized data and governance mechanisms of the small, seasonal transboundary rivers.[5] It is critical for the governance regime to factor in these along the India-Nepal border.

At present, the management of large transboundary rivers is the responsibility of the state and many other actors are involved in the process. There are hardly any mechanisms for the small and seasonal transboundary rivers. The present circumstances necessitate a cross-cutting collaborative effort for generating information about these small and seasonal transboundary rivers. Clearly, a detailed collaborative governance involving the varying constituents is needed in which civil society and other non-state actors at the national, state/province and district levels is needed.

It is crucial that non-state actors, both from Bihar and Nepal work out the collaborative mechanisms based on legal instruments, constitutional and administrative provisions, agreements, local concerns, and adapted practices for addressing the challenges Megh Pyne Abhiyan (2020). The present discourse on rivers needs to change and this ought to be done through a collective effort. Contributions from legislative, administrative, civil society, professionals, primary stakeholders are a prerequisite for developing a comprehensive strategy of collaboration.

11.7 Situation in Nepal

In Nepal, mitigation of water-induced disasters is accorded high priority. Prioritizing disasters management began in 1982 with the Natural Calamity (Relief) Act, and thereafter, the process accomplished many a milestone. Disaster management is led by the Central Natural Disaster Relief Committee, which is chaired by the Minister for Home Affairs and members include two other ministers, the secretaries of most ministries, various government officials, and representatives from the armed forces, police, scouts, and Nepal Red Cross Society (Shrestha and Bajracharya 2013). There

---

200, and 250 feet. The y-axis shows arsenic concentration in ppb. The red line shows the permissible limit of arsenic in water as per standards (i.e. 10 ppb).

[5] MPA's ongoing study across the Indo-Nepal border in the seven bordering districts of Bihar despite having identified 148 small and seasonal transboundary rivers indicate that there are many more that to be identified (Megh Pyne Abhiyan 2020).

is regional, district, and local natural disaster relief committees. The Department of Water Induced Disaster Prevention once responsible for implementing flood mitigation programmes no longer exists (Shrestha and Bajracharya 2013). The Department of Hydrology and Meteorology is responsible for issuing weather and flood forecasts. A number of NGOs are actively involved in flood risk management including community-based flood warning. A number of policy initiatives related to disaster management have been adopted since 1982 in Nepal. An overview of the same is shown in Table 11.4.

Since the official notification of the new constitution of Nepal in 2015, there have been several government initiatives towards strengthening the regulatory mechanisms for both generic and water-based disasters. The Constitution has endowed each of the country's 753 municipal governments with a jurisdiction and authority over the development sector. This has resulted in a new governance framework emphasizing the role of local governments in operationalizing disaster risk management strategies. In addition, in the 2015 Constitution, disaster risk reduction management (DRRM) is the common responsibility of all three tiers of government. From federal to local level, there are six disaster risk reduction committees/councils. The Prime Minister chairs the National Council (NC) for DRRM, while the Home Minister chairs the DRRM's National Executive Committee (NEC). Each provincial disaster management council is under the chairmanship of the Chief Minister of that province and a separate provincial executive committee is headed by the provincial internal (home) Minister. At local level, there is a local disaster management committee (LDMC) headed by the Mayor or Chair of the respective municipality/rural municipality as well as a district disaster management committee (DDMC) (Dixit et al. 2021).

The tasks of LDMCs include local-level planning, developing guidelines and criteria, reducing risks, responding to the needs of disaster-affected persons, camp operations, recovery, rehabilitation, and regulation of the non-governmental and private sectors. DDMCs work on response, the mobilization of security personnel, search-and-rescue and relief work, operation of shelters, and coordination with other bodies at district and Palika levels. The national platform for DRR (NPDRR) brings the Government of Nepal (GoN) and civil society organizations together to discuss issues related to disaster risk (Dixit et al. 2021).

The key question is whether these laws, policies, plans, and strategies are sensitive towards small and seasonal transboundary rivers between India-Nepal border. Further we need to examine whether issues concerning small and seasonal transboundary rivers have been provided adequate space to be acknowledged, deliberated and addressed.

11.8 Flood Management in India

In India, flood management is organized at the central, state, and local levels. The central level provides guidance to the states, which implement flood management at the field level. The Ministry of Home Affairs and the Department of Water Resources,

Table 11.4 Key laws, policies and plans pertinent to disaster management and water resources in Nepal

Policy instrument	Year	Salient features
Natural Calamity (Relief) Act	1982	• Milestone in systematic disaster management in Nepal • Provides a framework for disaster mitigation • Described the duties and responsibilities of the Ministry of Home Affairs, and not of other relevant disaster management agencies
Water Resources Act	1992	• Enables the formal creation of Water User Associations (WUAs) to govern water projects, prioritizes water allocations for drinking water and irrigation, requires permits for abstraction, outlines water quality protection principles, and establishes penalties for any violations of the Act
Water Resources Regulations	1993	• Establishes and outlines the structure and functions of District Water Resources Committees (DWRCs), including permit management and dispute resolution
National Report and Action Plan for Disaster Management	1996	• Contains a disaster preparedness plan and risk assessment for disaster management
Environment Protection Act	1996	• Has made provision of environmental impact assessment of the proposed activity before it is carried out
The Building Act	1998	• Has made provisions for the regulation of building construction works in order to protect building against earthquake, fire, and other natural calamities
Water Resources Strategy (WRS)	2002	• Defines water sector priorities and outlines actions, indicators, and timelines to address them. Adopts Integrated Water Resources Management (IWRM) principles for basin management, proposes river basin planning units, and calls for IWRM basin management plans
Tenth Plan	2002–2007	• Underlines the need to develop means for improved prevention, mitigation, and reduction of natural disasters
Multiple policies—Water Resource Strategy, 2002; Sustainable Agenda for Nepal, 2003; Disaster Risk Reduction Strategy, 2009; National Adaptation Programme of Action (NAPA) to Climate Change 2010 and Climate Change Policy, 2011	2002–2011	• Have adopted integrated water resource management and river basin management approaches with emphasis on community-based risk management

(continued)

Table 11.4 (continued)

Policy instrument	Year	Salient features
National Water Plan (NWP) (2002–2027)	2005	• Sets out detailed activities to be carried out to implement the Water Resource Strategy • Envisions the establishment of disaster prevention, warning/preparedness, and mitigation measures in at least 20 priority districts by 2010 and across the whole country by 2027 • Defines the long-term investment portfolio for water sector and assigns the Water and Energy Commission Secretariat (WECS) to develop and coordinate water policy and sectoral development. Restructured District Water Resources Committees (DWRCs) around IWRM principles and created Sub-Basin Committees, River Basin Authorities, District Water Assemblies, and inter-DWRC
Water-Induced Disaster Management Policy	2006	• Covers five areas: emergency protection; information dissemination for effective relief, rescue, and protection; awareness raising; warehouses for rescue and relief supplies; and a rehabilitation fund • To mitigate water-induced disasters and reduce loss of lives and property and to enhance institutional strengthening
Prime Minister Relief Fund Regulation	2007	• Has made provision of funding for rescue, assistance, medical treatment, aid to poor people, and aid for charity projects
The Constitution of Nepal	2015	• Has spelled out in its Directive Principles, Policies and Obligations of the State (Clause 51) about disaster management as to make advance warning, preparedness, rescue, relief, and rehabilitation in order to mitigate risks from natural disasters. The constitution has given the list of concurrent powers of federation, state and local level focusing on early preparedness for, rescue, relief and rehabilitation from, natural and human-made calamities for disaster management in Schedule-7, 8 and 9

(continued)

Table 11.4 (continued)

Policy instrument	Year	Salient features
Disaster Risk Reduction and Management Act	2017	• The prevailing law on reduction of disaster risks and management to protect human lives and properties of the public, private and individual, to preserve natural and cultural heritage, and to keep physical infrastructures safe from natural and non-natural disasters by effectively coordinating and managing all activities on reduction of disaster risk and management • Has made provision of effective disaster risk management throughout the disaster management cycle-preparedness, response and rehabilitation and mitigation • Replaces 1982 Act
Local Government Operation Act (LGOA)	2017	• Defines local government responsibilities for resource management, including forests, watersheds, wetlands, water use, and biodiversity • It recognizes that local people and local bodies are the most appropriate points of entry to meet the disaster management needs at the local level • Authorizes to undertake following functions with respect to DRR by local bodies – Local level policies, legislation, standards, plan implementation, monitoring and evaluation related to disaster management – Disaster preparedness and response plan, early warning system, search and rescue, advance store of relief materials, distribution and coordination at local level – Local embankment, river and landslide control, river management and evaluation Hazard mapping and identification of settlements at risk and transformation – Coordination between federal, state and local level and local community organizations and coordination with private sector in order to disaster management – Establishment of disaster management fund and operation and utilization of resources – Plan formulation, implementation, monitoring and evaluation for disaster risk reduction – Resettlement and rehabilitation after disaster – Data management and study and research about local level disaster – Development of local emergency work operation system and – Implementation of community based disaster management

(continued)

Table 11.4 (continued)

Policy instrument	Year	Salient features
Environment Protection Act	2019	• Defines environmental regulatory authorities, establishes an Environmental Fund, and requires environmental impact assessments
National Water Resources Policy	2020	• Integrates water policy, and all water uses and seeks to ease water-sharing disputes between provincial, district, and local governments. Establishes new federal, state, and local bodies for optimizing water use, development, management, and protection

Source Compiled from Shrestha and Bajracharya (2013), USAID and Sustainable Water Partnership (2021), and Nepal et al. (2018)

River Development and Ganga Rejuvenation in the Ministry of Jal Shakti are the main central-level agencies. In addition to the relevant ministries, commissions, and technical support agencies, created by the central government such as river boards for the major rivers such as the Ganga and Brahmaputra are involved also. A National Committee on Disaster Management was recently established under the Prime Minister to review disaster situations in the country (Shrestha and Bajracharya 2013). Central government formulated National Water Policy in 1987 which was subsequently reviewed and revised in the year 2002 and 2012. The National Water Policy has been sent to all States/UTs for appropriate action. As per available information, 16 States/UT have formulated and adopted their State Water Policies. Some of the actions taken by the Central Government in accordance with the principles indicated in the third National Water Policy (2012) are as follows (PIB 2020).

- The Inter-State River Water Disputes (Amendment) Bill, 2019 was introduced in Lok Sabha on July 2019 and was subsequently passed by Lok Sabha
- Ministry of Jal Shakti has prepared a draft National Water Framework Bill and draft River Basin Management Bill, 2018, which have been circulated to States/ UTs for comments
- Dam Safety Bill, 2019 was prepared and introduced in Lok Sabha on July, 2019. The bill was subsequently passed by Lok Sabha. The Rajya Sabha passed the Dam Safety Bill (2019) on December, 2021 paving the way for enactment of the Dam Safety Act in the country. The Act received the assent of the President of India on the December 13, 2021
- National Water Informatics Centre has been established under National Hydrology Project
- In 2019, the Ministry of Jal Shakti set up a committee of independent experts—led by Dr Mihir Shah to draft a new National Water Policy. The draft policy is presently being reviewed.

At the state level, the water discourse in Bihar alternates between flood and drought. However, flood takes precedence because it is an annual recurring phenomenon. MPA has categorized floods of Bihar into eight typologies as outlined in Table 11.5.

Table 11.5 MPA's typologies of floods in Bihar

Typologies of floods	Who are exposed	Character of the hazards
Waterlogged region	Adjacent to the embankments (Countryside)	Hazards predictable, sufficient time to seek refuge
Riverine flood	Between embankments of same river (Riverside)	Hazards predictable, insufficient time to seek refuge
Riverine flood + riverbank erosion	Adjacent to the embankments (Riverside)	Hazards unpredictable, with limited possibilities of safeguards
Riverine flood + riverbank erosion	Adjacent to river (No embankment)	Floods predictable, erosion frequent, with limited possibilities of safeguards
Riverine flood	Countryside (Post the breach of the embankment)	Floods unpredictable, with extremely limited time for safeguards
Flash floods	Between embankments of same river (Riverside)	Foreseeable, with short time frame for safeguards
Flash floods	Adjacent to river (No embankment)	Foreseeable, with short time frame for safeguards
Floods+ / Waterlogging+ / Inundation	Between embankments of two river systems	Foreseeable, moderate time frame for safeguards

Source Megh Pyne Abhiyan et al. (2018a, 2018b). Contextualizing Participatory Groundwater Management in Alluvial Flood Plains of Bihar: Overview, Processes, Findings and Way Forward

However, the flood management approach adopted by the Government of Bihar (GoB) is a generic one, rather than being typology specific. Thus, floods brought in by numerous small, seasonal transboundary rivers across the seven bordering districts remain out of the picture. To address challenges of small and seasonal transboundary rivers, a multi layered strategy needs to be devised. Beginning with the Water Resources Department (WRD), which can facilitate effective integration of knowledge, ideas, innovations and practices for the small and seasonal transboundary rivers at the regional level to address the concerns of the challenged ecosystem. Likewise, the thematic relevant departments for example—Disaster Management Department (DMD), Department of Agriculture (DoA), Department of Revenue and Land Reforms (DoRLR), Department of Animal and Fish Resources, Bihar State Disaster Management Department (BSDMA) and other departments can play a crucial role in addressing issues within their thematic jurisdiction. Thus, the competing and conflicting issues, and practices can be addressed in more a rational, strategic manner with in-situ consideration. The other government departments that can effectively contribute towards minimizing the impacts of recurring floods because in the small and seasonal transboundary rivers are—the Rural Development Department (RDD),

Public Health Engineering Department (PHED), Minor Water Resources Department (MWRD), State Pollution Control Board (SPCB), and Central Ground Water Board(CGWB)—Patna (Megh Pyne Abhiyan 2020).

Small, seasonal transboundary rivers are conspicuous by their absence in the national discourse as well. It remains a blind spot for all the formal institutions. For visioning the way forward for the small and seasonal transboundary rivers a mix of non-state actors such as local community representatives, rural extension workers, academics, local leaders (social and political), civil society members, thematic experts and networks must be involved equal representation of women and men. Themes like agriculture, non-farm economy, wetlands, sedimentation, floods, groundwater, water quality and protected water area have to be integrated in inclusive strategy for the seven bordering districts. Thus, the plan of action for flash flood-prone areas must integrate similar such themes generating in-depth information and knowledge base about the small and seasonal transboundary rivers in the bordering districts of Bihar and Nepal. Such actions must be followed by mapping of the river flow during the monsoon along with other themes of concern (Megh Pyne Abhiyan 2020).

11.9 Participatory Management of Transboundary Aquifers

A landscape impacted by regular flash flood from small transboundary rivers and affected by arsenic contamination in groundwater requires a participatory water governance framework. Such a framework can address water-based risk.

Participatory water governance for transboundary aquifers across Bihar-Nepal should aims to capture the pattern of groundwater's types, dynamics, in relation to various factors (natural, geographical, hydrogeological, social, economic, political, developmental, floods [riverine and flash flood]) required for identifying the groundwater related, challenges and opportunities.

Learnings from the action research undertaken by MPA on participatory groundwater management (PGWM) can be useful basis. The study was implemented in five flood typologies of Bihar, and flash flood prone typology was one. The study also involved transboundary aquifer. The PGWM study highlighted two critical issues. First, flood impact are perceived in a general sense. Those brought by flash flood occurring in the small transboundary rivers are not separated. Second, within the state and at the transboundary level, the shared aquifer with groundwater contamination has not received recognition.

This PGWM study will help in the understanding of the linkages between aquifers (local, regional or transboundary), arsenic contamination and flood prone areas drained by small and seasonal transboundary rivers. The analysis will help in understanding floods and related impacts from a new perspective.

In PGWM, the process of observing flood typology has gone through iterations of learning, exploring and documenting of groundwater system. In the process questions about the resource and responses have emerged. For instance, the use of participatory techniques of studying arsenic contamination at the village and panchayat has helped clarify doubts about contamination. It has triggered a process of thinking collectively about groundwater as a common pool resource. This collective effort of understanding can contribute to a positive transformation in managing groundwater by including insights from scientific inputs. This approach can also create an opportunity for PGWM to be scaled up as a governance framework in the transboundary context of small rivers (Megh Pyne Abhiyan et al. 2018a, 2018b).

For the people living in the arsenic contaminated districts along the Bihar-Nepal border access to safe contamination-free drinking water throughout the year is an urgent necessity. The people here are highly marginalised and responding to their welfare would require a collective localised approach. The PGWM framework incorporates elements that link groundwater with basic services like drinking water, sanitation and micro irrigation. A participatory framework can help highlighting the importance of contextual alternatives to uses of contaminated groundwater (Megh Pyne Abhiyan et al. 2018a, 2018b).

Developing a governance framework for managing transboundary aquifers should systematically evolve depending on undulating character of the region's aquifers, diversity of ecological and human-aspects. The experiences of the PGWM program in North Bihar can help facilitate this process to better understand the local context and peoples' interrelationship with groundwater. The communities are the most crucial link in any regional governance framework. PGWM can help in furthering an understanding of participation through the lens of culture, technology, conflicts, collaborations, and security. Thus, by linking it with pollution, power, politics and profits, PGWM can assist in developing a perspective about new governance framework involving the users, and custodians (Megh Pyne Abhiyan et al. 2018a, 2018b).

Ingraining transboundary groundwater, in aquifer settings calls for collaborative, participatory approach of management. It requires setting up of a prototype for contextualised safe resource usage (MPA et al. 2014). To contextualize transboundary aquifers governance with participatory framework, certain critical issues must be identified. They will help further our understanding of the extent and expanse of arsenic contamination and its impact in areas along the India-Nepal border.

Issues such as differential access to groundwater, hydrogeology, quality, socio-economic context, ecological and political perspectives needs to be better appreciated to help create a knowledge base. In addition, impact of low water quality on different uses and users along the aquifer region as well as groundwater and surface water interactions will help build foundation of plans. Thus actions in understanding the groundwater contamination, its impacts and ways of overcoming them can be taken while keeping in view the consequences of recurring flash flood at micro and meso levels.

Given the impacts of multiple hazards (recurring flash floods, arsenic contaminated groundwater) and climate change, a contextualised governance framework is imperative for future water security. The framework must keep the three broad categories of groundwater resource management in mind: quality assessment, supply, and demand management. The management of risk posed by flash floods needs to be mainstreamed in sectoral and cross-sectoral policies with consideration for impacts likely be made more serious by climate change. To operationalize such a framework for groundwater management, it will be important to bring in the principles such as:

- Seeing groundwater as a common pool resource
- Use an aquifer-based approach for management
- Consider different uses of groundwater; Aquifers, watersheds, and habitations as units of management
- Working on a long-term engagement
- Catalyze community action
- Integrate formal and people's knowledge
- Use social regulations to ensure enhanced water quality
- Tap into the local talent and capacitating them as para-hydrogeologists to help in improving the understanding of local aquifers using a combination of science and traditional knowledge (Arghyam 2015).

Such an approach will help communities and stakeholders manage the resource better. There are a series of steps that make the approach adaptive to community involvement. They are sensitization and capacity building; collection of baseline information; geological mapping; establishing a monitoring network (rain gauge, well inventory, water level data); aquifer delineation and mapping; conducting pumping tests; water quality analysis; socio-hydrological survey; and assessing groundwater balance. These steps can create local database as well as develop a cadre of para-professionals with capacity to collect relevant data, develop maps, and undertake water balance exercises. Paraprofessionals can work with communities to prepare water budgeting. Most importantly they will help in dissemination of information about the status of groundwater as part of a local management plan. The framework can be piloted in a region drained by small and seasonal transboundary rivers with flash floods, and arsenic contaminated groundwater. It can also help in addressing other local challenges peculiar to the region (MPA et al. 2019).

Building a repository of local knowledge about flash floods, their occurrences and impacts, incidence of arsenic contamination and exposure to it is imperative. A combination of science and people's knowledge should be used to build local hydrogeological information as a logical step. This should be followed by building of capacities of all stakeholders for comprehensive understanding of flash floods and the risks of drinking water that is contaminated with arsenic.

11.10 Conclusion

Governance framework for small and seasonal transboundary river and aquifers should create opportunities to collaborate, communicate, prepare, respond, and recover from the impacts of natural hazards. This process should include state and non-state actors (local community from India and Nepal, aid agencies, independent research institutions, civil society, philanthropist and advocacy groups). Each must play a constructive and critical role in bringing to the fore the underlying causes of vulnerability. The effort must be working for co-creating solutions.

This paper has highlighted the challenges of small and seasonal transboundary rivers and the transboundary aquifers. It has suggested an approach to be adopted to overcome the challenges. The governance framework to address the issues of small and seasonal transboundary rivers and the transboundary aquifers includes assessment of the limitations of the approach and risks of the existing system in use. And co-creation of a functional framework to respond to emerging challenges related to small and seasonal transboundary rivers and the transboundary aquifers is the need of the hour.

References

Arghyam (2015) Participatory groundwater management—a collective effort towards addressing India's water security. Bengaluru

Bhatia S et al (2014) High arsenic contamination in drinking water hand pumps in Khap Tola, Pashchim Champaran, Bihar, India. Front Environ Sci. Published 20 November 2014

Central Ground Water Board (2018) Ground water quality in shallow aquifers in India. Ministry of Water Resources RD & GR, Government of India

Dhiman SC, Jain SK (2010) An overview of transboundary of the Indian sub-continent. Bhujal News; Central Ground Water Board; Faridabad. http://cgwb.gov.in/Documents/Bhujalnews-25-01-02.pdf

Dixit A, Neupane S, Bhandari D, Aacharya BK (2021) Political economy of 2020 landslides, road construction and disaster risk reduction in Nepal. Oxford Policy Management, UKaid, Political and Institutions Facility

Ghimire P (2016) Chure conservation area: lessons for the management and use of natural and biological resources in Nepal. Case Study Report; South Asia Watch on Trade, Economics & Environment (SAWTEE); Nepal

Government of Bihar (2017) Cumulative Form IX—Flood Report Final 2017, Disaster Management Department

Government of Bihar (2019) Cumulative Form IX—Flood Report Final 2019, Disaster Management Department

Govt. of Nepal (2017) Nepal Flood 2017: post flood recovery needs assessment, National Planning Commission, Kathmandu. https://www.undp.org/sites/g/files/zskgke326/files/migration/np/PFRNA-Report.pdf

Goyal S (2020, July 07) Which states of India share boundaries with Nepal. Jagran Josh

International Federation of Red Cross & Red Crescent Societies (IFRC) (2019, July 17) Information bulletin Nepal: Monsoon floods & landslides. Nepal Red Cross Society. https://reliefweb.int/report/nepal/nepal-monsoon-floods-and-landslides-information-bulletin

Jamwal N (2015) How painting hand pumps red could save thousands from poisoning in North Bihar. Scroll. https://scroll.in/article/768615/how-painting-hand-pumps-red-could-save-thousands-from-poisoning-in-north-bihar

Megh Pyne Abhiyan (2017) Post disaster recovery: assessment of needs in moderate flood conditions. National Institute of Rural Development and Panchayati Raj (NIRDPR), Hyderabad

Megh Pyne Abhiyan (2020) Small and seasonal transboundary rivers between India and Nepal in Bihar. Internal Document; Delhi

Megh Pyne Abhiyan, Advanced Center for Water Resources Development and Management and Institute for Social and Environmental Transition-Nepal (2014) Background note for the preliminary study for possibilities of transboundary aquifers in Banderjhula village (Chitwan district, Nepal) and Kairi village (Pashchim Champaran district, India); Participatory Groundwater Management; New Delhi

Megh Pyne Abhiyan et al (2018a) Contextualising participatory groundwater management in alluvial flood plains of Bihar: overview, processes, findings and way forward. Program Report. New Delhi

Megh Pyne Abhiyan et al (2018b) Groundwater quality analysis of Naya Tola Bishambharpur, Nautan, Pashchim Chamaparan as part of the flood-resilient habitat pilot of hi-aware. Internal Document

Megh Pyne Abhiyan et al (2019). Concept note on flood resilient habitat—disaster and climate resilient water management framework—participatory groundwater management. New Delhi

NASC and UNICEF (2007) Report on blanket tube well testing in Sunsari, Bara, Dhanusha, Rupandehi, Kailali & Kanchanpur Districts-2007. A report prepared for NASC/UNICEF by Genesis Consultancy (P) Ltd. (unpublished), Kathmandu

NCVST (2009) Vulnerability through the eyes of vulnerable: climate change induced uncertainties and Nepal's development predicaments. Institute for Social and Environmental Transition-Nepal (ISET-N, Kathmandu) and Institute for Social and Environmental Transition (ISET, Boulder, Colorado) for Nepal Climate Vulnerability Study Team (NCVST) Kathmandu

Nepal P, Khanal NR, Prasad B, Sharma P (2018) Policies and institutions for disaster risk reduction in Nepal: a review. Geogr J Nepal 11:1–24

Press Information Bureau (2020) Draft National Water Policy. Ministry of Jal Shakti; Govt. of India, New Delhi. https://pib.gov.in/PressReleasePage.aspx?PRID=1607166

Sethi A (2017, September 10) Araria to Kishenganj, Bihar's decades-old flood tragedy has a worrying new trend. Hindustan Times. Jogbani Bihar. https://www.hindustantimes.com/india-news/changing-patterns-of-bihar-floods-a-cause-for-concern-for-the-state-authorities/story-YADcMZeZiAR7heywaucSoJ.html

Shrestha AB (2008) Resource manual on flash flood risk management—Module 2: non-structural measures. International Centre for Integrated Mountain Development, Kathmandu

Shrestha R (2012) Arsenic contamination of groundwater in Nepal: good public health intention gone bad. Inq J 4(09)

Shrestha AB, Bajracharya SR (eds) (2013) Case studies on flash flood risk management in the Himalayas: In support of specific flash flood policies. ICIMOD, Kathmandu. https://lib.icimod.org/record/27767

Shrestha A, Silwal S (2017) Water commons and transboundary rivers. Institute for Social and Environmental Transition (ISET)-Nepal and ActionAid Nepal

Tewary A (2020, June 12) 1 killed, 2 injured in firing along India-Nepal border in Bihar—national. The Hindu. https://www.thehindu.com/news/national/one-killed-2-injured-in-firing-along-India-Nepal-border-in-bihar/article31812741.ece

Thakur JK, Thakur RK, Ramanathan AL, Kumar M, Singh SK (2011) Arsenic contamination of groundwater in Nepal—an overview. Water 3:1–20

USAID and Sustainable Water Partnership (2021) Nepal Water Resources Profile

Aparna Unni is a water policy and governance professional from Tata Institute of Social Sciences, Mumbai. She has worked in both in the rural and urban contexts in different states and geographic terrains in India. She seeks to better understand water as a resource through different aspects—water security in rural areas, water quality, WASH in flood-prone terrains, recurring water-related disasters, sustainable sanitation, and community-management of water resources. Her contribution to the current paper comes from learnings, experiences and insights working in different flood typologies of North Bihar during her tenure with Megh Pyne Abhiyan.

Eklavya Prasad Eklavya Prasad is the social development professional from Tata Institute of Social Science, Mumbai. He is a founder member of Megh Pyne Abhiyan (MPA) working in the water-stressed rural and urban areas in East India focussing in the flood-prone region of Bihar, and water deficient and source contaminated areas of Jharkhand. Through MPA, Eklavya Prasad has been working on flood-related issues—flood-resilient habitation, typologies of floods in Bihar and appropriate interventions, mitigation norms and resettlement policy for people affected by coastal and river erosion, reconstruction and rehabilitation works in flood affected areas, post disaster assessments, and appropriate drinking water and sanitation intervention. Eklavya Prasad led a household-level flood loss assessment of the 2024 floods, covering 2,290 households across seven districts of Bihar. Additionally, he has been working on safe and secure drinking water processes and technologies in areas with fluoride and arsenic contamination in groundwater, hygienic and disaster resilient and ecologically sustainable sanitation technology and participatory groundwater management especially in climate challenged regions.

Ajaya Dixit pursues studying of challenges of water management, dams, and environmental impact analysis, climate change adaptation and resilience, transboundary water challenges, and environmental governance. Since the early 1990s, he has been involved in the study of climate change adaptation and resilience, flood, droughts, and earthquakes, political economic analysis, policy, and institutional studies. His work has also focused on community support, rural development and establishment of local financial institutions. He has particularly supported in capacity building and mentoring of young generation of professionals. Currently senior advisor at ISET-Nepal and Senior Fellow at SAWTEE, he is professor of practice at Kathmandu School of Law. Since 2020 he has been curating South Asia Nadi Sambad (SOANAS) with the aim of promoting dialogue across boundaries for healthy rivers and water.

Rakesh Mishra has been awarded MS degree in Agricultural, Environmental and Regional Economics by the Penn State University, USA where he studied under Fulbright scholarship program. He has altogether more than 25 years of experience of working with government and international organizations in Nepal. He has worked as Agricultural Economist for department of Irrigation, Government of Nepal at eastern Regional Directorate of Irrigation where he worked with various donor agencies like The World Bank, Asian Development Bank, DFID, JICA etc. He has experience of extended work in the field of development, disaster management, humanitarian assistance, and contextual/ political economy analysis. He has worked with partners at different level of government, donors and private sector in past 25 years.

Chapter 12
Conflicting Riparian Rights and Transboundary Water Governance of Small Rivers: A Case Study of Pandai, in Nepal and India

Shobhit Chepe and Jacquleen Joseph

12.1 Introduction

Water is an indispensable 'resource' for life on earth. While about three-fourths of the planet's surface is covered by water, freshwater available for human consumption is only a meagre fraction of the total. Subsequently, with an ever-increasing demand and diminishing availability, water is increasingly becoming 'a source of social tension' (Swain 2015, p. 443). Then, worrying as it may be, it comes as little surprise that water and wars are talked about in the same breath. In relation to water, wars imply the possibility of conflicts, ranging from verbal conflicts to violence (which may lead to actual wars). Scholars have argued over the notions of 'water wars' in the past (e.g. Gleick 1993; Homer-Dixon 1994; Lowi 1995; Toset et al. 2000; Wolf 2000, 2007; Gleditsch et al. 2006). The idea that the conflicts and wars of the future will be fought over water has not only gripped academia but also the general public. Ismail Serageldin, the former Vice President of the World Bank famously said in 1995—'Many of the wars in this century were about oil, but wars of the next century will be over water' (as quoted in New York Times, (Crossette 1995)). Similar contentions about water being a source of future conflicts, instead of oil, have been tossed around by several administrators such as the former-UN Secretary General Kofi Annan (Kasymov 2011). Owing to such claims, water has become a central theme in global security discourse.

While the idea of 'water wars' has been widely critiqued (and rightly so) (Yoffe et al. 2001; Barnaby 2009; Rahaman 2012), its capture of the public eye has given it

S. Chepe · J. Joseph (✉)
Mumbai, India
e-mail: jacquleenjoseph@tiss.ac.in

S. Chepe
e-mail: shobhitchepe@gmail.com

J. Andharia and P. Anand (eds.), *Shifting Lands, Flowing Waters*, Disaster Studies and Management, https://doi.org/10.1007/978-981-96-7691-0_12

some traction. To check the validity of the claim, Wolf (1998) analysed the International Crisis Behavior (ICB) dataset, collected by Jonathan Wilkenfeld and Michael Brecher. The author found that against 146 treaties, only 7 minor conflicts had occurred over water, and notably, no war could be attributed to water conflicts. Similarly, the 'basins at risk' thesis by Yoffe et al. (2001) also tried to put a rest to the 'water wars' agenda. Both theses downplay the agenda of wars over water, but do point to water being a potential threat multiplier in situations where other tensions already exist.

Rivers traversing geographic and political boundaries present such a situation, and possess the potential to incite conflicts between riparian states. This is especially true with rising concerns of pollution, and climate change (Bernauer and Siegfried 2012). The rivers shared between Israel and Jordan, India and Pakistan, Turkey and Syria are oft-cited examples where the rivers are seen as potential conflict creators. The transboundary rivers such as the Indus, Ganga, Nile, Jordan etc. too have seen disputes and conflicts in the past (Wolf 1998). However, as the criticism of the 'water wars' agenda goes, in these cases, water has only served as a catalyst to intensify conflict created by other sources such as territorial disputes, religious differences. (Brochmann and Gleditsch 2012; de Stefano et al. 2017; McCracken and Meyer 2018). Often these conflicts have been understood from a macro perspective of international relations, further complicating the understanding of the role of the river in the conflict.

While there are many small transboundary rivers, the study of conflicts has not focussed on them. In this paper we attempt to explore the factors contributing to water conflicts at the grassroot level in the context of small rivers flowing between two countries which in general share a cordial relationship. This would help provide an understanding of the key factors underlying water conflicts and how they are shaped by larger geo-political context. This chapter also discusses the case of Pandai by drawing from the international transboundary water governance framework to guide aspects of water sharing in small rivers.

12.2 Context

India and Nepal share numerous transboundary rivers that form a part of the Ganga basin. Apart from the major rivers such as Kosi, Gandak, and Mahakali, a number of smaller rivers, rivulets and streams flow across the hills and Terai plains at the India-Nepal Border. Having shared a friendly relationship, the two countries have chalked out bilateral agreements on the three major rivers mentioned. These agreements pertain only to the rivers themselves and not to any of their shared tributaries. The other rivers, large and small, remain out of the purview of the agreements between the countries. The Pandai is one such river, not covered by any agreement, that crosses from Nepal into India at the junction of the border at Thori (Nepal)-Bhikhna Thori (India).

Bhikhna Thori is a village located in the Jamuniya Panchayat of Gaunaha Block in West Champaran district of Bihar, India. It is a frontier village with Thori rural municipality being located at the other side in Parsa district in Nepal. It is situated at the edge of the Valmikinagar Tiger Reserve and was known because of an old railway line which has been abandoned since 2015. Bhikhna Thori is historically important because it was on the only road, connecting India and Nepal in the pre-British period. It is said that the village takes its name from Buddhist monks, locally called '*Bhikhshu' or 'Bhikkhu'*, who would use Bhikhna Thori as a sojourn on their pilgrimage from Tibet to India via Nepal.

Being on one of the only transit routes between India and Nepal, and the quickest one to Kathmandu and Tibet, helped Bhikhna Thori develop into a marketplace. Although nearby markets in Bettiah were considered bigger, the Bhikhna Thori market was an important one which catered to people's basic needs. However, the road was not a metalled one and mostly only bullock carts plied on it. During the British period, the King of Nepal and India's colonial British rulers met to develop the route. The British were the ones who established the railway line in Bhikhna Thori in 1917 to exploit the nearby forests and transport the timber. The railway line furthered the development of the village as a marketplace. After India's independence in 1947, this scenario changed. The state and central government considered Birgunj as a place with better prospect for developing a road to Nepal. This meant that the market at Narayanghat along the Birgunj highway blossomed while the Bhikhna Thori market started declining as more and more businessmen started settling towards Narayanghat. Since about the past 50 years, all traders who conducted business with Nepalese and Tibetan people have moved to Narayanghat. As a consequence, Bhikhna Thori's market faded leading to increased pressure on other livelihoods. It was compounded by other challenges related to the identity of forest dwellers.

The focus of this paper is the Pandai River, a small river that flows from the hills in Nepal into India. It also marks a small stretch of the India-Nepal border (Siddiqui 2017a, b). The river joins the Masan River, a tributary to the Burhi Gandak which eventually drains into the Ganga, making it a part of the larger Ganga basin. The governance of this shared river has become a problem with residents from both countries finding it difficult to amicably share the river's waters.

12.3 Transboundary Water Governance: National vs Local Priorities

Before delving into the specifics of Bhikhna Thori's crisis, it is important to understand some aspects of transboundary water governance. According to the Global Water Partnership water governance refers to a range of political, social, economic, and administrative systems put in place to develop and manage water resources, and the delivery of water services, at different levels of society (Rogers and Hall 2003, p. 16). When the waters cross political borders, they are termed as transboundary waters. They include the aquifers, lakes, and river basins, shared by two or more countries (UN-Water 2008). Article 1 of the Convention on the Protection and Use of Transboundary Watercourses and International Lakes 1992 defines transboundary waters as any surface or ground waters which mark, cross or are located on boundaries between two or more states (UNECE 2014). In that sense, transboundary water governance can be referred to the range of systems (as in the above definition) to develop and manage waters that traverse political borders.

Governance of transboundary waters can be complicated as each riparian shares hydrological, social and economic interdependencies and claims their right to water (UN-Water 2008). The ambition to control the water resources by one party may then lead to a conflict scenario with its neighbour (Biswas 2011; Uprety 2012). So, if a riparian community perceives a threat to their existing rights over the water, they will seek to manage that threat by denying or withdrawing the social license to operate (Camkin and Neto 2016). The actual scenario however, has been contrasting—there have been more cooperative events over water than there have been conflicting ones (Yoffe et al. 2001; Sinha 2016). Regardless of the cooperative events being greater in number, Earle and Bazilli (2013) comment that it is not sufficient to just count the number of such events without understanding the spirit and quality of such cooperation in essence, and without examining whether they fulfil the economic, social and environmental outcomes promised in the agreements.

Previous literature on transboundary water agreements of internationally shared water resources suggest that the state-centric approach of international relations is the primary approach to hold the negotiations (Furlong 2006; Akter 2016). Transboundary water negotiations are multi-level power games in which state representatives are the lynchpin (Warner 2008; as cited in Warner et al. 2014). However, Kranz and Vorwerk (2007) call attention to the fact that governance is not only ‘the government’ but also ‘the locals’, which includes civil societies, communities, and private sectors for self-government and local democracy. The state-centric ‘hydraulic mission’ approach, which focuses on ‘national’ interests, leaves the local populations behind (Milich and Varady 1999; Allan and Mirumachi 2010). These local populations are the ones who are directly dependent on and most affected by the water resources. The dichotomy of centre (state) vs periphery (communities) in the transboundary water governance permits the state to take the centre stage over the concerns and needs of communities from where the rivers flow. In this context of state-centric approach to water governance, it becomes pertinent to understand how international and national transboundary water governance frameworks address local needs and issues of riverine communities.

12.4 International Water Law

The international community, i.e. policy-makers and influencers, researchers, bureaucrats etc. has been making inroads into the governance of international watercourses for a few decades through various conventions and rules. But Stoa (2012) states that these institutions (conventions as well as the international community themselves) have not done much considering the size of the population that lives in transboundary basins. Different theories have existed on cooperation over international water bodies since the nineteenth century. These can be summed up into four major schools of thought:

1. Principle of Absolute Territorial Sovereignty
2. Principle of Absolute Territorial Integrity
3. Principle of Limited Territorial Sovereignty or Limited Territorial Integrity
4. Community of Co-riparian in the Waters of an International River

The first principle is based on the claim of right to unconditional use of water from an international river by a state in its territory. Also known as Harmon Doctrine, it holds no concern for the harm to other riparian by such use. It favours the upstream riparian. The second principle establishes the right of a downstream state to demand the continuation of the natural flow of a transboundary river into its territory from the upper riparian or riparian while imposing a duty on that state not to restrict such natural flow of waters to other lower riparian. It favours the downstream riparian. The third principle balances the rights to water of a riparian with duties of not restricting flow. It derives out of the first two principles and balances their extreme positions in favouring one riparian over the other. The fourth principle is an extension of the previous one and considers the river basin as an economic unit, and the rights to use are bestowed upon collective bodies formed of stakeholders from all riparian or divided among the riparian either by agreement or on the basis of proportionality. Only the third principle is a part of the modern customary international water law and has been codified into international law. But whether it balances the rights and duties, as mentioned above is a matter of debate in the international water law community (Salman, 2007a).

According to Dellapenna (2001, p. 266), customary international law 'consists of the practices of states undertaken out of a sense of legal obligation, that is out of a sense that the practice is required by law'. He further states that it is complex and evolves through a pattern of how states behave when one takes actions that affect the other. Customary rules are generally adhered to as they find root in widely followed practices. Put simply, they are general practices that are accepted as law. This also allows for an easy codification of customary international law into treaties, conventions and international law. However, the author also sheds light on the fact that it is 'ill-defined' and 'uncertain', but still legitimises the claims of international actors. This claim is echoed by Zeitoun and Warner (2006), who refer to Dellapenna and add that this legitimisation might increase the bargaining power of such states for the continued exploitation of the water resource. Then, there is a need to scrutinize these customs in light of the power relations and social practices during which they were established (Wegerich and Olsson 2010).

For legal purposes, the use of international rivers and lakes have been classified into navigational and non-navigational uses. Navigational use of water have attracted early attention in terms of international law than the non-navigational use. Rules on navigational 6 use of water started emerging as early as the beginning of the nineteenth century (Salman 2007b). Whereas, the rules on non-navigational use have been comparatively recent, mostly a creation of the industrial revolution. Disputes regarding non-navigational use of freshwater were rare before the industrial era (Dellapenna 2001). Three major international law frameworks have emerged in the last half a century.

1. Helsinki Rules

One of the earliest international set of rules for shared watercourses is the 'Helsinki Rules on the Uses of the Waters of International Rivers'. These rules came into existence in 1966. Although its name might suggest applicability only to rivers, the rules apply on any international water body including groundwater connected to the surface water. The rules apply to non-navigational uses of water (Salman 2007a). They call for an 'equitable' and 'reasonable' sharing of international waters. Importantly, as Caflisch (1998) notes, the rules have no official standing or legal obligation.

2. UN Watercourses Convention

The most significant effort from the international community came thirty years later in the form of the United Nations Convention on the Law of the Non-navigational Uses of International Watercourses (UNWC) (Stoa 2012). Adopted by the UN General Assembly in May 1997, the convention takes a broad definition of international watercourses like the Helsinki Rules and also calls to attention the inter-linkage between groundwater and surface water systems (McCaffrey 1998). However, this definition excludes 'confined aquifers', which are not connected to a surface water body. As of 2019, the convention has entered into force with 36 parties voting in favour of it and 16 of them becoming signatories to it (United Nations 2019).

3. Berlin Rules

Following the creation of the UNWC and the development of international environmental law, the International Law Association (ILA) convened to develop a revised set of rules for international fresh watercourses. Adopted in 2004 as the 'Berlin Rules on Water Resources', these were also planned as a revision of the Helsinki Rules. As a departure from the Helsinki Rules and the UNWC, the Berlin rules also apply to national water bodies. The rules go to a higher level of detail and are quite extensive.

The international law framework lays a strong emphasis on equitable and reasonable utilization, and the obligation to cause no harm. However, the three sets of rules are, in themselves, state centric and do not look at the local interests. While in theory the rules would support the riverine people by providing them access to participation in water governance, with the exception of what the Helsinki rules emphasize, they do not specify what this would entail (Ebbesson 1998, p. 59; in Woodhouse 2003, p. 165). Further, there has been no evidence which sets a precedence for the governance of small rivers through the international frameworks. This may be attributed to the limited signatories to such agreements, thus limiting the pool of available choices. However, the unit of governance may also be an important reason as why international frameworks may not cater to the governance of small rivers.

Water, specifically the sharing of it, has become a sensitive and political issue. Despite the earliest of the above frameworks being formulated more than half-a-decade ago, most countries have chosen not to be a party to the convention/rules. For instance, majority of the countries in the 'global south' are not a party to the UNWC. However, this is not to say that the countries that have chosen not to accept the international frameworks do not follow the principles that constitute these frameworks. Bilateral International water agreements have existed between countries along with the overarching international frameworks. These have often been built on the foundations of a cordial bilateral relationship, like in the case of India and Nepal. The next section outlines the India-Nepal relationship and the bilateral transboundary water governance frameworks.

12.5 India-Nepal Relationship

India and Nepal share a unique and unparalleled relationship (Upreti 2003). This finds its roots in not only the geographical conditions that the nations share but also 'kinship, common religious and social values, and common civilizational factors' (Baral 1992, p. 816). Upreti (2003) also comments on the informal nature of the relationship between the two nations. These informal ties and popular perceptions between the citizens mean that understanding the bilateral relations only through formal diplomatic relationships are difficult. Communities along the over 1850 km of the open border have interacted with each other since time immemorial. The open borders have also added to the dynamics of the relationship by creating social and economic interdependencies.

India-Nepal Water Relationship: Of the 1850 km long India-Nepal border, 640 km is defined by rivers (Sinha 2016). Cooperation on water resources between India and Nepal traces its beginnings to British India with the 1920 Sarada Barrage Project. The then rulers of the two nations agreed upon the construction of an irrigation project consisting of a barrage and system of canals upon the Sarada River. The agreement enumerated the exchange of land and the sharing of water.

The Kosi agreement of 1954 was the first agreement between the two countries after the independence of India in 1947. The Government of India and His Majesty's Government of Nepal agreed upon the construction of headworks at Hanuman Nagar at the India-Nepal border. Located just upstream of the border in Nepal, the project was envisaged as a multipurpose one with flood and erosion control, irrigation and hydroelectric benefits. It also called for the setting up of communication systems—roads, telegraph, telephone and radio—to support the project. The various clauses of the agreement also cover other aspects associated with the project and the river such as right to fishing, land revenue, compensation for land, arbitration etc.

Researching India-Nepal water diplomacy through the Kosi project, Bisht (2008) reports that compensation was an issue of friction between the parties. In addition, the project has been marred with controversy due to its provisions which are deemed inequitable in Nepal. The agreement makes no mention of what irrigation benefits are to be accrued through the project, neither to India nor to Nepal. There also have been grievances on the use of water and power, navigation rights, fishing rights and administration of the project area. However, the clause of land ownership and hence, sovereignty was the most controversial one. It stated that the lands acquired, and the water rights associated with it would be owned by the Union (of India) (Dhungel, 2009). Considering all the hullabaloo generated by the original agreement, the Kosi agreement was amended in 1966. Nepal's sovereignty was ensured by converting the ownership of the land to a lease restricted to 199 years through the amendment. An alternative to the barrage has been suggested by India in the face of Sapt Kosi High Dam, but its social and environmental feasibility has been under the scanner (Bisht 2008).

The Gandak agreement was the third event of cooperation on water resources between India and Nepal, the second after India's independence. Similar to the Kosi Project, the Gandak agreement detailed the clauses for the construction of a barrage at Valmikinagar where the river enters India. It was signed on December 4, 1959. Implemented as the Gandak Irrigation and Power Project, the agreement specifies the provisions for the construction of headworks, canals and necessary communication infrastructure. It also outlines the water and power-sharing between the parties as well as river training works such as afflux bunds, spurs and embankments. The agreement was seen as nonreciprocal and inequitable in Nepal and also contrary to Nepal's national interests. This prompted a revision in the agreement in 1964 wherein one clause to establish Nepal's right to use of waters of the Gandak was modified along with another clause which talked about water allocation in the dry season being removed. Additionally, the operation of the head regulator of the Don Canal was handed over to Nepal.

Dhungel (2009, p. 30) calls the 24-year period after the revision of the Kosi treaty in 1966 till 1990 as a 'Period of Suspicions and Misunderstanding'. Although the sides discussed projects over Karnali and Sarada, nothing was achieved till the 1991 agreement over the Tanakpur Barrage project. This was followed by a treaty over the Mahakali River in 1996, (known as Sarada river in India), regarding the Sarada Barrage, Tanakpur Barrage and Pancheshwar project. The 1920 Sarada Agreement was to be subsumed under the treaty. The Tanakpur barrage had been operational in 1990/91, before the signing of this agreement. It was constructed in the Indian territory downstream of the Sarada Barrage but caused erosion of 36 hectares of land in Nepal. Sarada Barrage's design life had been surpassed by then. It also ran the danger of the Mahakali River passing the barrage through its left bank. With this in mind, the Indian government requested their Nepalese counterpart to permit connecting the barrage's left afflux bund to the highlands in Nepalese territory. The same was granted in 1991 under a Memorandum of Understanding (MoU) between the two governments. The author (*ibid.*) notes that under this MoU, India also agreed to share 10 MW of power with Nepal. An intervening period of five years between the signing of the MoU and the 1996 Mahakali treaty saw protests in Nepal over the dealing with India. It also caused the fall of the Koirala government in Nepal and India's reputation in the international community was blemished by its unilateral decision to build the Tanakpur barrage. So, in 1996 when the new government was formed in Nepal, both parties saw it as an opportunity to carve out a new deal over water cooperation. Thus, the Mahakali treaty was signed. Its scope covered agreement sharing of water and energy from Tanakpur barrage, water from Sarada barrage and a link road from Tanakpur barrage to the East-West Highway at Mahendra Nagar in Nepal. It also included the provisions for the preparation of the Detailed Project Report (DPR) for the Pancheshwar Project, which was envisaged as a bi-national project.

In terms of other instances of water cooperation, projects such as the Arun-III, Kosi High Dam, and projects over the medium-sized rivers such as Kamala and Bagmati are in the pipeline. However, they have been held back by Nepal's reluctance to go ahead with them. There is also cooperation over flooding and inundation, and navigation, which are being considered by the governments.

The water resources policy of both countries plays an important role in how they approach transboundary water negotiations. India's national water policy, for instance, although accepting of the basin approach, states that water cooperation should be carried out bilaterally. Practicability and ease of implementation are provided as the rationale for this. The policy also demands setting up of institutional mechanisms for implementation at the centre (Clause 13.1 and 13.2) (Ministry of Water Resources, India 2012). According to Vishwanath (2018), India has shown a preference for short-term bilateral treaties largely aimed at preventing conflict. This means that India avoids basin-wide agreements on transboundary rivers. In India, water is a state subject. It means that state governments can decide the direction they want to take over water resources development. Moreover, it also implies that they can influence transboundary water negotiations, which create impediments in the policymaking process. Although Nepal's National Water Plan (NWP) of 2005, Water Resources Act of 1992, and the Water Resources Strategy of 1993 do not discuss transboundary water-related issues, the NWP 2005 focuses on integrated water resources management approach. The focus in both countries is largely on water resources development promoting national interest.

12.6 Pandai—The Case for Small Rivers

A wide array of literature looks at river basins from the point of view of 'national interests' (Rahaman 2012; Zhang 2016; Petersen-Perlman et al. 2017). However, the 'local interests at the border' lens remains largely unexplored. This mirrors the international transboundary water resources discourse which has largely remained state centric and built around 'national interests'. There are for example, a number of scholars calling for the transboundary governance of the Ganga-Brahmaputra-Meghna basin, which is an important concern for research as well as practice. However, there is little work on the smaller rivers which largely remain out of the purview of the current research and practice. And while arguments may be made for taking a basin-wide or a watershed view of the river basins, these too have catered to a 'national interests' agenda.

The hydro diplomacy between the two nations fails to engage with smaller rivers such as the Pandai. Further, there has not been much research around water sharing of these rivers. Siddiqui, Chohan and Sharma (2017a, b) are some of the few authors who have explored this subject. Their work focussed on rivers such as Pandai, Khado, the conflicts and other issues associated with them. With respect to Pandai, their work has tackled the equitable water distribution conundrum in the transboundary context. In summary, literature on transboundary water governance as well as international water law has focussed predominantly on 'national interests' over major rivers, leaving out the smaller rivers where interests are largely 'local' in nature.

Siddiqui et al.'s (2017a, b) work is based on a series of transboundary consultations conducted in the Kosi and Gandak basins. Called as 'Gram Shivir' (Village Camps), the consultations were conducted with riverine communities along the border in India and Nepal. Their consultation on the Pandai river documented damming of the river, the riparian claims, land and water rights associated with transfer of land between the two countries, and so on.

The Pandai river is the major source of water for all purposes for Bhikhna Thori, Thori, and neighbouring habitations on both sides of the borders as the local terrain is rocky and groundwater extraction is difficult and expensive. Conflicts over the river are nothing new to the people of Bhikhna Thori. The unavailability of water and the diverging riparian beliefs are primary reasons associated with the conflicts. The river is marked by high variation in annual flow. It runs dry in the months of summer, but wreaks havoc through floods and causes erosion in the monsoons. Additionally, the locals from Nepal in Thori and beyond assert territorial sovereignty on the water as the upper riparian. There have been instances where they have diverted water upstream and left little to flow into India. Bhikhna Thori locals, on the other hand, claim territorial integrity and chronological use over the water. They allege that even when a water distribution arrangement existed, the Nepalese have now started claiming water which is theirs (Bhikhna Thori's) rightfully. A water resources officer from Nepal called such blockading of streams and rivers as unfair according to the principles of international law. This has fuelled disputes amongst the border communities, as seen earlier. According to an Indian SSB officer from Thori post, the conflicts have been going on for almost 20 years now, particularly surfacing during the months of March-April when the flow in Pandai River begins to dwindle.

12.7 Increasing Demand for Water and Related Conflict

The traditional water sharing arrangement that existed in the past has now declined and no new system is in place to share the water—neither community based nor government established. The decline of the traditional system coupled with increasing population in the region has aided conflicts. The river flows through many streams. There are four channels which are considered major. Traditionally, out of these, two were used by Nepal while the other two flowed into India, where they were utilised downstream. Some versions of the account point to three major streams out of which one was used by Nepal while two flowed downstream. But all accounts establish that upstream demand for water was low. Bhikhna Thori residents claim that since there was scant population upstream, there was little use of the water by the upstream communities. Agriculture was not practiced as it is now, and the major water use was for rearing livestock. This is corroborated by academic work too (Weiner 1973; Thapliyal 1999; Portnov et al. 2007; Kansakar 2012; Jha 2013). Since the population in Terai, where Thori is located, was sparse the water use in the upstream was minimal. Portnov et al. (2007) argue that the Nepal government turned the Terai from a densely forested region to an agricultural region (cited from Schwartz 2000). They posit how this led to a rise in population density in Terai. Additionally, Weiner (1973) comments how the increase in rice and maize production in Terai was directly a consequence of this effort to populate Terai. Many settlers were squatters who slashed or burnt the forest to find land to settle. The author also points to the irreversible damage this might have caused in terms of decline of forest and increase in erosion and flooding. For Bhikhna Thori, this meant that the water usage upstream increased. Consequently, the traditional arrangement of water sharing on the river could no longer suffice the requirements of the upstream.

The boom in agriculture stimulated by the population growth and availability of fertile land drove the increased usage of water in upstream communities of Thori and above who diverted the erstwhile free streams that flowed into India for their use. This flagged off a conflict with the downstream community in Bhikhna Thori who felt their riparian rights were gradually being compromised. The upstream residents, on their part, claimed territorial sovereignty over water. They saw the water as 'their own' and did not see a breach of downstream riparian rights. The downstream residents do not deny the right of the upstream ones, but point to their own territorial integrity and historical use of water. They contend that they still have the right to the water being diverted upstream. Yet, in the absence of the traditional arrangement or, for the matter, any new one, the differences in riparian rights standpoint do not allow for the conflict to be settled. The diversion is created through the construction of earthen dams on the Pandai during the summers. This allows the upstream residents to irrigate their paddy crop, which is highly water intensive. However, the same diversion causes a water scarcity in Bhikhna Thori and downstream which leads to low productivity. Further, the situation for the downstream communities is worsened during the monsoons when these dams are dismantled and water gushes down leading to destructive flooding and erosion. The erosion, locally called as *katav*, can be dangerous—engulfing almost 30–35 houses in its spate on one occasion.

Yet another related problem is the construction of a bridge over the water channel by the Nepal administration. The bridge was designed in such a way that it would impede the flow of the streams to Bhikhna Thori. This would worsen the water security of Bhikhna Thori and create a problem for its residents. As the construction of the bridge had reached halfway, the West Champaran district administration intervened on request of the Bhikhna Thori locals and proposed a solution for ensuring the flow of water. Inserting an RCC pipe through the culvert would provide a path for the water to flow, but the Nepalese engineers and administration felt that this would make the structure weak. This led to an impasse due to which the culvert stands constructed halfway and has led to a few motor accidents. But the larger issue is the perception among Nepal citizens that the Indian authorities are impeding their development works and thus compromising their territorial sovereignty. They feel that they should be able to carry out such development interventions in their own territory. However, this would be seen as against the international customary law of the obligation to not cause harm.

12.8 Bhikhna Thori—Development in a Tiger Reserve Area

It is not just the flooding and erosion during the monsoons, and the scarcity of water in the lean periods that affects the Bhikhna Thori locals. The daily lives of Bhikhna Thori locals are impacted on account of being a border village located in a Tiger Reserve. The Forest department's regulations in India mean that it is difficult to create infrastructure such as roads, railways, electric lines etc. because these interfere with the wildlife. The forest is also the reason that the old railway has been abandoned and a new line has not found clearance. In addition, since electric lines cannot be laid till Bhikhna Thori through the forest, the villagers have to depend on whatever solar energy is made available to them through government funded solar panels and solar water pumps. However, this does not suffice the needs of the locals because it is not available for long periods of the year during the monsoons and the winters. Pumping water through borewells is also expensive and difficult due to the rocky terrain. Coupled with lack of power, this makes irrigation through groundwater highly inconvenient and therefore increases the local residents' dependence on Pandai's water.

Further, the Forest department has also laid claim over the village as a part of the reserve and are looking to vacate the village. This has created a dispute between the forest authorities and the locals. The latter allege that the department has produced fake maps to show that Bhikhna Thori was a part of the forest. Being a remote border village at the edge of the forest meant that initial documentation by the government was quite poor. However, the locals have been able to produce maps from the British era which establish Bhikhna Thori as a revenue village and thus not part of the reserve. They have also been able to get documents and land ownership certificates over the past two decades to prove their residence in the village and its status as a village separate from the sanctuary. Further, the fact that the government has established some primary facilities in the panchayat has aided their claim. However, the department has also been relentless in its pursuit in claiming the village as a part of the reserve.

Forest regulations and local politics too have acted against local interests and have banned stone mining in the river, one of the predominant livelihoods of the indigenous Tharu tribals. The Tharus have a long history of mining stone from the river. The lease exclusively belonged to the nearby Shikarpur estate, an erstwhile royal estate. Later the stone mining was brought under the license system and leased to a politically influential person. This led to the exploitation of the local mining labour for about

25 years. It was a large operation and employed as many as 10,000 labours. Later, as governments changed, a local (state) politician close to the then Chief Minister of Bihar obtained the lease. After a spat between the two, the chief minister ordered a ban on mining and later reduced the lease from 200 acre per person to 2 acre per person. However, Shikarpur estate which still owned the rights intervened and approached the courts. The courts ruled in favour of the estate and reset the lease to 200 acre per person. Following this judgement, the state government turned to the Supreme Court, and the matter has remained unresolved. The stone mining business has closed, without offering any alternative, impacting the livelihood of the people dependent on it. This has in turn forced people to migrate to cities in search of employment as labour. The closure of the mining business also created an environmental problem. Stone mining was carried out in the middle of the channel which increased the channel depth and created space for the river to flow through the middle. After the ban, stone kept accumulating into the middle of the bed and caused the river to overflow its banks and cause widespread riverbank erosion. Subsequent governments have not restored the lease to the community even after promising it. Instead they have leased out the mining to Jharkhand based contractors who the locals see as stone mafia. This has resulted in a rise in the cost of stone as the contractor has increased the prices.

Then there are other problems such as the forest department demanding a toll tax from the locals to enter into the village. The department has set up a check post along the road passing through the tiger reserve, and charges even the local vehicles at the entry. Not only does it impact the daily commuters, it presents major problems when there are weddings or other events in the village. There is also a ban on the entry of tractors and other large vehicles in the area, as the department sees them as disturbances to the wildlife. Moreover, despite being located at the edge of the reserve and in the hills, there are no special accommodation facilities created for tourists. The condition of the road is poor and there is very limited telecommunication facility available. The forest department has not allowed the establishment of communication towers. Thus, despite having a tremendous potential for tourism, locals do not receive any benefits. They are rather troubled by the department's policies and see them as responsible for 'negligible development' in the village.

12.9 Conclusion

Development failure due to forest department's regulations has added to the vulnerability of the residents of Bhikhna Thori who face exposure to flash floods and riverbank erosion. The government policies and the transboundary water conflict impact their daily lives and make them highly vulnerable to the flooding and water scarcity. The lack of development has also left them with reduced capacity to cope with the hazards they face. Topped with the decline of institutions, their agriculture activities have been hampered, leaving their livelihoods weakened. The conflict has also disrupted the previously harmonious relation between the communities across the border. Although they are not hostile to each other, fault lines have become evident in the recent past. While there existed, a traditional water sharing agreement that was seen as acceptable to both parties, it no longer fits the present demography and its requirement. It has thus been rendered defunct without a viable alternative in place.

The contestations highlighted in the case of the Pandai river, reflect the complexities that are often involved in the operationalization of laws or agreements on transboundary water governance at the level of the riverine communities. Upstream and downstream communities present their case over the rights to the use of Pandai's waters. While they may not take the principle of territory and sovereignty to the absolute scale, the contestations do present a clear picture of divergent riparian beliefs being shaped by the international discourse, as seen in the claims of the people on both sides of the border. Drawing from the international transboundary water law frameworks (Helsinki rules, UNWC, and Berlin rules), the arguments of the upstream and downstream residents build upon the principles of equitable and reasonable utilization, do no harm, acquired rights, non-recognition of unilateral action by one riparian, among others. Riverine communities on both sides of the border draw from the same principles to counter each other's right claims. For example, the do no harm argument is not just in terms of foreclosure of future uses for Nepal but includes other forms of physical harm emerging from floods, water scarcity, and environmental degradation, that are often used to counter each other's claims. The inconclusive nature of debates around the principles of international water law and the hierarchy within them, even among international scholars adds to the complexity of the issue in hand.

Considering India and Nepal's transboundary water policy, it might be said that an agreement on the small rivers seems difficult. However, the diplomacy has only focused on individual rivers till now and shows no indication of a change in the future. This means a larger basin-wide agreement approach which encompasses all the major and minor rivers in the primary river basin also seems unlikely. Though such basin-wide agreement further strengthens the state-centric approaches, it is deemed to be necessary to acknowledge the issues of all riverine communities. As indicated above, while the approach may seem unlikely, Pandai's woes can be reduced by taking up a whole of basin approach along with local level engagement over smaller rivers. This may be done by strengthening the local or community level mechanisms for water governance. The communities have shown, through traditional arrangements, that they can engage in effective transboundary water governance. However, some of these traditional institutions of water governance have started eroding. The growing population pressures and changing practices (in water governance and otherwise) have further weakened the institutions, rendering them defunct. Moreover, a failure to replace the local level arrangements compounds the problems, creating fertile grounds for conflict to grow. Therefore, there is an urgent need to reinvigorate these institutions. Community level engagement along with formal government cooperation may perhaps facilitate better transboundary cooperation and provide relief to riverine communities living along the small rivers like Pandai.

Community engagement or public participation is emphasized by both the Berlin rules and UNWC. However, the requirement of participation at the international level in UCWC is only limited to that of the states. Woodhouse (2003) notes that though there is no condition for public participation at the international level, the convention does call for it at a domestic level to conduct activities such as the environment needs assessment within the state. The Berlin rules include a section on 'Public Participation and Access to Information' (ILA 2004, p. 24). They ask the states to ensure that people likely to be affected (under their jurisdiction) are included in decision-making on water management. And also, that the affected public is given 'reasonable opportunity' to voice their concerns and ideas about the water in question. Other than the three frameworks detailed thus far, the first comprehensive declaration which addressed the requirement for public participation in decision-making in water management was made in 1992 in Dublin. The Dublin International Conference on Water and the Environment established four guiding principles for managing freshwater resources. The second guiding principle is solely concerned with public participation. The principle stipulates that, 'Water development and management should be based on a participatory approach, involving users, planners and policymakers at all levels'. Another instrument for public engagement in water governance was the Rio Declaration on Environment and Development. Principle 10 of the Rio Declaration reconfirms the need for public involvement by stating that environmental issues are best handled with the participation of all concerned citizens, at the relevant level (United Nations Conference on Environment and Development, Rio de Janeiro 1992). Though the international water law frameworks increasingly emphasize public participation they do not specify what participation would entail. As Woodhouse (2003) notes international agreements cannot be specific in what participation would entail. He quotes Ebbesson (1998) who says, 'It would be contradictory in international treaty to agree upon the precise manner in which public participation is to be given form in all contexts and legal systems' (p. 59). Thus, the conception of public participation is either limited to the neo-liberal conception of water users or totally missing in the current bilateral transboundary water governance frameworks between India and Nepal. The state-centric frameworks need to comprehend how the agreements signed for so-called mutual benefits affect the riparian communities. In the context of India and Nepal they are excluded from the processes of decision-making and from decisions which affect them negatively.

The emerging international perspectives to move beyond environmentalism, to integrate social and human development concerns, the need to reconsider downstream hegemony, and integrate public participation, within the international water law framework, underlines the hope for a better transboundary water governance mechanism between the two countries, especially sharing smaller rivers; although not in the immediate future.

References

Akter M (2016) Conceptualizing environmental governance on the GBM basin. Bdung J Glob South 3(1):1–17. Springer International Publishing. https://doi.org/10.1186/s40728-015-0025-7

Allan JA, Mirumachi N (2010) Why negotiate? Asymmetric endowments, asymmetric power and the invisible nexus of water, trade and power that brings apparent water security. In: Earle A, Jägerskog A, Öjendal J (eds) Transboundary water management – Principles and practice. Earthscan, London, pp 13–26

Baral LR (1992) India-Nepal relations- Continuity and change. Asian Surv 32(9):815–829. Available at: https://about.jstor.org/terms

Barnaby W (2009) Do nations go to war over water? Nature 458(7236):282–283. https://doi.org/10.1038/458282a

Bernauer T, Siegfried T (2012) Climate change and international water conflict in Central Asia. J Peace Res 49(1):227–239. SAGE Publications Ltd. https://doi.org/10.1177/0022343311425843

Bisht M (2008) Revisiting the Kosi agreement: lessons for Indo-Nepal water diplomacy. IDSA Comment Inst Def Stud Anal India 22

Biswas AK (2011) Cooperation or conflict in transboundary water management: case study of South Asia. Hydrol Sci J 56(4):662–670. Taylor & Francis

Brochmann M, Gleditsch NP (2012) Shared rivers and conflict – A reconsideration. Polit Geogr 31(8):519–527. https://doi.org/10.1016/j.polgeo.2012.11.001

Caflisch L (1998) Regulation of the uses of international watercourses. In: Salman SM, Boisson de Chazournes L (eds) International watercourses: enhancing cooperation and managing conflict. Proceedings of a world bank seminar. The World Bank, pp 3–16

Camkin J, Neto S (2016) Roles, rights, and responsibilities in water governance. World Aff 179(3):82–112. https://doi.org/10.1177/0043820017690944

Crossette B (1995) Severe water crisis ahead for poorest nations in next 2 decades. N Y Times 10:13. Available at: https://www.nytimes.com/1995/08/10/world/severe-water-crisis-ahead-for-poorest-nations-in-next-2-decades.html

Dellapenna JW (2001) The customary international law of transboundary fresh waters. Int J Glob Environ Issues 1(3–4):264–305

De Stefano L, Petersen-Perlman J, Sproles E, Eynard J, Wolf A (2017) Assessment of transboundary river basins for potential hydro-political tensions. Glob Environ Chang 45:35–46

Dhungel DN (2009) Historical eye view. In: Dhungel D, Pun S (eds) The Nepal-India water relationship: challenges. Springer, Netherlands, pp 11–68

Earle A, Bazilli S (2013) A gendered critique of transboundary water management. Fem Rev 103(1):99–119. Nature Publishing Group. https://doi.org/10.1057/fr.2012.24

Ebbesson J (1998) The notion of public participation in international environmental law. Yearb Int Environ Law 8(1):51. Oxford Publishing Limited (England)

Furlong K (2006) Hidden theories, troubled waters: international relations, the "territorial trap", and the Southern African development community's transboundary waters. Polit Geogr 25:438–458

Gleditsch NP et al (2006) Conflicts over shared rivers: Resource scarcity or fuzzy boundaries?. Polit Geogr 25(4):361–382. Elsevier

Gleick PH (1993) Water and conflict: fresh water resources and international security. Int Secur 18(1):79–112. The MIT Press

Homer-Dixon TF (1994) Environmental scarcities and violent conflict: evidence from cases. Int Secur 19(1):5–40. The MIT Press. https://doi.org/10.2307/2539147

International Law Association [ILA] (2004) Berlin rules on water resources. Report of the seventy-first conference (Berlin). International Law Association, London, pp 334–421

Jha HB (2013) Nepal's border relations with India and China. Eurasia Bord Rev 4(1):63–75

Kansakar VBS (2012) History of population migration in Nepal

Kasymov S (2011) Water resource disputes: conflict and cooperation in drainage basins. Int J World Peace 28(3):81–110. http://www.jstor.org/stable/23266720

Kranz N, Vorwerk A (2007) Public participation in transboundary water management. In: Amsterdam conference on the human dimensions of global environmental change

Lowi MR (1995) Rivers of conflict, rivers of peace. J Int Aff 49(1):123–144. Temporary Publisher. Available at: http://www.jstor.org/stable/24357445

McCaffrey S (1998) The UN convention on the law of the non-navigational uses of international watercourses: prospects and pitfalls. In: Salman SM, Boisson de Chazournes L (eds) International watercourses: enhancing cooperation and managing conflict. Proceedings of a world bank seminar. The World Bank, pp 17–28

McCracken M, Meyer C (2018) Monitoring of transboundary water cooperation: review of sustainable development goal indicator 6.5.2 methodology. J Hydrol 563:1–12

Milich L, Varady RG (1999) Openness, sustainability, and public participation: new designs for transboundary river basin institutions. J Environ Dev 8(3):258–306. SAGE Publications Inc.https://doi.org/10.1177/107049659900800304

Ministry of Water Resources (India) (2012) National Water Policy (2012). Government of India. https://doi.org/10.1061/(ASCE)1052-3928(1988)114:4(408)

Petersen-Perlman JD, Veilleux JC, Wolf AT (2017) International water conflict and cooperation: challenges and opportunities. Water Int 42(2):105–120. Routledge. https://doi.org/10.1080/02508060.2017.1276041

Portnov BA, Adhikari M, Schwartz M (2007) Urban growth in Nepal: does location matter?. Urban Stud 44(5–6):915–937. SAGE Publications Ltd. https://doi.org/10.1080/00420980701258118

Rahaman M (2012) Water Wars in 21st century: speculation or reality? Int J Sustain Soc 4:3–10. https://doi.org/10.1504/IJSSOC.2012.044658

Rogers P, Hall AW (2003) Effective water governance, vol 7. Stockholm

Salman SMA (2007a) The Helsinki rules, the UN Watercourses convention and the Berlin rules: perspectives on international water law. Int J Water Resour Dev 23(4):625–640. https://doi.org/10.1080/07900620701488562

Salman SMA (2007b) The United Nations watercourses convention ten years later: why has its entry into force proven difficult?. Water Int 32(1):1–15. Taylor & Francis.

Schwartz M (2000) The Beginnings of Israeli resettlement assistance to Nepal: 1960–1965. M. Schwartz and AP Hare, foreign experts and unsustainable development: applying israeli experience to Zambia, Nigeria and Nepal, pp 129–151

Siddiqui S (2017) Villages along the India-Nepal border are fighting over water from the Pandai river, Scroll.in. Available at: https://scroll.in/article/836365/villages-along-the-india-nepal-border-are-fighting-over-water-from-the-pandai-river. Accessed 26 March 2020

Siddiqui S, Chohan S, Sharma PM (2017a) Dehydrated villages Himalayan uncertainties: challenges for equitable water allocations in the marginalized rivers shared by India and Nepal.

Siddiqui S, Chohan S, Sharma PM (2017b) Inundated villages-drowned futures.

Sinha UK (2016) Riverine neighbourhood: hydro-politics in South Asia. Pentagon Press, New Delhi

Stoa RB (2012) Harmonizing international water law: current challenges and future prospects. In: Legal instruments for the implementation of sustainable development, II. Available at: http://ssrn.com/abstract=2417031

Swain A (2015) Water wars. In: Wright JD (ed) International encyclopedia of the social & behavioral sciences: second edition, vol 25, 2nd edn. Elsevier, pp 443–447. https://doi.org/10.1016/B978-0-08-097086-8.91087-0

Thapliyal S (1999) Movement of population between India and Nepal: emerging challenges. Strat Anal 23(5):777–789. Routledge. https://doi.org/10.1080/09700169908455083

Toset HPW, Gleditsch NP, Hegre H (2000) Shared rivers and interstate conflict. Polit Geogr 19(8):971–996. https://doi.org/10.1016/S0962-6298(00)00038-X

UNECE (2014) Convention on the protection and use of transboundary watercourses and international lakes (amended). Available at: http://www.unece.org/fileadmin/DAM/env/documents/2013/wat/ECE_MP.WAT_41.pdf

United Nations (2019) Status of treaties: convention on the law of the non-navigational uses of international watercourses, United Nations treaty collection. New York. Available at: https://treaties.un.org/Pages/ViewDetails.aspx?src=TREATY&mtdsg_no=XXVII-12&chapter=27&lang=en

United Nations Conference on Environment and Development (1992) Report of the United Nations conference on environment and development: Rio de Janeiro, 3–14 june 1992; official records. New York: United Nations

UN-Water (2008) Transboundary waters - Sharing waters, sharing responsabilities, UN-water thematic paper. Available at: www.unwater.org

Upreti BC (2003) India-Nepal relations: dynamics, issues and problems. South Asian Surv 10(2):257–274

Uprety K (2012) Transboundary water governance: lessons for South Asia. GWF Discussion Paper 1233, Canberra, Austalia. Available at: http://www.globalwaterforum.org/2012/09/03/transboundary-water-

Vishwanath A (2018) Paddling upstream transboundary water politics in South Asia. Carnegie India.

Warner J (2008) Contested hydrohegemony: hydraulic control and security in Turkey. Water Altern 1(2):271–288

Warner J, Zeitoun M, Mirumachi N (2014) How “soft” power shapes transboundary water interaction. In: Grafton RQ et al (eds) Global water: issues and insights. ANU Press, Canberra, Austalia, pp 50–56

Wegerich K, Olsson O (2010) Late developers and the inequity of “equitable utilization” and the harm of “do no harm.” Water Int 35(6):707–717. https://doi.org/10.1080/02508060.2010.533345

Weiner M (1973) The political demography of Nepal. Asian Surv 13(6):617–630. University of California Press. https://doi.org/10.2307/2642967

Wolf AT (1998) Conflict and cooperation along international waterways. Water Policy 1(2):251–265

Wolf AT (2000) Hydrostrategic territory in the Jordan basin: water, war, and Arab–Israeli peace negotiations. In: Water in the Middle East: a geography of peace. University of Texas Press, Austin, USA, pp 63–120

Wolf AT (2007) Shared waters: conflict and cooperation. Annu Rev Environ Resour 32:241–269. Annual Reviews

Woodhouse M (2003) Is public participation a rule of the law of international watercourses? Nat Resour J 43(1):137–183

Yoffe S, Wolf AT, Giordano M (2001) Conflict and cooperation over international freshwater resources: indicators and findings of the Basins at RISR1. J Am Water Resour Assoc 39:1109–1126. Available at: http://tbw.geo.orst.edu/research/basins_at_risk/bar/BAR_chapter4.pdf%5Cnpapers2://publication/uuid/638D6B51-420B-4AC3-BC9C-B810A7C148A7

Zeitoun M, Warner J (2006) Hydro-hegemony–a framework for analysis of trans-boundary water conflicts. Water policy 8(5):435–460. IWA Publishing

Zhang H (2016) Sino-Indian water disputes: the coming water wars?. WIREs Water 3(2):155–166. John Wiley & Sons, Ltd. https://doi.org/10.1002/wat2.1123

Shobhit Chepe is a Disaster Risk Reduction professional with experience in both programming and research. He holds an M.Sc. in Disasters, Adaptation, and Development from King's College London and has a background in civil engineering. He has led the water governance portfolio at Oxfam India, including work on community-led advocacy on flood warnings and water quality in transboundary river basins. He has also worked on humanitarian responses during Covid-19 pandemic and floods in Odisha (2022). As a researcher, he supported the study of civic engagement in transboundary water governance in the Gandak River basin between India and Nepal at the JTSDS, Tata Institute of Social Sciences, Mumbai. Currently, he is supporting the DRR efforts of the Government of Maharashtra, India as a consultant with Palladium Consulting India Pvt. Ltd. His professional interests include exploring social aspects of disasters, climate, and water.

Jacquleen Joseph is Professor and Dean at the Jamsetji Tata School of Disaster Studies (JTSDS), Tata Institute of Social Sciences, Mumbai, India. She has a Masters' degree in Social Work, followed by an M.Phil. and a Ph.D. in Psychiatric Social Work from the National Institute of Mental Health and Neuro Sciences (NIMHANS), Bangalore, India. She has been associated with JTSDS since its inception. Her research, field action, and teaching centre around psychosocial care and support in disaster and humanitarian contexts. She is actively engaged in long-term research and field action in the psychosocial domain in several disasters, starting with Tsunami, 2004; to the Nepal Earthquake 2015; COVID-19 pandemic and the 2023 Balasore Train Disaster in Odisha.

Some of the significant recent research and mobility grants that she has secured in her areas of research interest are- The Disaster Resilience Leadership Research and Fellowship Project (BMG research grants 2017–19), Transboundary Water Governance and Capacity Building for Civic Engagement (ICIMOD-HUC Research Networking Grant, 2021), and Disaster Risk Creation in Urban Resettlement Processes (Independent Research Fund Denmark, 2021–25), among others. She has several publications in peer reviewed international journals of high repute.

Chapter 13
Navigating a Transboundary Terrain in a Delta: Water Governance, Lives and Livelihoods in the Sundarban Region of India and Bangladesh

Bateswar Das and Meghna Guha

13.1 Introduction

It is well known that deltas have a unique ecosystem and are dynamic landforms at the land—water boundary, involving intricate mazes of rivers and small waterways, wetlands, estuaries and coastal barrier islands. Delta regions across the world experience vulnerabilities of different kinds and although the nature and form of vulnerabilities differ, deltas in Asia are considered most at risk as they face the most serious challenges in terms of human vulnerability caused by a combination of relative sea level rise and population pressures and sometimes poor delta governance (Driel et al. 2015). Globally, 60% of transboundary basins lack formal cooperative arrangements and the existing treaties along this line have proved to be weak in areas of mandate, resources, design and the enforcement mechanisms. The situation is even worse for the transboundary aquifers with only six internally accepted aquifer agreements (Transboundary Water Governance 2016).

Deltas are fertile areas which make them attractive for people to live and work and yet extremely fragile and vulnerable. Worldwide delta systems are under dramatic threat from sea level rise, cyclones, river flooding, storm surges, rapid urbanization, agricultural overuse and pollution, salinization, sediment starvation, coastal erosion, and natural and man-made subsidence (Foufoula-Georgiou 2013). This paper focuses on the Sundarban delta region that poses unique challenges of governance. The Sundarbans (21°32' to 22°40'N and 88°05' to 89°51'E) is a cluster of low-lying

B. Das
Kolkata, India
e-mail: bateswar.das.u2@elms.hokudai.ac.jp

M. Guha (✉)
Mumbai, India
e-mail: meghna.guha@leadschool.in

J. Andharia and P. Anand (eds.), *Shifting Lands, Flowing Waters*, Disaster Studies and Management, https://doi.org/10.1007/978-981-96-7691-0_13

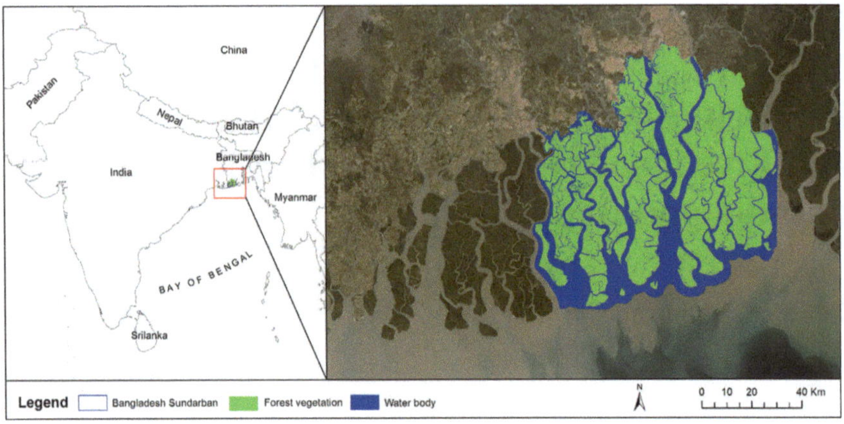

Fig. 13.1 Sundarban Mangrove region between India and Bangladesh. *Source* Kanan et al. (2023)

islands in the Bay of Bengal and is home to the world's largest mangrove forest. Comprising a total area of 10,000 sq km (WWF n.d.) this vast marshy clustered island spreads across two neighboring countries, viz, India and Bangladesh (Fig. 13.1).

While 40% (Sundarbans n.d.) of the land lies in Indian territory, around 60% falls under Bangladesh. This saline, swampy land constitutes the lower part of the Padma (Ganga)- Brahmaputra River Basin in the south eastern part of West Bengal (North eastern state of India) and lower Meghna River Basin in southern Bangladesh. It stretches in a west-east direction along the Bay of Bengal from the eastern edge of River Hugli in India to the western edge of River Meghna in Bangladesh for about 260 km (Pletcher 2020); it reaches nearly 80 km (Pletcher 2020) inward into the mainland. It is crisscrossed by several tidal rivers in both the countries making the land even more saline. The Sundarbans is internationally recognized by the United Nations Educational, Scientific, and Cultural Organization (UNESCO) as a World Heritage Site since 1987 in Bangladesh, and heritage status was accorded to the Indian region in 1997. However, to what extent such recognition actually protects the ecology as well as the communities of the region, continues to remain a matter of question and debate.

Crisscrossed by several rivers across the international boundary between India and Bangladesh, the Sundarbans constitutes an important case study for transboundary water governance. However, management of the transboundary waters is becoming increasingly difficult in the face of complex social and environmental changes. Further, transboundary water governance and related issues cannot be studied in isolation, separate from the challenges posed by climate change and other developments in the region. It is well acknowledged that climate change has serious impacts on water with abrupt variability in precipitation, increasing flood and drought risks, rendering the local population vulnerable on several counts. This paper explores these aspects of transboundary water governance in the context of Sundarbans and related socio-cultural impacts on the population inhabiting the region.

13.2 Tensions Around Transboundary Water Governance in South Asia

Globally, over the years, transboundary governance especially among riparian states has posed a significant challenge. The policies and agreements substantially depend on international collaborations and play a significant role in bilateral and multilateral ties. However, the inclusion of local riparian communities in decision making, or inclusive development has been ignored by the national and international leadership. This further leads to incongruous impacts on livelihoods, socio-cultural sustainability, local governance, and ecosystem practices of the transboundary region. The neglect of local traditional knowledge by the state in transboundary approaches increases the likelihood of multi-dimensional vulnerability, enhancing environmental migration, poverty and the negative impact on indigenous ways of living, and intense conflicts regarding water sharing.

Many would argue that the emerging risks and increasing vulnerabilities are due to a lack of cooperation among the riparian countries in the context of transboundary water conflicts worldwide. One of the critical concerns in transboundary water governance in Asia pertains to the Mekong basin. China, a major political power, is involved in building dams to fulfill the energy and water requirements for growing industries in the region, and a similar approach is followed by comparatively small but industrially strong Vietnam. On the other hand, economically poor countries like Laos and Cambodia are far behind in harnessing water resources for their development. Though Laos, Cambodia, Thailand, and Vietnam have agreements in place with respect to water governance and Mekong River Commission (MRC), the emerging problem of unequal water resource distribution for fishing, and industrialization further creates adverse impacts on the livelihoods of native inhabitants, and put pressure on various developmental aspects (Hirsch et al. 2006). In Africa, groundwater distribution and its increasing demand in the agri-sector due to arid and semi-arid climate is a common phenomenon. UNESCO has found more than 38 groundwater blocks which are situated in transboundary regions and every riparian country makes claims for the utilization of aquifers. The lack of strong trans-national cooperation and governance creates hostility in the region, and influential countries take advantage of their power to extract more water, compared to poorer riparian states. The shrinking of Lake Victoria and Chad are among the major concerns due to imprudent transboundary governance in the region which push some low-income countries into further poverty considering the nature of existing vulnerabilities in the region (Scheumann 2006).

In South Asia, observing the long rivalry between Nepal, India, Bangladesh, and Pakistan over water distribution, organizations such as the World Bank Group (2014) began to intervene in water governance of the Ganga–Brahmaputra-Meghna (GBM) basin since mid-2000s with one of the most ambitious projects called the South Asia Water Initiative (SAWI). Yet neither SAWI nor any international policy of more developed countries such as Australia and UK have been able to improve water governance in this region (Hanasz 2019). The South Asian region has always been considered strategically important worldwide since significant players like China and India are considered more aggressive in expanding their water resources utilization strategy. For that, many times, both countries have seen military conflict. Bangladesh has always accused India of being hostile in the context of the GBM basin water sharing, despite the Ganga Water Treaty that has been in place since 1996. India has issues with Nepal for restricting water flow in the upper riparian region. Conflicts between India-Pakistan regarding Indus water share pose significant barriers for collaborative transboundary water governance in the area. The 2016 report of the UN Environment Programme and associated organizations assessed the GBM basin as the most vulnerable delta in the world based on more than 15 key indicators including water quantity, governance, socio-economic vulnerability and ecological impact (Vishwanath 2018).

Spread over 10,000 kms, the UNESCO has recognized that biodiversity enriched Ramsar site, Sundarbans in India and Bangladesh, is vulnerable to climate change. It is a coastal land, comprising a cluster of several islands, crisscrossed by numerous tidal rivers that experience frequent cyclonic storms (the most recent one being cyclone Amphan which made landfall on 20th May, 2020), cyclone induced floods as well as recurring tidal and riverine floods. The threat of climate change is however a consequence of both natural and human factors (Dasgupta et al. 2020, xi). Since the Sundarbans fall in a transboundary region, the challenges of governance are compounded. A detailed analysis of some of the key indicators of vulnerability of the Sundarbans region is presented in the next section.

13.3 Transboundary Challenges in the Sundarban

13.3.1 The Unique Topography of Sundarban in India and Bangladesh and Nature of Conflicts

Although the Sundarban delta fall under the jurisdiction of two sovereign states-India and Bangladesh. Its ecosystem is contiguous with the same flora, fauna, water, and land forms. The vulnerability of the population is also similar. The landmass of Sundarbans undergoes continual change with the actions of the tidal rivers like Matla (one of the largest rivers), and the Bidyadhari. Over time, the river Ganga is also meandering eastwards resulting in increased sedimentation and reducing inflow of freshwater in the region (Yash 2021).

Bangladesh and India share a close relationship in terms of bilateral ties. The two countries are very closely related to each other in terms of economy, environment and a common culture with the two countries having a shared history. Historically, prior to 1947 Bangladesh was a part of India. Further after the partition of India and Pakistan, the region was called East Pakistan. India played a very important role in securing the independence of East Pakistan from West Pakistan and the creation of Bangladesh in 1971. However, the contemporary situation has become more complex as both India and Bangladesh experience some degree of tension ranging from water disputes, illegal migration of Bangladeshis into India as well as due to religious strife.

Water related disputes are prominently visible in the northern part of the state of West Bengal in India and Bangladesh. These pertain to the distribution of the water of River Teesta, where the conflict has been ongoing for over a decade. Bangladesh has expressed concern over India's policies on Teesta. In the most recent virtual summit held in 2020 the two countries discussed ways to boost trade, investments and transportation links, diplomatically avoiding the water issue of Teesta (DW 2021). However, the region experiences transboundary conflicts and challenges. The vast open border in a delta with no distinct water division between the two countries, often results in fishermen facing legal action for straying into international waters. Secondly, the unique species of Sundarbans like the Royal Bengal Tiger, birds of kingfisher groups, swamps, gray-headed fishing eagle, masked fin-foot and so on, are under threat from pirates and hunters. Finally, there is both human and cattle trafficking, as the means of livelihoods are limited and the large majority of the population remains poor and vulnerable.

13.3.2 Vulnerability of the Population

About 4.5 and 7.5 million people inhabit the Indian and Bangladesh parts of the Sundarban, respectively (Gopal and Chauhan 2018). The people living in the region are mainly dependent on fishing, collecting honey, bee wax, tanning, and other forest products. This population happens to be among the poorest in South Asia and extremely vulnerable (Andharia 2020). Reduction of freshwater is affecting the human population as well as the flora and fauna of the place. While these are some of the natural factors, human activities also play a very significant role in destroying the fragile ecosystem of Sundarbans. A large proportion of the population is dependent on agriculture and fishing for their livelihood. One of the most significant impacts of climate change here is the increased soil salinity.

Further, construction of embankments along the rivers to protect the land from periodic flooding during monsoons, construction of upstream dams, overexploitation of the forest products, especially the mangroves, the clearing of forests for agricultural purposes and urban and industrial pollution have significantly affected the landscape of the region. These have an adverse impact on the topography, water supply, sedimentation and hydrology of the region.

Climate change induced threats have serious implications on both the forest-dependent livelihoods of the people inhabiting the region, as well as the regional economy (Roy & Guha, 2017). A rapidly growing population deals with developmental constraints like lack of proper transportation, modern energy services, proper healthcare, and education. However, now global climate change is making the situation even worse for both human beings as well as the wildlife (Danda 2010). Several climatic related (Dasgupta et al. 2020, xi) threats such as rise of sea level, increase of atmospheric CO2, higher frequency of storms and precipitation, and increased air

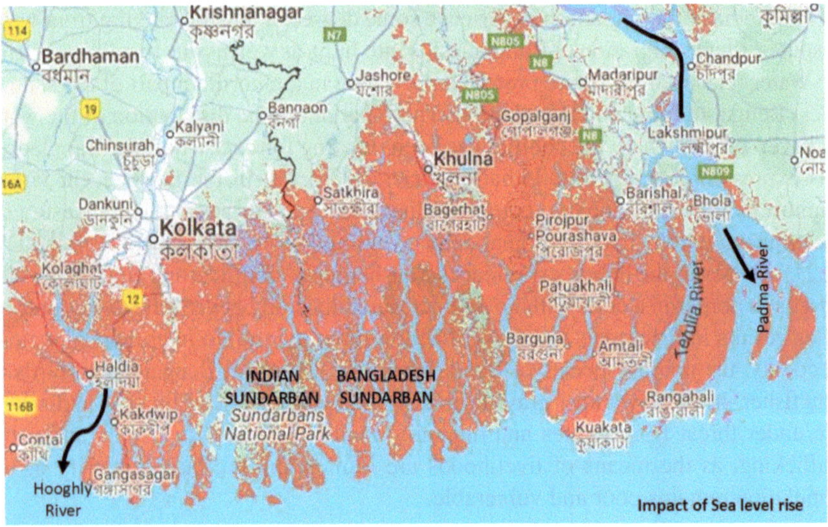

Fig. 13.2 Areas in red would be below water level by 2050. *Source* Sea Level Rise Map and Coastal Flood Tool, Climate Central (2021)[1]

and water temperatures are affecting vulnerable communities inhabiting Sundarbans on both sides of the border. According to Climate Central's recent research (2021), the sea level rise is so visible that by the year 2050 (Fig. 13.2), a great portion of the Sundarbans would be below the annual flood level.

This implies that a great proportion of the human population is exposed to high-risk conditions and the land may even get submerged completely in the next few decades (Sardar & Samadder, 2023).

The overall topography is also highly vulnerable due to inefficient disaster risk management and weak governance in the region. The inefficiency may be on account of a system of patronage between local and state powers where the relationship between communities and government is defined by doles given in form of relief packages or monetary support majorly. There are several instances of neglecting the communities in decision making and instead patronizing them with cash in the aftermath of disaster events. The Aila super cyclone in the Sundarban region saw a huge political mess both in India and Bangladesh and affected people were made even more vulnerable due to weak governance (Tenhunen et al. 2023). A strong case needs to be made for community-based disaster risk reduction interventions in this region that is frequently ravaged by cyclones.

[1] Available at: Climate Central | Threats from Antarctic ice loss.

13.3.3 Cooperation Between India and Bangladesh on Climate Change Issues: Environmental Diplomacy

Since the two countries share environmental concerns there is considerable scope for environmental diplomacy. The governments of India and Bangladesh signed a Memorandum of Understanding (MoU) in 2011 on "Conservation of Sundarban" and it was executed under the aegis of the Ministry of Environment, Forest and Climate Change of the two countries. A Joint Working Group (JWG) was also formed to look into the implementation of the actions listed in the MoU. Two actions in the MoU were related to climate change, one of which was undertaking research to develop a common and shared understanding of impacts of climate change and the other was related to the capacity building of field level forest officials on climate change adaptation (Danda 2019). Both countries recognized the need for monitoring and conserving the Sundarban, a home to a large variety of rare species of flora and fauna and one that also acts as a barrier protecting the mainland of both the countries from flooding, tidal waves and cyclones. The two countries recognized that they must adopt strategies for joint management and joint monitoring of resources. They agree that the Sundarbans' ecosystem is greatly influenced by the human population and settlement inhabiting the region and they delineated the human settlements of this region in both the countries for better management of livelihood problems, damages by flooding, and related disasters, man-animal conflict and pollution and monitor depletion of the resources. The two countries also agreed to conduct an exercise to identify and catalog the diversity of flora and fauna and their spatial distribution in Sundarbans. Another important aspect of the MoU was that both the parties agreed to carry out research to develop a common shared understanding of the impacts of climate change and develop adaptation strategies that can be implemented (Ministry of External Affairs, GOI 2011). However, it was observed that there was hardly any capacity at the field level for climate change adaptation and no joint training on the subject was conducted in the Sundarban region of the two countries. The level of achievement was minimal because the JWG met only once between 2011 and 2016. It took almost five years to convene the first meeting. Further, the MoU of 2011 fails to engage with the core climate change issues threatening the population on both sides of the Sundarbans. Sea level rise, coastal flooding, loss of land, and severe cyclones are merely mentioned on paper. However, in reality these are lagging far behind what was expected (Danda 2019). Current environmental concerns neglect climate change, which in turn renders the local communities increasingly vulnerable, severely impacting their livelihood (Nath 2011). Therefore, it is imperative that the two countries pay more attention to it and consider the creation of an effective collaborative process for mitigation measures based on science and local wisdom, as a transboundary effort. This transboundary ecosystem and the population residing here continues to suffer from the adverse effects of climate change-related issues, which hampers lives and livelihoods and impacts the culture of the communities. In other words, a more robust transboundary agreement is the need of the hour in order to protect the land from further degradation and to protect communities living in the region.

13.3.4 *Localized Transboundary Conflict and Impacts on the Inhabitants*

Given that the natives of Sundarbans have low literacy and minimal legal awareness about the nuances that govern a transboundary region, fishermen inadvertently foray into international waters and experience criminal penalization. Therefore, people's participation in policy making, generating awareness through workshops about the significance of a biosphere reserve and a tiger reserve, is an urgent need that both the countries need to respond to.

According to the Wild Life Protection Act (WLPA), 1972 and Sundarban Tiger Reserve (STR) Act, 1973, motorized fishing boats are not allowed in buffer zones. Prior to that fishermen were freely entering transboundary areas. Since the 1980s restrictions came into force, the Forest Department of West Bengal started to issue Boat License Certificates (BLC) and special permits were required to operate non-motorized boat in the buffer region of two countries. Scholars (Mistri and Das 2015) have argued that often government notices are issued in English and the local people do not understand them and remain oblivious to what the national and international regulations are. Moreover, communication has primarily been top-down. There is little or no effort to understand the needs of the locals which also includes scheduled tribes. The fisher communities face a host of problems including lack of resources (Fig. 13.3) and are routinely targeted by local government institutions. Often, they are arrested for minor transgressions arising out of ignorance. Their licenses are then canceled and their boats confiscated and released on payment of bribes. These practices are reflective of corruption within the governance machinery and has a direct impact on the livelihoods of the struggling fisher community, which experiences systematic dispossession. They lose their jobs, their earnings and become vulnerable and are then compelled to migrate out from the region in search of jobs.

Besides that, the Bangladesh Border Guard (BGB) and Bangladesh Coast Guard (BCG) are often accused of arresting the Indian fishermen who cross the international water divide under the Protection and Conservation of Fish Act, 1950.[2] For example, the Government of Bangladesh is very cautious to regulate the breeding of Hilsa fish. Therefore, they put a ban on fishing in this region for a month every year under the above law. The fishermen then have to look for other sources of livelihoods. As studies reflect, this is one of the most common phenomena which occurs due to lack of governance based on local needs and weak communication strategies between these South Asian countries. Murshed-e-Jahan et al. (2014, p. 68) argue,

> Conflict over access rights also occurs when the fishers of bordering nations (Myanmar and India) enter Bangladesh's territorial waters or vice-versa, and become involved in conflict

[2] See Bhandari (2019).

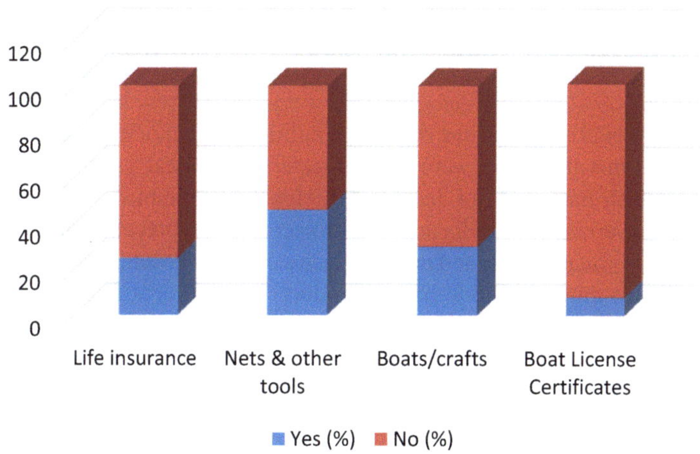

Fig. 13.3 Problems faced by fisher communities in Sunderbans. *Source* Survey Report on Case Study (Mistri and Das 2015)

with local fishers. This type of transboundary conflict comes to the fore when the border security force of the neighboring nation seizes boats and nets and arrests fishers, claiming that they entered territorial waters illegally. These incidents are made more frequent because of unresolved issues of boundary demarcation at sea.

13.3.5 Issues of Industrial Developments Affecting the Water Ecosystem and Community Resistance

The industrial development initiatives of both the countries near the bank of Bhagirathi-Hugli and Padma River led to community resistance over in Sundarban region. The Government of both sides tried to enforce rules that no industrialization shall take place within 10 kms of Sundarban, however multiple Cement factories, leather industry and processing units were built near this ecologically critical area (ECA) of Sundarban. Mongla port area in Bangladesh is the hub of more than 150 industries of such category which was built amidst concerns of ecological crisis that leads to water pollution in Sundarban region, making it unsustainable for aquatic species and local communities (Pole 2018).

The Bangladesh-India Friendship Power Company Ltd. (BIFPCL) led initiative of Rampal coal power plant raised tension in the Sundarban region of both countries considering significant water-air pollution, loss of ecosystem. The UNESCO led team evaluated the project of coal power plant which was identified as harmful despite the push for the coal power plant. Local communities and activists resisted the Government for a long time and the sole reason was the unique ecosystem of Sundarban as an irreplaceable asset.[3]

[3] See Eco-Business (2016).

13.3.6 *Issues in the Protection of Royal Bengal Tigers Amidst Increasing Commercial Trade Through Waterways*

The British colonial state was one of the first entities to venture into the Sundarbans. They believed that the tigers were all man-eaters. The local government in 1782 under the British magistrate of Jessore Mr. Henckell promoted *shikar* (hunting). In 1972 the Government of India adopted Project Tiger to prevent hunting but even today the hunters attack tigers and unique species of birds. Poaching and commercial transportation of timber and non-timber products are prevalent in the dense region of Sundarbans which is mainly done through water channels and have been a major challenge for transboundary water governance for both the countries (Chakrabarti 2009, Mahmood et al. 2021). In 2018, a Union Govt. of India report titled "Status of Tigers in the Sundarbans' Landscape, Bangladesh and India" (Wildlife Institute of India & Govt. of India 2018, p. 22) reflects upon the challenges of transboundary water governance and protection of wildlife in this region against increasing illegal human activities:

> The encounter rate of human sign and sightings were higher near the edges of the forest and overall higher in Bangladesh Sundarbans which is further exacerbated by the usage of river channels for transportation of commercial vehicles.

Barges and ships of cargo usually move through the creeks and river channels mainly in the Bangladesh side and create barriers in the movement of tigers. One of the largest populations of tigers are found here, and has been facing increasing threat. The report calls for regulation of transboundary collaboration and regulation of cargo in order to protect the movement of tigers (Thakur 2020).

Piracy is also common from the sea route. Pirates are active mainly in the Bangladesh side of the Sundarbans (Rahman 2022). Several pirate gangs have been identified by the Bangladeshi government. Such gangs have been known to operate a very lucrative business with impunity in the region, causing losses to the exchequer. All this poses serious hurdles for the local communities who face suspicion and harassment at the hands of the authorities.

13.4 Conclusion

The Ganges-Brahmaputra-Meghna Delta (GBMD) is one of the world's largest (~100,000 km^2) and most dynamic deltas draining land in Bangladesh, Bhutan, China, India and Nepal. As stated earlier, the delta covers most of Bangladesh and part of West Bengal, India, with many of the 147 million people (in the year 2000) living in the delta, under extreme poverty and facing multiple challenges. The case study of Sundarban is an example in the context of global transboundary conflicts that raises difficult and challenging questions around development, governance and policy making.

Worldwide delta systems are under threat from sea level rise, cyclones, river flooding, storm surges, rapid urbanization, agricultural overuse and pollution, salinization, sediment starvation, coastal erosion, and natural and man-made subsidence. Following Foufoula-Georgiou (2013) comprehensive global delta sustainability initiative that promotes awareness, science integration, data and knowledge sharing, and development of decision support tools for an effective dialogue between scientists, managers and policy makers is recommended. This is the only way that regions like Sundarbans which are regarded as "hot spots of vulnerability and change" can become again "seedbeds of sustainability and resilience."

Acknowledgements The authors would like to gratefully acknowledge the anonymous reviewer and Prof. Andharia for their valuable suggestions to improve the quality of the paper.

References

Andharia J (2020) A web of vulnerabilities: ecofragility, poor livelihoods and cyclones in Sundarban Region, India. In: Andharia J (ed) Disaster studies. Springer Nature, Singapore

Bhandari P (2019) Bangladesh claims Indian fishermen defied the ban on catching Hilsa fish and BSF men intruded. National Herald, October 18, available at https://www.nationalheraldindia.com/international/bangladesh-claims-indian-fishermen-defied-the-ban-on-catching-hilsa-fish-and-bsf-men-intruded

Chakrabarti R (2009). Local people and the Global Tiger: an environmental history of the Sundarbans. Glob Environ 2(3):72–95. https://doi.org/10.3197/ge.2009.020304

Danda A (ed) (2010) Sundarbans: future imperfect climate adaptation report. WWF India

Danda AA (2019) Environmental security in the Sundarban in the current climate change era: strengthening India-Bangladesh cooperation. Observer Research Foundation. https://www.orfonline.org/research/environmental-security-in-the-sundarban-in-the-current-climate-chane-era-strengthening-india-bangladesh-cooperation-57191/

Dasgupta S, Wheeler D, Sobhan MI, Bandopadhyay S, Nishat A, Paul T (2020) Coping with climate change in Sundarbans. In: International development in focus, World Bank Group. Google Scholar

Driel WFV, Bucx T, Makaske A, van de Guchte C, van der Sluis T, Biemans H, Ellen GJ, van Gent M, Prinsen G, Adriaanse B (2015) Vulnerability assessment of deltas in transboundary river basins. In: Delta Alliance contribution to the Transboundary Water Assessment Program, River Basins Assessment. Delta Alliance Report 9. Delta Alliance International, Wageningen – Delft, The Netherlands

DW (2021) Why Bangladesh-India friendship is under pressure. DW Made for minds. Retrieved 2021 from https://www.dw.com/en/india-bangladesh-ties/a-56473457

Eco-Business (2016) UNESCO investigates environmental impact of Sundarbans coal plant. Eco-Business, April 1, available at https://www.eco-business.com/news/unesco-investigates-environmental-impact-of-sundarbans-coal-plant/

Foufoula-Georgiou E (2013) A vision for a coordinated international effort on delta sustainability. Conference Proceedings

Gopal B, Chauhan M (2018) The transboundary Sundarbans Mangroves (India and Bangladesh). The wetland book. https://link.springer.com/referenceworkentry/10.1007/978-94-007-4001-3_26

Hanasz P (2019) Transboundary water governance and international actors in South Asia: the Ganges-Brahmaputra-Meghna Basin (Earthscan Studies in Water Resource Management), 1st edn. Routledge

Hirsch P, Jensen KM, Boer B, Carrard N, FitzGerald S, Lyster R (2006) National interests and transboundary water governance in the Mekong. Mekong Resource Centre; Danish International Development Assistance; The University of Sydney

Kanan AH, Pirotti F, Masiero M, Rahman MM (2023) Mapping inundation from sea level rise and its interaction with land cover in the Sundarbans mangrove forest. Clim Chang 176(8). https://doi.org/10.1007/s10584-023-03574-5

Mahmood H, Ahmed M, Islam T, Uddin MZ, Ahmed ZU, Saha C (2021) Paradigm shift in the management of the Sundarbans mangrove forest of Bangladesh: issues and challenges. Trees For People 5:100094. https://doi.org/10.1016/j.tfp.2021.100094

Ministry of External Affairs, GOI (2011) MOU between India and Bangladesh on Conservation of the Sundarban. Ministry of External Affairs, Government of India. Retrieved 2021 from https://mea.gov.in/bilateral-documents.htm?dtl/5141/MOU+between+India+and+Bangladesh+on+Conservation+of+the+Sundarban

Mistri A, Das B (2015) Environmental legislations and livelihood conflicts of fisherman in Sundarban. Asian Profile, India

Murshed-e-Jahan K, Belton B, Viswanathan K (2014) Communication strategies for managing coastal fisheries conflicts in Bangladesh. Elsevier, pp 65–73

Nath S (2011) India- Bangladesh: saving the Sundarbans. Gateway House. Retrieved 2021 from https://www.gatewayhouse.in/india-bangladesh-saving-sundarbans/

Pletcher K (2020) Sundarbans. Sundarbans. Retrieved 2021 from https://www.britannica.com/place/Sundarbans

Pole T (2018) Rapid ndustrialization poses pollution risk to the Sundarbans. South Asia Journal. https://southasiajournal.net/rapid-industrialisation-poses-pollution-risk-to-the-sundarbans/

Rahman, S. A. (2022, January 17). The untold stories of pirate-free Sundarbans. Eurasia Review. https://www.eurasiareview.com/17012022-the-untold-stories-of-pirate-free-sundarbans/

Roy C, Guha I (2017) Economics of Climate Change in the Indian Sundarbans. Global Business Review. https://journals.sagepub.com/description/GBR

Sardar P, Samadder SR (2023) Long-term ecological vulnerability assessment of Indian Sundarban region under present and future climatic conditions under CMIP6 model. Eco Inform 76:102140. https://doi.org/10.1016/j.ecoinf.2023.102140

Scheumann W (2006) EconStor: transboundary water management in Africa: challenges for development cooperation. EconStor. https://www.econstor.eu/handle/10419/199146

Tenhunen S, Uddin MJ, Roy D (2023) After cyclone Aila: politics of climate change in Sundarbans. Contemp South Asia 31(2):222–235. https://doi.org/10.1080/09584935.2023.2203903

Thakur J (2020) India, Bangladesh to use Sundarbans-waterways carefully so as not to disturb movement of tigers: Report. Hindustan Times. https://www.hindustantimes.com/india-news/india-bangladesh-to-use-sundarbans-waterways-carefully-so-as-not-to-disturb-movement-of-tigers-report/story-dFY8bseoRxvGvu38rUFDzL.htm

World Bank Group. (2014). Building resilience for sustainable development of the Sundarbans: Strategy report (English). Washington, DC: World Bank Group. https://documents.worldbank.org/curated/en/879351468259748207

Transboundary Water Governance. (2016). Transboundary Water Governance. https://sustainabledevelopment.un.org/content/documents/hlpwater/04-TransbounWaterGovernace.pdf

Vishwanath A (2018) Paddling upstream: transboundary water politics in South Asia. Carnegie India. https://carnegieindia.org/2018/10/10/paddling-upstream-transboundary-water-politics-in-south-asia-pub-77298

Wildlife Institute of India & Govt. of India (2018) Status of Tigers in the Sundarban Landscape Bangladesh and India. Govt. of India. https://www.wii.gov.in/images/images/documents/publications/status_tigers_sundarban_landscape_india_bangladesh.pdf

Yash AGPBCKAMNS (2021) A story of River Ganga: reconstructing the Sundarbans. Digital.ecomagazine.com. http://digital.ecomagazine.com/publication/?i=724539&article_id=4133545&view=articleBrowser

Bateswar Das is Ph.D. Scholar in Hokkaido University, Graduate School of Environment Science (Japan) and currently working on urban flood risk and ecosystem trade-off in Kolkata (India). He formerly worked as a Research Associate cum Programme Co-ordinator at the Gujarat Institute of Disaster Management leading the capacity development activities to multi-layer government and non-governmental stakeholders including public health professionals and played a pivotal role in course curriculum development. Previously, he contributed to the JTSDS (Tata Institute of Social Sciences, Mumbai) led project focused on extreme heat and coping strategies, granted by GDPC. He is trained and professionally involved in the areas of nature and human-induced disaster risk management, sustainable development and urban governance, risk informed development applying geospatial technologies, and future oriented capacity building.

Meghna Guha is currently a Curriculum Developer for Social Science at LEAD Group, where she designs holistic, future-ready curricula aimed at cultivating twenty-first century skills among students. She is a former Teach for India fellow, where she taught in a low-income private school and focused on the holistic development of underprivileged children, with a strong emphasis on leadership and equity in education. Her work is grounded in community engagement and resilience-building to empower children to navigate complex social realities. Prior to this, she served as a government consultant with the Department of School Education, Government of Tamil Nadu, contributing to large-scale education reform initiatives. Her professional interests lie at the intersection of curriculum design, equity in education, and social transformation.

MIX
Papier aus verantwortungsvollen Quellen
Paper from responsible sources
FSC® C105338

If you have any concerns about our products,
you can contact us on
ProductSafety@springernature.com

In case Publisher is established outside the EU,
the EU authorized representative is:
Springer Nature Customer Service Center GmbH
Europaplatz 3, 69115 Heidelberg, Germany

Printed by Libri Plureos GmbH
in Hamburg, Germany